Computational Chemistry Insights into Molecular Interactions

Computational Chemistry Insights into Molecular Interactions

Guest Editor

Wim Buijs

MDPI

Basel • Beijing • Wuhan • Barcelona • Belgrade • Novi Sad • Cluj • Manchester

Guest Editor
Wim Buijs
Delft University of Technology
(Retired)
Delft
The Netherlands

Editorial Office
MDPI AG
Grosspeteranlage 5
4052 Basel, Switzerland

This is a reprint of the Special Issue, published open access by the journal *Molecules* (ISSN 1420-3049), freely accessible at: https://www.mdpi.com/journal/molecules/special_issues/I8ZP7087RF.

For citation purposes, cite each article independently as indicated on the article page online and as indicated below:

Lastname, A.A.; Lastname, B.B. Article Title. *Journal Name* **Year**, *Volume Number*, Page Range.

ISBN 978-3-7258-5271-0 (Hbk)
ISBN 978-3-7258-5272-7 (PDF)
https://doi.org/10.3390/books978-3-7258-5272-7

Contents

About the Editor

Wim Buijs

Wim Buijs was educated at Leiden University as an organic chemist with a broad interest in society problems. His PhD work on the relation between structure, biotransformation and mutagenic activity of a series of halogenated hydrocarbons is a witness to this. After his PhD work he joined the chemical industry. He worked as an industrial chemist for 30 years and fulfilled various roles in plant optimization research and new process development for large- and small-scale processes. Molecular modeling has become an indispensable part of mechanistic research and serves as a visual and quantitative language for bridging the communication gap with chemical and mechanical engineers. At the same time, he always stayed in touch with academic research both to explore new scientific approaches and share industrial experiences. His professorships at the Davy Faraday Lab of the Royal Institution in London, UK, the University of Sydney, Australia, and Delft University of Technology, Netherlands, are typical examples, as well as his still-ongoing list of publications on subjects like industrial catalysis (transition metal complexes, metals, and porous materials), nanomaterials, and scaling in oil and gas. Currently, CO_2 capture and cannabinoids are among his prominent research subjects, while molecular modeling plays a key role in his research activities. Since Wim Buijs has retired from Delft University of Technology, he has become an independent researcher.

molecules

MDPI

Article

A Holistic View of the Interactions between Electron-Deficient Systems: Clustering of Beryllium and Magnesium Hydrides and Halides

Otilia Mó [1], M. Merced Montero-Campillo [1,*], Manuel Yáñez [1], Ibon Alkorta [2,*] and José Elguero [2]

1 Departamento de Química, Módulo 13, Facultad de Ciencias, and Institute of Advanced Chemical Sciences (IAdChem), Universidad Autónoma de Madrid, Campus de Excelencia UAM-CSIC, Cantoblanco, 28049 Madrid, Spain; otilia.mo@uam.es (O.M.); manuel.yanez@uam.es (M.Y.)

2 Instituto de Química Médica, IQM-CSIC, Juan de la Cierva, 3, 28006 Madrid, Spain; iqmbe17@iqm.csic.es

* Correspondence: mm.montero@uam.es (M.M.M.-C.); ibon@iqm.csic.es (I.A.); Tel.: +34-91-497-3462 (M.M.M.-C.); +34-91-258-706-75 (I.A.)

Abstract: In the search for common bonding patterns in pure and mixed clusters of beryllium and magnesium derivatives, the most stable dimers and trimers involving BeX_2 and MgX_2 (X = H, F, Cl) have been studied in the gas phase using B3LYP and M06-2X DFT methods and the G4 ab initio composite procedure. To obtain some insight into their structure, stability, and bonding characteristics, we have used two different energy decomposition formalisms, namely MBIE and LMO-EDA, in parallel with the analysis of the electron density with the help of QTAIM, ELF, NCIPLOT, and AdNDP approaches. Some interesting differences are already observed in the dimers, where the stability sequence observed for the hydrides differs entirely from that of the fluorides and chlorides. Trimers also show some peculiarities associated with the presence of compact trigonal cyclic structures that compete in stability with the more conventional hexagonal and linear forms. As observed for dimers, the stability of the trimers changes significantly from hydrides to fluorides or chlorides. Although some of these clusters were previously explored in the literature, the novelty of this work is to provide a holistic approach to the entire series of compounds by using chemical bonding tools, allowing us to understand the stability trends in detail and providing insights for a significant number of new, unexplored structures.

Keywords: Be and Mg hydrides and halides; clusters; bonding; rehybridization; stability trends; multicenter bond

1. Introduction

Electron-deficient compounds very often exhibit peculiar bonding characteristics. The most paradigmatic example is diborane, in which the two BH_3 monomers share two hydrogen atoms. As a result of the formation of this quite singular bond, three centers share a pair of electrons as in any typical two-center covalent bond [1,2]. These bonding arrangements, usually named three-center two-electron (3c-2e) bonds, are also found in dialane, the corresponding aluminum hydride dimer. On top of these singular bonding patterns, electron-deficient compounds often present elusive aggregates. A typical example is the corresponding aluminum hydride, whose dimer was predicted to be stable by ab initio calculations [3,4] in the late 1980s but would not be characterized experimentally for the first time until the beginning of the XXI century [5]. In contrast to the stability of diborane, diborane halides, namely B_2F_6 and B_2Cl_6, which were supposedly not stable in the gas phase [6], were very recently found to be weakly bound, as revealed by high-level ab initio calculations. The low stabilization enthalpies are due to the fact that interaction between monomers is mainly dispersion [7]. In the same paper, a systematic study of dimers and trimers involving BX_3 and AlX_3 (X = H, F, Cl) in the gas phase showed that besides dispersion, the rehybridization of the electron-deficient element and its ability to reach pentacoordination are key factors to understand their structure, stability, and bonding characteristics. These findings prompted us to explore the behavior of similar dimers and trimers involving electron-deficient elements from group IIA of the periodic table. Accordingly, in the present paper, we present a theoretical examination of dimers and trimers involving BeX_2 and MgX_2 (X = H, F, Cl) in the gas phase.

Not surprisingly, BeH_2 and its dimer have received a lot of attention. The first two studies we are aware of were on Be_2H_4 and Be_3H_6 [8] at the Hartree–Fock level, with an estimation of electron correlation effects, and on Be_2H_4 and Mg_2H_4 using non-empirical approaches, the latter one including also the mixed $BeMgH_4$ dimer [9]. A rather complete survey of the dimers, trimers, and tetramers of BeH_2 and MgH_2 using high-level ab initio calculations was published in 2005 [10]. More recently, an analysis of the properties of Be_2H_4 and other beryllium hydride oligomers and their spectral characteristics have been reported [11], but the Be_3H_6 trimer was not included in this survey [11]. In 2011, a study of the interactions of BeH_2 and Mg_2H_4 with H_2 included information on the structure and energetics of these two clusters, revealing their ability for hydrogen storage purposes [12]. The Raman spectrum of Be_2H_4 (D_{2h}) was predicted [13] using a new code from variational configuration interaction theory to allow the calculation of such spectra in a pure ab initio fashion [14]. The magnesium hydride dimer was characterized by matrix infrared spectroscopy, which led to the conclusion that Mg_2H_4 is a dibridged molecule analogous to dialane [15]. Some attention was also paid to polymers of BeH_2 [16,17]. The information is scarcer for the BeF_2, MgF_2, $BeCl_2$, and $MgCl_2$ dimers. It can be reduced to Hartree–Fock calculations on Be_2F_4, Mg_2F_4, and BeF_2MgF_2 clusters [18], calculations based on polarized models for MgF_2 and $MgCl_2$ [19], SCF calculations on the dimers of BeF_2, $BeCl_2$, MgF_2, and $MgCl_2$ [20], and electron diffraction experiments on the $MgCl_2$ dimer [21]. For the specific case of the $BeCl_2$ dimer, gas-phase electron diffraction (GED) and mass spectrometric (MS) experiments allowed us to obtain its structure assuming a D_{2h} symmetry, and its force constants and frequencies were also estimated [22]. Some complexes involving Be_2Cl_4 and phosphorus-containing compounds have been very recently synthesized and structurally characterized, evidencing the Lewis acid behavior of beryllium in the dimer [23]. However, we are not aware of theoretical studies on trimers involving BeX_2 (X = H, F, Cl) derivatives, with the only exception of the BeH_2 trimer from [8,10]. The magnesium fluorides and chlorides have instead received most of the attention. To our knowledge, the first survey on the structure and energetics of $(MgF_2)_n$ clusters up to n = 24 was published in 1995 [24]. Years later, the structure of MgF_2 clusters, from dimers to decamers, was studied [25] using a stochastic optimization technique, namely a genetic algorithm (GA) in association with density functional theory. The structures of Mg_3Cl_6 clusters, optimized at the MP2/6-311G* level, were reported in a paper that presents a systematic study of 168 isomers of $(MgCl_2)_n$, where n = 1–4 [26]. It is worth noting that, with the exception of the $BeMgH_4$ [9] and $BeMgF_4$ [18] dimers, there is no information on mixed dimers or trimers involving BeX_2 and MgX_2 (X = H, F, Cl) compounds. The aim of this paper is to present a systematic theoretical study not only of the corresponding homodimers and homotrimers but also of the mixed dimers and trimers. Although, as indicated above, previous studies have been reported for some of these systems, most of them were focused exclusively on structures and energetics. In our study, we will focus our attention on the bonding characteristics of these clusters to precisely explain their structures and relative stabilities. We wonder what changes are observed in the bonding of the dimers when the two metals involved are not identical and how the nature of the substituent may affect bonding and relative stability trends. A similar scenario arises when dealing with trimers. What are the effects on bonding and stability when the dimer interacts with a monomer of the same nature as those forming the dimer? What occurs if the third component is a monomer different from those forming the dimer? How do these effects depend on the nature of the substituent? These are typical questions we will try to answer in this publication.

2. Results and Discussion

Since we are using high-level ab initio calculations and DFT approaches, we first compare the stabilization energies obtained with both methods. The correlation between the G4 and M06-2X values is very good (see Figure S1 of the Supplementary Materials), whereas the correlation with B3LYP functional results is worse (see Figure S2 of the Supplementary Materials). It should be noted that when dispersion contributions are included in our B3LYP calculations using the D3BJ empirical dispersion term, the total stabilization enthalpy becomes more negative from 8 to 20 $kJ \cdot mol^{-1}$, improving the agreement with the G4 outcomes. However, the effects of including dispersion on the geometry of the cluster are negligibly small, to the point that if the B3LYP + D3BJ geometry is used in the standard G4 formalism (that does not include dispersion corrections in the geometry optimization procedure), the changes in the final G4 energy are typically smaller than 0.4 $kJ \cdot mol^{-1}$. Hence, the conclusion is that the G4 formalism is reliable for investigating this type of complexes, in contrast to the results obtained when dealing with B and/or Al derivatives, where the dispersion effects in some of the complexes are a key factor for a proper description [7]. Additionally, in a few cases that will be commented on later, the G4 formalism predicts stationary points that, if obtained, including dispersion corrections, would collapse to the global minimum. In view of this, for the sake of simplicity, we decided to make our discussion using the M06-2X calculated values, knowing that the general conclusions are fully in line with the G4 values. All investigated clusters are closed-shell systems with stable wavefunctions.

2.1. Dimers

Standard graphical programs visualize the clusters under investigation with bonding connections based on default internuclear distances, which may lead to an ambiguous description of the systems depending on the software used. Therefore, instead, we will use the molecular graphs of the clusters in all figures in such a way that we can simultaneously have information on the shape and networking of bonds stabilizing the clusters. Following this univocal criterion, we summarized our results for the dimers in Figure 1.

Quite unexpectedly, we found that the stabilization enthalpies of the hydrides are practically equal for the three dimers, both the homodimers containing Be and Mg and the heterodimer. This is not the case, however, for the corresponding halides, in which the magnesium homodimers are clearly more stable than the beryllium homodimers and the heterodimers. It can also be observed that the largest stabilization enthalpies are those of the fluorides, being those of the chlorides not very different from those of the hydrides. Another very interesting result is that, independently of the nature of the substituent, the stabilization enthalpy of the heterodimers is very close (for the chlorides equal) to the average of the stabilization enthalpies of the corresponding homodimers. The latter observation was already suggested by Kirillov et al. for the dimer hydrides [9].

The rationalization of these findings required a detailed analysis of the bonding of these clusters. The first question we want to address is the nature of the bonds in these clusters and how they depend on the substituents. To quantify these two questions, we have carried out an LMO-EDA analysis whose results are shown in Table S1 of the Supplementary Materials. The first conspicuous fact is that the ionic character of the bonds is significant already in the hydrides, where the percentages of electrostatic and exchange components are similar, the former being slightly higher than the latter. For fluorides, the electrostatic component becomes dominant, while for chlorides, the situation is again similar to that of hydrides. This finding is fully consistent with the values of the Laplacian of the electron density, which is, in all cases, positive. However, whereas for the hydrides, the values are rather small and slightly larger for the chlorides, for fluorides, the values are from 3 to 6 times larger. The same trends are observed for the corresponding energy density values (see Figure S3 of the Supplementary Materials). The larger electrostatic weight in the fluorides is consistent with the much larger natural positive charge of Be and Mg when bonded to F and Cl (+1.7 and +1.8) with respect to the ones in the hydrides (+1.1

and +1.2). Even though the charges are similar for fluorides and chlorides, the weight of the electrostatic component is smaller in the latter due to the larger size of the substituent, which results in longer bond lengths. Another important difference between hydrides and halides, which is related to the higher covalent character of the bonds of the former, is that the three hydride dimers are stabilized by the formation of 3c-2e bonds. These bonds are not present in the halides due to the dominant electrostatic character of the interaction, as illustrated by the AdNDP analysis shown in Table S2 of the Supplementary Materials.

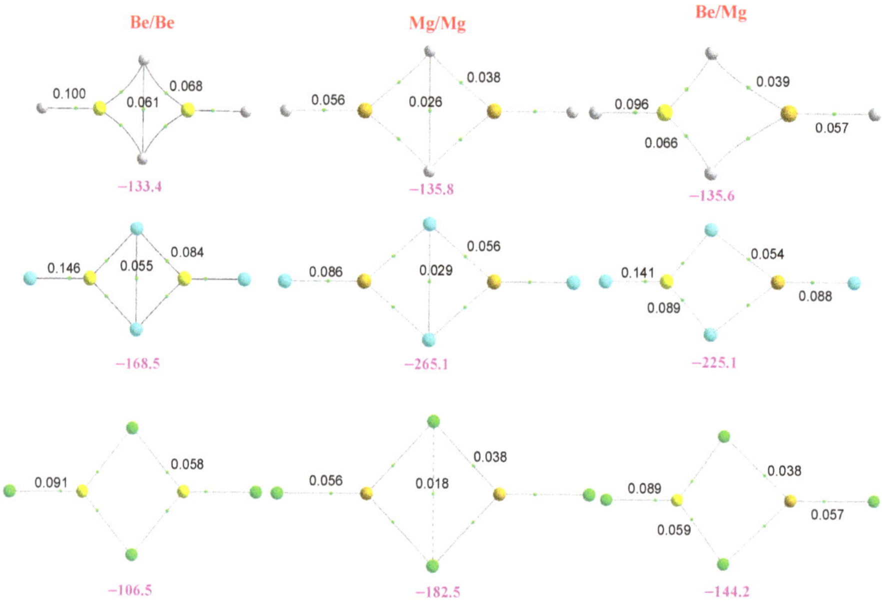

Figure 1. Molecular graphs of the homo and heterodimers involving BeX_2 and MgX_2 (X = H, F, Cl) monomers. The electron densities at the bond critical points (BCPs) are in a.u. The numbers in magenta are the stabilization enthalpies in kJ·mol^{-1}. Atomic colors code: Be (yellow), Mg (orange), H (white), F (blue), Cl (green).

This result is consistent with the ELF plots (see Figure 2). It can be seen that the hydrides present trisynaptic Be-H-Be, Mg-H-Mg, and Be-H-Mg basins with a population of two electrons. In contrast, only disynaptic ones are found for fluorides, reflecting the polarization of the lone pairs of fluorine by the positive charge of the metal. The same is found for chlorides, though the polarization of the chlorine lone pairs is clearly larger, as well as the corresponding populations.

It is also worth noting that, as shown in Table 1, the dispersion contributions in these systems are not negligible, accounting for 6 to 9% of the total stabilization energy, explaining why the B3LYP+D3BJ energies are 9 to 20 kJ·mol^{-1} lower than the B3LYP ones.

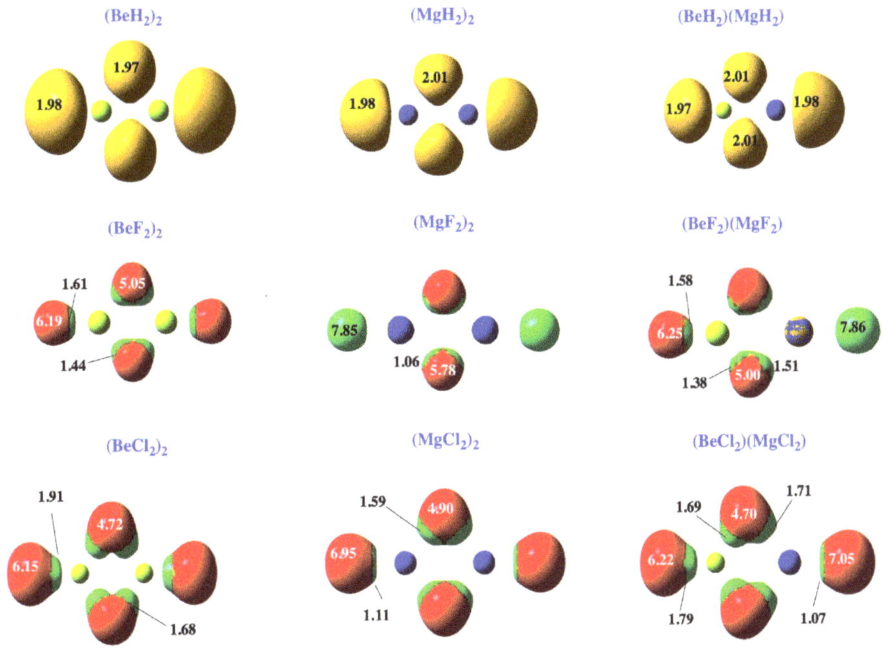

Figure 2. ELF (0.8) disynaptic and trisynaptic basins for the hydride, fluoride, and chloride homodimers and heterodimers. Basins involving H atoms are colored in yellow, and core basins involving Mg appear in dark blue. For the halides, disynaptic basins between the metal atom and fluorine atom appear in green, while lone pairs belonging to halogen atoms are colored in red. Populations are shown in atomic units (e).

Table 1. MBIE analysis of the binary complexes formed by BeX_2 and MgX_2 (X = H, F, Cl). All values in $kJ\cdot mol^{-1}$.

Binary Complex	$E_R(A)$	$E_R(B)$	$\Delta^2E(AB)$	E_{total}
Hydrides				
$BeBeH_4$	58.7	58.7	−261.4	−144.1
$MgMgH_4$	36.3	36.3	−216.6	−143.9
$BeMgH_4$	66.3	31.6	−243.2	−45.2
Fluorides				
$BeBeF_4$	84.1	84.1	−340.8	−172.6
$MgMgF_4$	42.3	42.3	−353.1	−268.5
$BeMgF_4$	85.8	42.6	−357.5	−229.1
Chlorides				
$BeBeCl_4$	80.5	80.5	−269.8	−108.9
$MgMgCl_4$	44.2	44.2	−273.6	−185.3
$BeMgCl_4$	88.0	40.0	−274.8	−146.8

The stability trends within each dimer family still remain to be explained. For this purpose, it is very useful to carry out a MBIE energy decomposition analysis, which is shown in Table 1.

Starting from the hydrides, it is clear that although the stabilization energies E_{total} are practically equal for the three dimers, the stabilizing two-center energy contributions, $\Delta^2 E(AB)$, are not, being the largest one that of the $BeBeH_4$ binary complex. In contrast, the monomer distortion E_R in the $BeBeH_4$ cluster is almost twice that of $MgMgH_4$. Indeed, the largest contribution to this term comes from the rehybridization undergone by the metal from *sp* to sp^n ($n \approx 2$), but this energy cost is much larger for BeH_2 than for MgH_2 (80 vs. 50 kJ·mol^{-1}) [27]. Therefore, although the attractive two-center term is larger for Be_2H_4, the larger monomer deformation energy compensates for the difference. The rehybridization cost also explains the trends for the halides. In this case, the attractive two-center contributions are rather similar, reflecting the electrostatic character of the interaction, but again, the rehybridization cost is much higher for BeF_2 than for MgF_2. Accordingly, the stabilization energy of the $MgMgF_4$ cluster is larger than that of the $BeBeF_4$ analog. Similar arguments explain the trends for the chlorides.

Finally, the stabilization enthalpy of the mixed dimers is very close to the mean of the enthalpies of the two homodimers, reflecting that the intrinsic characteristics of each monomer are essentially conserved in the dimer. This is corroborated by the different analyses. The AIM shows that the electron density at the BCP of the BeX_2 subunit practically does not change when going from the BeX_2-BeX_2 dimer to the BeX_2-MgX_2 one, and the same can be said as far as the MgX_2 subunit is concerned. The same conclusion is reached when looking at the characteristics and populations of the ELF basins, the components of the MBIE decomposition analysis, or the Wiberg bond indexes [28] (see Table S3 of the Supplementary Materials).

2.2. Homotrimers

The two most stable trimers of BeH_2 are shown in the first row of Figure 3. The most stable one corresponds to a linear aggregate, whereas the second is a cyclic structure labeled **A**.

Both ternary clusters can be seen as the result of the attachment of a BeH_2 monomer to the Be_2H_4 dimer, in the first case along the Be-Be axis and in the second case perpendicular to it. Accordingly, they present a very different bonding pattern, though the energy gap between them is rather small (9 kJ·mol^{-1}). This is the result, as we will discuss below, of subtle differences between the different energy components. If the stabilization enthalpy of the dimer (see Figure 1) is compared with those of the two trimers (see Figure 3), it is evident that the trimerization is followed by some kind of cooperativity since both trimers' stabilization energies are more than twice the stabilization energy of the dimer. As shown in Figure 3, the central Be atom is tetracoordinated, and according to both the AdNDP and ELF analyses, it is involved in two Be-H-Be bonds with each of the terminal Be atoms. Note, however, that the electron densities at the corresponding BCPs are greater than in the dimer, indicating stronger bonding interactions. Consistently, the ELF analysis finds trisynaptic basins in the trimer, similar to those in the dimer (see second row of Figure 3). These trisynaptic basins also have a population very close to 2 e but within a smaller volume (122 vs. 129 au^3), whereas the volume of the disynaptic Be-H basins of both terminal groups remains unchanged. This contraction of the trisynaptic basins in the trimer is reflected in a shortening (0.03 Å) of the distance between the central Be atom and the terminal ones with respect to the dimer, reflecting a reinforcement of the interaction. This is also coherent with the MBIE partition energy shown in Table 2 compared to that in Table 1. The distortion energy of the terminal Be atoms are equal in the dimer and the trimer, whereas that of the central Be atom becomes about 40 kJ·mol^{-1} greater as beryllium undergoes a change of hybridization from sp^2 to sp^3 to become tetracoordinated. Consistently, the two-center contributions in the trimer are almost identical to those in the dimer, but the additional three-center term leads to its enhanced stabilization.

Table 2. MBIE analysis of the homotrimers formed by BeX_2 and MgX_2 (X = H, F). All values in kJ·mol^{-1}.

Ternary Complex	$E_R(A)$	$E_R(B)$	$E_R(C)$	$\Delta^2E(AB)$	$\Delta^2E(AC)$	$\Delta^2E(BC)$	$\Delta^3E(ABC)$	E_{total}
$BeBeBeH_6$ (linear)	59.2	102.6	59.2	−262.0	7.8	−262.0	−30.9	−326.2
$BeBeBeH_6$ (cyclic, **A**)	87.4	8.0	86.7	−202.1	−115.3	−116.0	−66.8	−318.1
$BeBeBeH_6$ (hexagonal)	82.9	82.9	82.9	−140.1	−140.1	−140.1	−57.0	−228.7
$MgMgMgH_6$ (linear)	37.2	66.8	37.2	−215.2	0.3	−215.2	−10.4	−299.3
$MgMgMgH_6$ (cyclic, **A**)	58.4	4.2	58.2	−185.5	−99.6	−99.9	−30.6	−294.8
$MgMgMgH_6$ (cyclic, **B**)	28.1	44.5	54.7	−124.6	−54.0	−188.6	−42.0	−281.9
$MgMgMgH_6$ (hexagonal)	42.9	42.9	42.9	−100.1	−100.1	−100.1	−99. 8	−271.4
$MgMgMgH_6$ (hexagonal, non-planar)	51.6	46.7	28.5	−177.8	−77.2	−67.8	−69.3	−265.3
$BeBeBeF_6$ (linear)	84.0	167.3	84.0	−345.2	−2.1	−345.2	−2.0	−359.3
$BeBeBeF_6$ (hexagonal)	95.1	95.1	95.1	−175.1	−175.1	−175.1	−114.4	−354.4
$BeBeBeF_6$ (cyclic, **A**)	162.2	29.1	162.4	−358.3	−146.2	−146.5	80.1	−217.3
$BeBeBeF_6$ (cyclic, **B**)	78.2	124.8	181.4	−139.7	−168.9	−338.5	−40.3	−303.0
$MgMgMgF_6$ (linear)	42.4	82.1	42.5	−356.3	0.3	−356.3	1.2	−546.1
$MgMgMgF_6$ (cyclic, **A**)	92.3	16.0	92.4	−351.9	−214.4	−214.4	48.2	−531.6
$MgMgMgF_6$ (cyclic, **B**)	39.5	56.2	63.1	−250.2	−133.0	−307.4	6.8	−525.0
$MgMgMgF_6$ (hexagonal)	47.5	47.6	47.6	−188.5	−188.5	−188.5	−65.8	−488.8

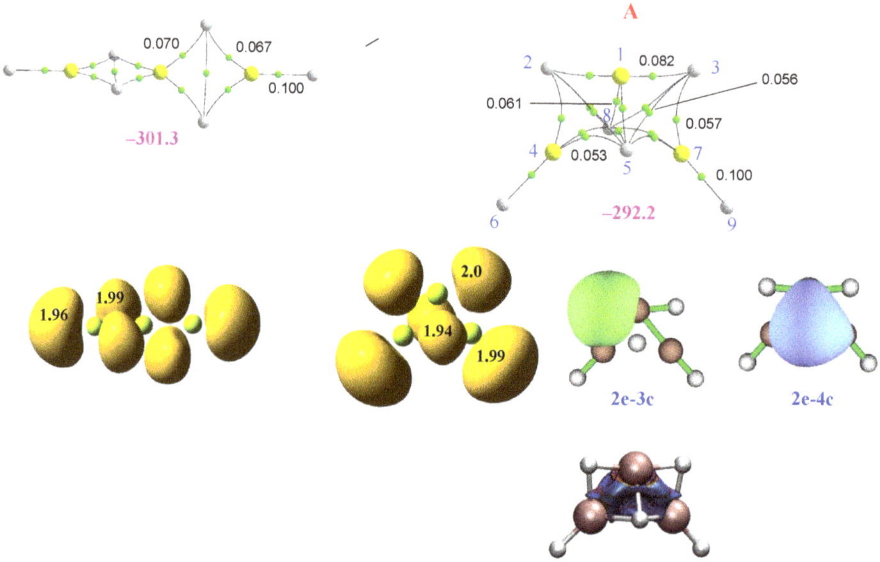

Figure 3. Molecular graphs and ELF plots of the homotrimers involving BeH_2. The electron densities at the bond critical points (BCPs) are in a.u. The numbers in magenta are the stabilization enthalpies in kJ·mol^{-1}. The second row shows the ELF (0.8) trisynaptic basins and their populations, as well as the 3c-2e and 4c-2e orbitals involved in the bonding of cycle **A** obtained by means of the AdNPD approach. In the third row, the 3D representation of non-covalent interactions obtained with NCIPLOT (s = 0.3); color code: red (strongly repulsive), green (weakly attractive and weakly repulsive), blue (strongly attractive). A less stable hexagonal trimer (not shown in the figure) is also a local minimum of the potential energy surface.

The bonding pattern of the cyclic minimum **A** is rather different. In this case, the three Be bonds are tetracoordinated, and non-covalent H···H interactions between the negatively charged hydrogens are also detected. The presence of these interactions implies a certain increase in the dispersion contributions to the stabilization energy, which in the cyclic trimer are significantly larger (−198 kJ·mol^{-1}) than in the linear cluster (−148 kJ·mol^{-1}). NCIPLOT shows (see third row of Figure 3) that, due to these non-covalent interactions, there is a strongly attractive and quite homogeneous interstitial density between the metals and the hydrogens, in line with the concentration of BCPs found in this area when using AIM. It can also be observed that the dimer subunit is, in this case, significantly distorted, with curved Be-H-Be bond paths, whereas the distortion of the third BeH_2 subunit is very small, with an almost linear arrangement. This subunit also appears connected to the dimer through the hydrogen atoms of the former. The AdNDP description provides some interesting additional information, showing that the connectivity between 1Be (see numbering in Figure 3) and the dimer subunit takes place through 1Be-2H-4Be and 1Be-3H-7Be 3c-2e bonds and through 1Be-5H-7Be-8H and 1Be-4Be-5H-8H 4c-2e bonds (both kinds depicted in the second row of Figure 3). The ELF description is not strictly identical since all the basins are trisynaptic, though the ones involving 4Be-5(8)H-7Be are more compact (volume 97.5 au^3) than those involving 1Be-2H-4Be and 1Be-3H-7Be (volume 141.4 au^3). The MBIE decomposition analysis shows (see Table 2) that the two-center term between the Be atoms of the dimer subunit is smaller in absolute value than in the linear trimer (−202.1 vs. −262.0 kJ·mol^{-1}), but the interaction of these two Be atoms with the third Be atom is more than double in cycle **A** (−66.8 vs. −30.9 kJ·mol^{-1}), reflecting the formation of the aforementioned 3c-2e bonds. Still, the overall attractive components in **A** are 47 kJ·mol^{-1} above the linear ones. Nevertheless, this difference reduces to only 9 kJ·mol^{-1} in the stabilization energy due to the E_R deformation energies. Indeed, the E_R distortion values of the dimer subunit of cluster **A** are higher than in the same subunits of the linear complex (87 vs. kJ·mol^{-1}). In contrast, the E_R value for the 1Be is much smaller (8 kJ·mol^{-1}) than that of the central Be atom of the linear trimer (102 kJ·mol^{-1}). Accordingly, the overall destabilization energies in the linear isomer are 38 kJ·mol^{-1} greater than in the cyclic one, reducing the gap between their stabilization energies in this amount.

The two main conclusions are that (a) it is not enough to look at the strength of the binding interactions but also at the cost of distortion they entail, and (b) the presence of weak non-covalent interactions indicates that a good description of the geometries of these cyclic clusters requires to account for dispersion, which is not contemplated in the geometry optimization of the standard G4 formalism.

For the MgH_2 trimers, the scenario, as illustrated in Figure 4, is a little more complicated, with five (only four shown in the figure) low-energy conformers instead of two. Note that the linear structure is still the global minimum.

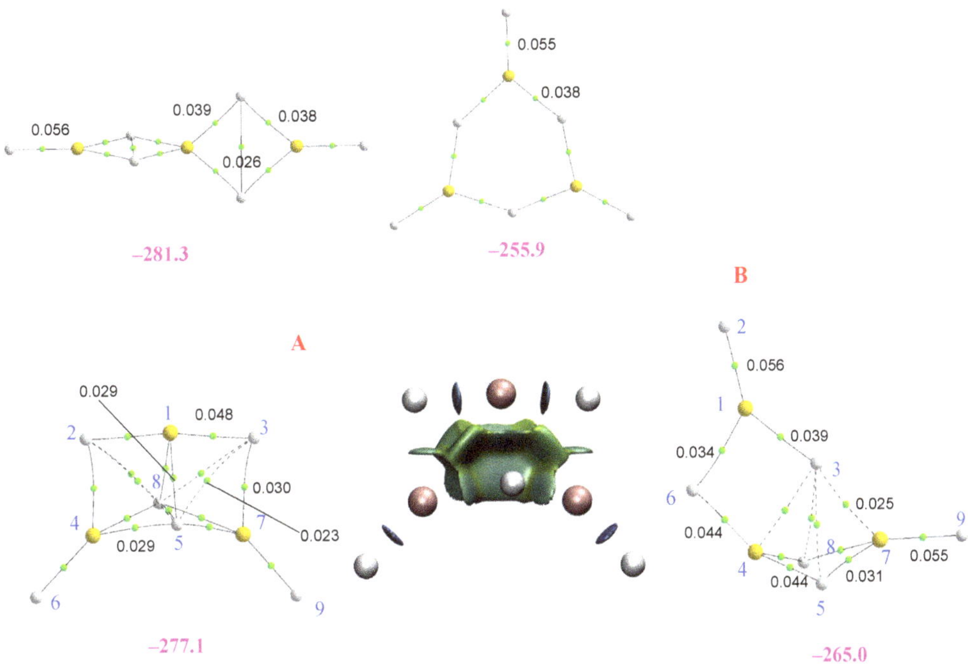

Figure 4. Molecular graphs of the homotrimers involving MgH_2. The electron densities at the bond critical points (BCPs) are in a.u. The numbers in magenta are the stabilization enthalpies in kJ·mol^{-1}. For cluster **A,** the 3D representation of non-covalent interactions obtained with NCIPLOT (s = 0.3); color code: red (strongly repulsive), green (weakly attractive and weakly repulsive), blue (strongly attractive), is included.

From the four cyclic local minima of the potential energy surface, the most stable one is again cycle **A**. This structure is followed in stability by another cyclic structure, **B**, that, as with the previous one, can be seen as the result of the interaction of a MgH_2 monomer with the MgH_2-MgH_2 dimer. Cycles **A** and **B** are distinguished by the *cis* or *trans* arrangement of hydrogens 6 and 9 in the dimer, with the result that in **B,** the 1Mg atom is only tricoordinated. At this point, it should be noted, as mentioned previously, that for the BeH_2 trimers, only **A** was found to be stable, as any attempt to find **B** led to the linear trimer. The third cyclic isomer in terms of stability is a planar hexagonal structure, though another non-planar conformer (not shown in Figure 4) was found to be also a local minimum, but 7 kJ·mol^{-1} less stable. The bonding analysis discussed above for the BeH_2 trimers (linear and **A)** can be extended to the MgH_2 ones, and for similar reasons, again, the linear tautomer is slightly more stable than the trimer **A**. Specifically, the linear $MgMgMgH_6$ trimer is 20 kJ·mol^{-1} less stable than its Be-containing analogue. For this structure, in line with what was discussed for the dimers, the Δ^2E attractive interactions and the E_R repulsive ones are smaller in $MgMgMgH_6$ than in the $BeBeBeH_6$ analog (see Table 2), resulting in a smaller stabilizing enthalpy in the former. A comparison of Figures 3 and 4 shows that the structure of cluster **A** for Mg is less compact than the homologous Be-containing isomer due to the longer interatomic distances (see Table S4 of the Supplementary Materials). Consistently, some of the non-covalent interactions are weaker. NCIPLOT shows that, due to its smaller compactness, the attractive homogeneous interstitial density between the Mg atoms and the hydrogens is less strongly attractive than in the Be analog. Going from cluster **A** to **B**, as expected, the overall attractive contributions (Δ^2E and Δ^3E) decrease by about 6 kJ·mol^{-1}, whereas the E_R terms increase by about the same amount, explaining why complex **B** is only 12 kJ·mol^{-1} less stable than conformer **A**. The lower stability of the hexagonal cycle just reflects the decrease in the Δ^2E attractive terms because no 3c-2e bonds are formed in this case, which is only partially compensated by an increase of the Δ^3E contribution due to the hexagonal arrangement of this complex.

The stability trends of the different conformers change dramatically when moving to the halides, though the linear trimer is still the global minimum. Let us discuss the fluorides shown in Figure 5 in more detail.

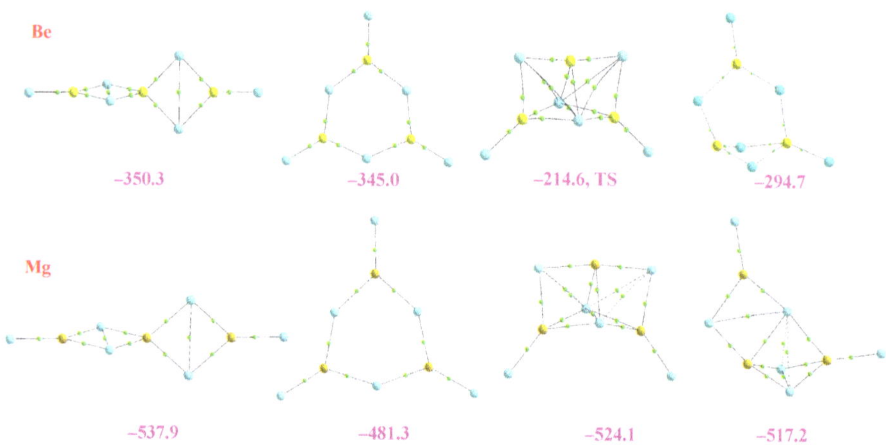

Figure 5. Molecular graphs of the homotrimers involving BeF_2 and MgF_2. The numbers in magenta are the stabilization enthalpies in kJ·mol^{-1}.

For the BeF_2 trimers, the linear arrangement is the global minimum, followed by the hexagonal cycle, only 5 kJ·mol^{-1} less stable, and by cycle **B,** which is 50 kJ·mol^{-1} further less stable. In this case, cycle **A** is a stationary point with one imaginary frequency. The same clusters are found when Be is replaced by Mg, but in this case, all of them are local minima of the potential energy surface. However, the energy trends are totally different, with cycle **A** being the second most stable after the linear trimer. As in the dimers, the dominant electrostatic character in fluorides renders the linear trimer more stable than the homologous containing BeH_2 (−350.3 vs. −301.3 kJ·mol^{-1}). Conversely, since this dominant electrostatic character prevents the formation of 3c-2e bonds, cycle **A** is found to be significantly less stable than its homologous hydride (−214.6 vs. −277.1 kJ·mol^{-1}). Cycle **A** is indeed much less stable than the linear trimer due to its compactness, which results in a close vicinity of F atoms reflected by the repulsive terms in the LMO-EDA results of Table S5. The consequence is that for the fluorides, **A** is a TS that leads to the linear cluster. The MBIE analysis (see Table 2) also shows that the hexagonal cycle is marginally less stable than the linear trimer. This is caused by the very small difference between the attractive ($\Delta^2E + \Delta^3E$) and the repulsive (E_R) terms, which is slightly greater (4 kJ·mol^{-1}) than in the former, so those species can be considered practically degenerate. The main difference when dealing with the Mg-containing systems is the larger size of Mg and the much longer bonds, which contribute to significantly stabilizing cycles **A** and **B**. These cycles are less compact, closing the gap with respect to the linear global minimum. It should be mentioned that although the MgF_2 trimers we found coincide with those in the literature [25], this is not the case as far as their stabilization energies are concerned since our G4 and M06-2X calculations both predict a different stability order (see Table S6 of the Supplementary Materials), likely due to the effect of dispersion contributions only included in our calculations.

The structures and stabilities of the homotrimer chlorides are presented in Figure S4 of the Supplementary Materials. As was the case for the hydrides and fluorides, the most stable trimer is the linear one. However, there are some differences with respect to fluorides in what concerns the relative stabilities of the other minima. For BeF_2 homotrimers, the hexagonal trimer is planar, whereas in the corresponding chloride, it is not. On top of that, the fluoride is 5 kJ·mol^{-1} less stable than the linear conformer, whereas for the chloride, this gap becomes ten times larger. Also, for Be trimers, cycle **A** is found to be a TS and the least stable stationary point, showing once more the significative effect of the repulsion in these compact systems when the substituent is voluminous as Cl. Conversely, for Mg, where these interactions are much weaker due to the much larger interatomic distances (see Table S4 of the Supplementary Materials), cycle **A** is not only a minimum but close in energy to the global minimum.

2.3. Heterotrimers

The conformational richness when dealing with heterotrimers is very high. Starting from the linear clusters, the number multiplies by three on going from homo to heterodimers. This can be easily understood if we remember that a linear trimer can be seen as the result of the interaction of a dimer with a monomer along the axis. Hence, if we take the homodimers X_2Be-BeX_2, its interaction can be exclusively with MgX_2, leading to a unique $BeX_2BeX_2MgX_2$ arrangement. However, if the interaction involves the heterodimer, X_2Be-MgX_2, the interaction can take place with any of the two monomers. The interaction with BeX_2 will yield the same trimer as before if the attachment takes place on the BeX_2 or to a new $BeX_2MgX_2BeX_2$ conformer if this attachment takes place on the MgX_2 side. If the monomer involved is MgX_2, two new clusters, $MgX_2BeX_2MgX_2$ and $BeX_2BeX_2MgX_2$, would be produced. All these possibilities, with their corresponding stabilization enthalpies, are shown in Figure 6. As expected from our discussion on the dimers, fluorides are significantly more stable than hydrides, whereas chlorides are only slightly more stable or unstable than hydrides, depending on the case. A second conspicuous fact is that this stability depends on the nature of the central atom and that the stabilities observed for the hydrides reverse on going to fluorides and chlorides. Indeed, for the hydrides, the most stable linear clusters of each kind are those in which the central atom is Be, whereas in fluorides and chlorides, the most stable are systematically those in which the central atom is Mg.

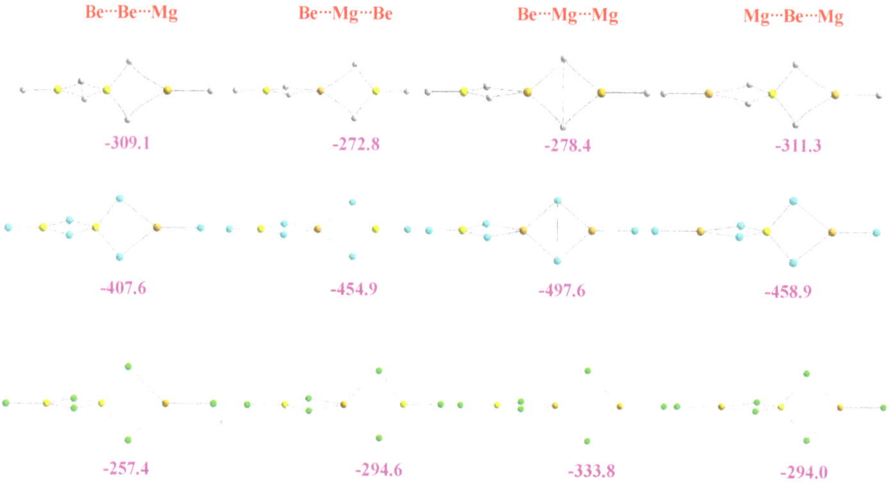

Figure 6. Molecular graphs for the linear heterotrimers, showing their stabilization enthalpies in kJ·mol^{-1}.

This finding is the result of mainly two features: the distortion energies of the central atom and the three-body contributions to the total energy. As it can be seen in Table S7 of the Supplementary Materials, for the hydrides, E_R is almost twice when the central atom is Be instead of Mg. This effect is more than compensated by the Δ^2E terms, more negative in the first case, but mainly by the Δ^3E term, more than three times larger in the former than in the latter thanks to stronger 3c-2e bonds in H_2Be-BeH_2 than in BeH_2-MgH_2. When moving to the halides, the Δ^3E contribution is marginal because, in these systems, no 3c-2e bonds are formed. It only remains, as a key factor, the much higher E_R value when the central atom is Be, leading to less stable clusters than those where the central atom is Mg.

Concerning the cyclic trimers, one option is the formation of hexagonal trimers, also found for the homotrimers. However, these structures were only found to be stable for hydrides and fluorides, but in both cases, more than 50 kJ·mol^{-1} is less stable than the corresponding linear trimers (see Figure S5 of the Supplementary Materials). More interesting are the cycles similar to the clusters **A** and **B** described in the homotrimers section. When dealing with heterotrimers, the different cycles that can be envisaged amount to five instead of two, as shown in Figure 7.

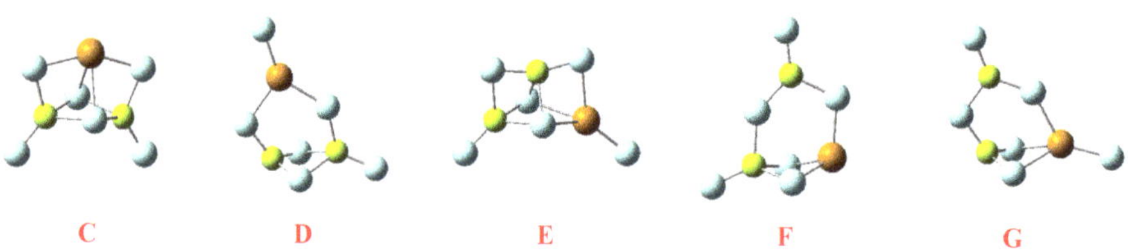

Figure 7. Different kinds of cyclic heterotrimers formed by the interaction of a dimer with a monomer that approaches the former in a direction perpendicular to the dimer axis.

Cycles **C** and **D** result from the interaction of a homodimer with a different monomer. They differ in the *cis* (**C**) or *trans* (**D**) arrangement of the terminal substituents of the homodimer. On the other hand, cycles **E**, **F**, and **G** arise from the interaction between a heterodimer and a certain monomer. Again, if the substituents of the dimer are *cis*, cycle **E** is formed. If the substituents are *trans*, there are two possibilities, cycles **F** and **G**. The important matter is that in some specific cases, these cycles become the global minimum of the potential energy surface. Let us start with the hydrides. The molecular graphs of these clusters are shown in Table 3, together with their stabilization enthalpies and their relative enthalpies with respect to the corresponding linear trimers in Figure 6.

Table 3. Molecular graphs of cycles **C**–**G** of $BeH_2BH_2MgH_2$ and $BeH_2MgH_2MgH_2$ and their stabilization enthalpies (bold numbers). In blue, their relative stabilities with respect to the corresponding linear trimer are also provided. In red, the two cases in which the cycle is the global minimum. All values in $kJ{\cdot}mol^{-1}$.

	BeBeMgH$_6$		**BeMgMgH$_6$**	
	Molecular graph	Stab. enth./ Rel. Stab.	Molecular graph	Stab. enth./ Rel. Stab.
Trimers	BeBeMg		BeMgMg	
C	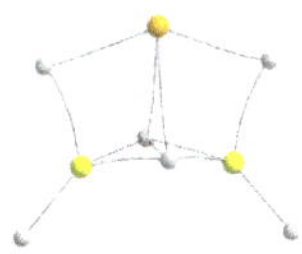	**−253.6** 55.5	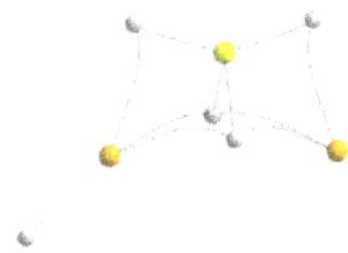	**−319.5** −8.2
D	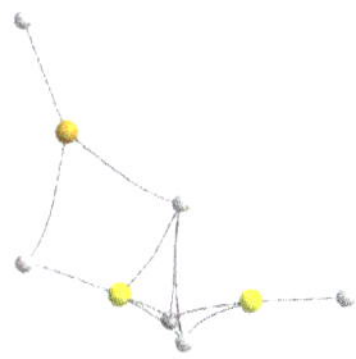	**−299.8** 9.3	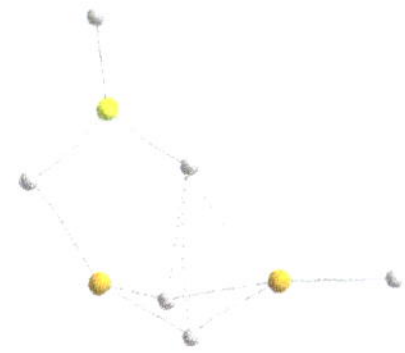	**−248.4** 62.9
E	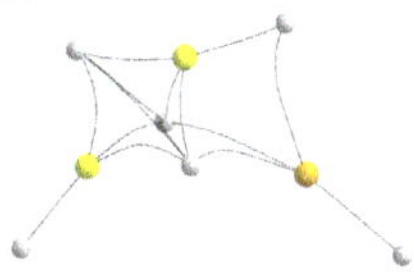	**−313.7** −4.6	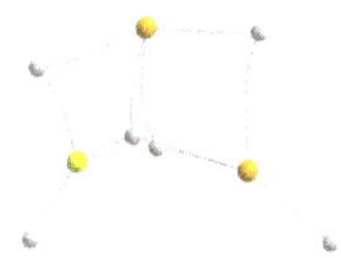	**−271.3** 40.0
F	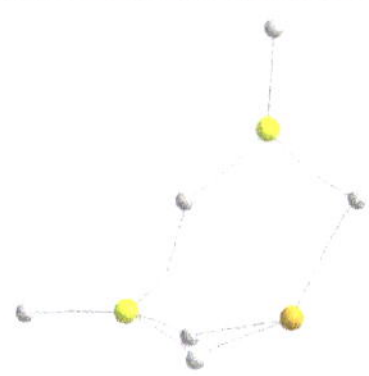	**−221.7** 87.4	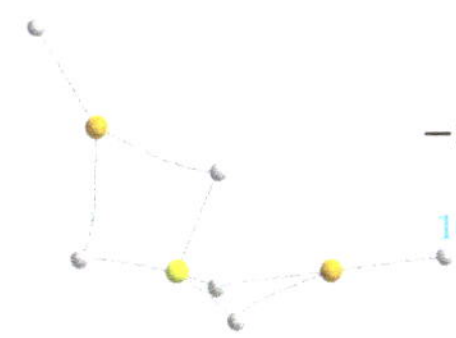	**−208.1** 103.2

The first conspicuous fact is that cycle **G** does not exist because it collapses to the linear global minimum. The second and most important is that some of them are rather stable, to the point that one within each family is predicted to be the global minimum of the potential energy surface (negative relative stabilities in red). This is a result of subtle differences, related again essentially to the distortion energy of the monomers and the Δ^3E terms. In cycle **E** (see Table S8 of the Supplementary Materials), the E_R contributions are smaller than in the linear trimer because the BeH_2 moiety at the top of the cycle is almost linear. As expected, the three Δ^2E in cycle **E** are negative, whereas in the linear trimer, only those of the central unit with the other two are negative. The overall balance is still favorable to the linear trimer, but this is compensated by a larger Δ^3E contribution, which renders cycle **E** the most stable. For the case of the $BeMgMgH_6$ heterotrimers, the situation is slightly different. Now, the E_R contributions are very similar in both cycle **C** and the linear trimer (see Table S8 of the Supplementary Materials) because of a larger distortion of the Be derivative at the top in cycle **C**. Since the Δ^2E contributions are globally similar, even though in the linear trimer, as expected, only two are significantly large, the enhanced stability of cycle **C** comes essentially from the Δ^3E term. Hence, once more, the two key factors are the E_R and the Δ^3E components, and when the former does not contribute significantly, it is only the Δ^3E term that is behind the stability differences.

When moving to fluorides and chlorides, the scenario changes completely with respect to the relative stability of cycles **C**–**G** (see Table S9 of the Supplementary Materials). Indeed, neither in the fluorides nor in the chlorides do the cyclic trimers compete in stability with the linear ones. This reflects the fact already mentioned in previous sections: the dominant electrostatic character of the interactions avoids the formation of 3c-2e bonds and, accordingly, since these cycles are rather compact structures, the repulsive interactions between the substituents increase significantly, and the Δ^3E contributions become highly unstabilizing (see Table S10 of the Supplementary Materials).

3. Computational Details

The first step of our analysis was a screening of the different dimers and trimers that can be possible minima of the corresponding potential energy surface. From them, we chose the most stable form for each of the different isomers, linear or cyclic. It should be mentioned that for the particular cases of BeH_2, MgH_2, MgF_2, and $MgCl_2$ trimers, our most stable structures are the same as those previously reported in the literature [10,24–26].

The structure and final energies of the clusters under investigation were obtained using both ab initio and density functional theory methods. In the first case, we have used the composite Gaussian-4 (G4) formalism [29]. In this method, the different ab initio calculations are carried out on geometries optimized with the B3LYP DFT approach together with a 6-31G(2df,p) basis set expansion. Final energies are calculated by combining different methods that properly account for the electron correlation effects, namely Møller–Plesset (MPn) perturbation theory up to the fourth order and CCSD(T) coupled-cluster theory. A final correction should be added to this: an estimation of the Hartree–Fock energy limit (HFlimit) together with two high-level empirical corrections. The result is that final energies are accurate up to a CCSD(T,full)/G3LargeXP + HF limit level, with an average absolute deviation [29] of 3.47 $kJ \cdot mol^{-1}$.

For our DFT calculations, we have chosen the B3LYP method, which is the one used in the G4 formalism for the geometry optimizations. To specifically check whether dispersion effects are significant for the systems under investigation, we have also added to the B3LYP method the D3BJ empirical dispersion term proposed by Grimme, including the Becke–Johnson damping correction [30]. The second DFT method chosen was the M06-2X, as this functional also provides much better quality descriptions of dispersion-dominated systems than standard functionals [31]. Moreover, M06-2X yields values that correlate very well with the MP2 ones, in particular when an extended basis set is used [32–35]. Very recently, we have found it to provide results in good agreement with G4 calculations when dealing with clusters involving electron-deficient systems [7]. For all these DFT calculations, a rather flexible aug-cc-pVTZ basis set has been used.

In particular, when dealing with trimers, a good understanding of the nature of bonds in this kind of cluster requires knowing the contribution of one-, two- and three-body terms, as well as the weight of the electrostatic, exchange, and dispersion contributions to the total binding energy.

For the first goal, we have used the many-body interaction energy (MBIE) formalism [36,37], which, for a ternary complex, allows the decomposition of the total binding energy ΔE (Equation (1)) into one- (Equation (2)), two- (Equation (3)), and three-body interactions (Equation (4)), as follows:

$$\Delta E = E(ABC) - \sum_{i=A}^{C} E_m(i) = \sum_{i=A}^{C} E_R(i) + \sum_{i=A}^{B}\sum_{j>i}^{C} \Delta^2E(i,j) + \Delta^3E(ABC) \quad (1)$$

$$E_R(i) = E(i) - E_m(i) \quad (2)$$

$$\Delta^2 E(ij) = E(ij) - [E(i) - E(j)] \quad (3)$$

$$\Delta^3 E(ABC) = E(ABC) - [E(A) + E(B) + E(C)] - [\Delta^2 E(AB) + \Delta^2 E(AC) + \Delta^2 E(BC)] \quad (4)$$

The value of $E_R(i)$, which measures the energy associated with the monomer distortion when it is part of the trimer, is the difference between $E_m(i)$, the energy of the i-monomer in its equilibrium geometry, and $E(i)$, the energy of the i-monomer within the geometry of the ABC complex. $\Delta^2 E(ij)$ and $\Delta^3 E(ABC)$ are the two- and three-body interaction energies computed at the corresponding geometries in the complex. For the second objective, we have employed the LMO-EDA [38] decomposition analysis based on the generalized Kohn–Sham (GKS) and localized molecular orbitals, which permits us to write the total interaction energy as the sum of electrostatic, exchange, repulsion, polarization, and dispersion contributions (Equation (5)).

$$E_{int} = E_{elec} + E_{exc} + E_{rep} + E_{pol} + E_{disp} \quad (5)$$

It should be mentioned that the distortion energy present in the MBIE analysis is not included in the LMO-EDA energy decomposition procedure. The LMO-EDA calculations were carried out by using the GAMESS code (version 2012-R1) [39].

To detect the new bonds that stabilize each cluster and to have reliable information on their nature and strength, we have carried out a topological analysis of the molecular electron density, ρ(r), by means of the Atoms in Molecules (AIM) approach [40], using the AIMAll (Version 19.10.12) code [41]. In this way, we could locate its first-order saddle points, the so-called bond critical points (BCPs), and obtain the corresponding molecular paths that provide information on the bonds stabilizing the system. In fact, since an inspection of the cluster structures in terms of interatomic distances and angles does not necessarily lead to knowing which bonds have been formed, the discussion of the cluster structures will be carried out using the corresponding molecular graphs. In all cases, the AIM analysis was followed by NBO (Natural Bond Orbitals) [42] calculations to complement the bonding information, in particular on what concerns the formation of (3c-2e) bonds. These calculations have been carried out with the NBO 5.G code [43]. In this respect, another useful technique to detect the existence of multicenter bonds is the AdNDP analysis [44], which was used for dimers and trimers. For those systems in which non-covalent interactions might be present, the information provided by the AIM method is nicely complemented by the NCIPLOT approach [45], which allows finding regions of low reduced density gradient (s) and low-density values typically associated with these non-covalent interactions.

Further information on the bonding of these clusters may be obtained by using the Electron Localization Function (ELF) formalism [46], which permits the location of the areas in which the electrons of the system are distributed in monosynaptic and disynaptic (or polysynaptic) basins. These regions are characterized by a low value of the excess local kinetic energy, thus identifying highly localized electrons. Monosynaptic basins are typically associated with core and electron lone pairs, whereas disynaptic (or polysynaptic) basins are associated with two-center (or more than two) bonding interactions.

4. Conclusions

Considering the unpaired electron available for bonding in H and halogens, one might expect that clustering between pure and mixed beryllium and magnesium clusters could follow relatively similar stabilities and bonding patterns for these substituents. However, a global picture of the dimers and trimers of this family offers a much more complex panorama. This work has dealt with the reasons behind the observed trends.

We have seen that magnesium homodimers are significantly more stable than beryllium homodimers and heterodimers and that the largest stabilization enthalpies are those of the fluorides, far from hydrides and chlorides, which are more alike. Thanks to LMO-EDA and NBO/AdNDP, we have shown that the electrostatic component is, as expected, much larger in halides than in hydrides, but chlorine has a lower contribution because of its larger size and cannot form multicenter two-electron bonds as hydrides do. The MBIE analysis helped to reveal the importance of the deformation (rehybridization) energy cost, which is much larger for beryllium than for magnesium.

Regarding the hydride homotrimers, we have understood why cyclic structures such as cyclic **A** can compete in stability with the linear structure. As the MBIE showed, in line with observations through topological tools, not only the strength of binding interactions must be taken into account but also the distortion involved to form the structures, together with the presence of non-covalent interactions in cycle **A** that have to be properly included in the theoretical treatment. Instead, beryllium fluoride clusters prefer the linear conformation, avoiding the compactness of cyclic structures, whereas magnesium, with larger bonds, presents cyclic structures **A** and **B** closer in energy to the linear one.

Finally, we have also paid attention to hydride heterotrimers, locating linear and several cyclic structures, where the many-body interactions allowed us to explain the preferences for central Be or Mg atoms in the linear structures. MBIE can also rationalize the subtle differences in the cyclic cases and explain why, for the $BeBeMgH_6$ and $BeMgMgH_6$ hydrides, the compact cycles, as **C** and **E**, become the global minima, respectively. However, halogen heterotrimers cannot compete in their cyclic forms with the linear ones in the absence of multicenter two-electron bonds and the predominance of three-body unstabilizing terms. As a final comment, we would like to point out that the study of the interactions between the $BeCl_2$ dimers with phosphorus-containing compounds reported in ref. [23] suggests that the clusters investigated here may exhibit a rather interesting reactivity, whose study could be a good benchmark to analyze it in terms of HOMO-LUMO interactions, as proposed in a very recent publication [47].

Supplementary Materials: The following supporting information can be downloaded at: https://www.mdpi.com/article/10.3390/molecules28227507/s1. Figure S1. Correlation between G4 and M06-2X interaction enthalpies. Figure S2. Correlation between the interaction enthalpies calculated at the G4 and B3LYP levels of theory. Table S1. LMO-EDA analysis for $BeBeX_4$, $MgMgX_4$ and $BeMgX_4$ (X = H, F, Cl) dimers. Figure S3. Molecular graphs of the homo and heterodimers involving BeX_2 and MgX_2 (X = H, F, Cl) monomers, showing the electron density, its Laplacian and the energy density at the bond critical points (BCPs). Table S2. AdNPD orbital list for the $BeBeX_4$, $MgMgX_4$ and $BeMgX_4$ dimers. Table S3. Wiberg bond indexes for Be_2X_4, Mg_2X_4, $BeMgX_4$ (X = H, F, Cl). Table S4. Interatomic distances in the cycles **A** of Be_2H_4 and Mg_2H_4. Table S5. LMO-EDA analysis for the $BeBeBeF_6$ trimers. Figure S4. Bond paths and stabilization enthalpies for the $BeCl_2$ and $MgCl_2$ homotrimers. Table S6. Relative stabilities for the MgF_2 trimers obtained by different theoretical approaches. Table S7. MBIE analysis of linear heterotrimer complexes formed by BeX_2 and MgX_2 (X = H, F, Cl). Figure S5. Bond paths for the stable hexagonal heterotrimers $BeCl_2$ and $MgCl_2$ homotrimers, showing their stabilization and relative enthalpies with respect to the corresponding linear heterotrimer in $kJ \cdot mol^{-1}$. Table S8. MBIE analysis of $BeBeMgH_6$ and $BeMgMgH_6$ heterotrimer complexes. Table S9. Molecular graphs of cycles **C-G** of $BeX_2BX_2MgX_2$ and $BeX_2MgX_2Mgx_2$ (X = F, Cl) and their stabilization enthalpies. Table S10. MBIE analysis of heterotrimer complexes for fluorides and chlorides.

Author Contributions: Conceptualization, I.A., M.M.M.-C., J.E., O.M. and M.Y.; methodology, I.A., M.M.M.-C. and M.Y.; software, I.A. and M.M.M.-C.; validation, I.A., M.M.M.-C. and M.Y.; formal analysis, I.A., M.M.M.-C., J.E., O.M. and M.Y.; investigation, I.A., M.M.M.-C., J.E., O.M. and M.Y.; resources, I.A. and M.Y.; writing—original draft preparation, M.Y.; writing—review and editing, I.A., M.M.M.-C., J.E., O.M. and M.Y.; visualization, I.A., M.M.M.-C., J.E., O.M. and M.Y.; supervision, I.A., M.M.M.-C., J.E., O.M. and M.Y.; project administration, I.A. and M.Y.; funding acquisition, I.A., M.Y. and O.M. All authors have read and agreed to the published version of the manuscript.

Funding: This research was funded by Projects PID2021-125207NB-C31, PID2021-125207NB-C32, and PID2019-110091GB-I00 from the Ministerio de Ciencia e Innovación (MICINN) of Spain.

Institutional Review Board Statement: Not applicable.

Informed Consent Statement: Not applicable.

Data Availability Statement: The authors confirm that the data supporting the findings of this study are available within the article and its Supplementary Materials.

Acknowledgments: We acknowledge the help provided by Al Mokhtar Lamsabhi to carry out the screening of possible conformers. Computational time at Centro de Computación Científica (CCC) of Universidad Autónoma de Madrid is also acknowledged.

Conflicts of Interest: The authors declare no conflict of interest.

References and Note

1. Lipscomb, W.N. Advances in Theoretical Studies of Boron Hydrides and Carboranes. In *Boron Hydride Chemistry*; Muetterties, E.L., Ed.; Academic Press: Itaca, NY, USA, 1975; pp. 39–78.
2. Mayer, I. Bond orders in 3-center bonds—An analytical investigation into the electronic-structure of diborane and the 3-center 4-electron bonds of hypervalent sulfur. *J. Mol. Struct. Theochem* **1989**, *55*, 43–52. [CrossRef]
3. Liang, C.X.; Davy, R.D.; Schaefer, H.F. Infrared-spectra of the unknown dialane (Al_2H_6) and recently observed digallane (Ga_2H_6) molecules. *Chem. Phys. Lett.* **1989**, *159*, 393–398. [CrossRef]
4. Lammertsma, K.; Leszczynski, J. Ab initio study on dialane(6) and digallane(6). *J. Phys. Chem.* **1990**, *94*, 2806–2809. [CrossRef]
5. Andrews, L.; Wang, X.F. The infrared spectrum of Al_2H_6 in solid hydrogen. *Science* **2003**, *299*, 2049–2052. [CrossRef]
6. Nori-Shargh, D.; Yahyaei, H.; Mousavi, S.N.; Maasoomi, A.; Kayi, H. Natural bond orbital, nuclear magnetic resonance analysis and hybrid-density functional theory study of sigma-aromaticity in Al_2F_6, Al_2Cl_6, Al_2Br_6 and Al_2I_6. *J. Mol. Model.* **2013**, *19*, 2549–2557. [CrossRef] [PubMed]
7. Mó, O.; Montero-Campillo, M.M.; Yáñez, M.; Alkorta, I.; Elguero, J. Dispersion, Rehybridization, and Pentacoordination: Keys to Understand Clustering of Boron and Aluminum Hydrides and Halides. *J. Phys. Chem. A* **2023**, *127*, 5860–5871. [CrossRef]
8. Ahlrichs, R. Ab initio Calculations on Small Hydrides Including Electron Correlation. *Theoret. Chim. Acta* **1970**, *17*, 348–361. [CrossRef]
9. Kirillov, Y.B.; Boldyrev, A.I.; Klimenko, N.M.; Charkin, O.P. An ab initio calculation of the structure and stability of the complex hydrides MgBeH- and Mg_2H-. *J. Struct. Chem.* **1983**, *24*, 134–136. [CrossRef]
10. Chen, Y.L.; Huang, C.H.; Hu, W.P. Theoretical study on the small clusters of LiH, NaH, BeH_2, and MgH_2. *J. Phys. Chem. A* **2005**, *109*, 9627–9636. [CrossRef] [PubMed]
11. Lingam, C.B.; Babu, K.R.; Tewari, S.P.; Vaitheeswaran, G. Quantum chemical studies on beryllium hydride oligomers. *Comput. Theor. Chem.* **2011**, *963*, 371–377. [CrossRef]
12. Alkorta, I.; Elguero, J.; Solimannejad, M.; Grabowski, S.J. Dihydrogen Bonding vs. Metal-σ Interaction in Complexes between H_2 and Metal Hydride. *J. Phys. Chem. A* **2011**, *115*, 201–210. [CrossRef]
13. Erfort, S.; Tschope, M.; Rauhut, G. Efficient and automated quantum chemical calculation of rovibrational nonresonant Raman spectra. *J. Chem. Phys.* **2022**, *156*, 124102. [CrossRef]
14. Erfort, S.; Tschoepe, M.; Rauhut, G. Toward a fully automated calculation of rovibrational infrared intensities for semi-rigid polyatomic molecules. *J. Chem. Phys.* **2020**, *152*, 244104. [CrossRef] [PubMed]
15. Wang, X.F.; Andrews, L. Infrared spectra of magnesium hydride molecules, complexes, and solid magnesium dihydride. *J. Phys. Chem. A* **2004**, *108*, 11511–11520. [CrossRef]
16. Abdurahman, A. Ab initio studies of static dipole polarizability of the polymeric beryllium hydride chain. *J. Phys. Chem. A* **2003**, *107*, 11547–11552. [CrossRef]
17. Wang, X.F.; Andrews, L. One-dimensional BeH_2 polymers: Infrared spectra and theoretical calculations. *Inorg. Chem.* **2005**, *44*, 610–614. [CrossRef]
18. Ramondo, F.; Bencivenni, L.; Spoliti, M. Ab initio study on the Be_2F_4, Mg_2F_4 dimers, on the mixed dimers $BeMgF_4$ and $LiNaF_2$ and on the Li_2BeF_4, $LiBCl_4$ and $LiAlCl_4$ ion-pairs. *J. Mol. Struct. Theochem* **1992**, *96*, 171–184. [CrossRef]
19. Gigli, G. On the structure of the alkaline earth dihalides dimers. *J. Chem. Phys.* **1990**, *93*, 5224–5233. [CrossRef]
20. Ystenes, B.K. Quantum chemical studies of molecular difluorides and dichlorides of beryllium and magnesium. *Spectrochim. Acta A Mol. Biomol. Spectrosc.* **1998**, *54*, 855–868. [CrossRef]
21. Molnar, J.; Marsden, C.J.; Hargittai, M. Molecular-structures and force-fields of monomeric and dimeric magnesium dichloride from electron-diffraction and quantum-chemical calculations. *J. Phys. Chem.* **1995**, *99*, 9062–9071. [CrossRef]
22. Girichev, A.G.; Giricheva, N.I.; Vogt, N.; Girichev, G.V.; Vogt, J. Structural investigation of molecules in the vapour over beryllium dichloride using electron diffraction and mass spectrometric data. *J. Mol. Struct.* **1996**, *384*, 175–182. [CrossRef]
23. Buchner, M.R.; Spang, N.; Ivlev, S.I. Hydrolysis and oxidation products of phosphine adducts to beryllium chloride. *Z. Naturforsch. B J. Chem. Sci.* **2022**, *77*, 381–390. [CrossRef]
24. Eichkorn, K.; Schneider, U.; Ahlrichs, R. An ab-initio investigation of structure and energetics of clusters Mg_nCl_{2n}. *J. Chem. Phys.* **1995**, *102*, 7557–7563. [CrossRef]
25. Neogi, S.G.; Chaudhury, P. Structure, electronic properties and vibrational spectra of $(MgF_2)n$ clusters through a combination of genetic algorithm and DFT-based approach. *Mol. Phys.* **2015**, *113*, 3729–3739. [CrossRef]

26. Luhtanen, T.N.P.; Linnolahti, M.; Laine, A.; Pakkanen, T.A. Structural characteristics of small magnesium dichloride clusters: A systematic theoretical study. *J. Phys. Chem. B* **2004**, *108*, 3989–3995. [CrossRef]
27. *Values calculated at the M06-2X/aug-cc-pVTZ level of theory* (this work).
28. Wiberg, K.B. Application of pople-santry-segal cndo method to cyclopropylcarbinyl and cyclobutyl cation and to bicyclobutane. *Tetrahedron* **1968**, *24*, 1083–1088. [CrossRef]
29. Curtiss, L.A.; Redfern, P.C.; Raghavachari, K. Gaussian-4 theory. *J. Chem. Phys.* **2007**, *126*, 084108. [CrossRef]
30. Grimme, S.; Ehrlich, S.; Goerigk, L. Effect of the Damping Function in Dispersion Corrected Density Functional Theory. *J. Comp. Chem.* **2011**, *32*, 1456–1465. [CrossRef]
31. Grimme, S.; Hansen, A.; Brandenburg, J.G.; Bannwarth, C. Dispersion-Corrected Mean-Field Electronic Structure Methods. *Chem. Rev.* **2016**, *116*, 5105–5154. [CrossRef]
32. Walker, M.; Harvey, A.J.A.; Sen, A.; Dessent, C.E.H. Performance of M06, M06-2X, and M06-HF Density Functionals for Conformationally Flexible Anionic Clusters: M06 Functionals Perform Better than B3LYP for a Model System with Dispersion and Ionic Hydrogen-Bonding Interactions. *J. Phys. Chem. A* **2013**, *117*, 12590–12600. [CrossRef]
33. Castro-Alvarez, A.; Cameros, H.; Sanchez, D.; Vilarrasa, J. Importance of the Electron Correlation and Dispersion Corrections in Calculations Involving Enamines, Hemiaminals, and Aminals. Comparison of B3LYP, M06-2X, MP2, and CCSD Results with Experimental Data. *J. Org. Chem.* **2015**, *80*, 11977–11985. [CrossRef] [PubMed]
34. Lopez-Lopez, J.A.; Ayala, R. Assessment of the performance of commonly used DFT functionals vs. MP2 in the study of IL-Water, IL-Ethanol and IL-$(H_2O)(3)$ clusters. *J. Mol. Liq.* **2016**, *220*, 970–982. [CrossRef]
35. Stortz, C.A.; Sarotti, A.M. Exhaustive exploration of the conformational landscape of mono- and disubstituted five-membered rings by DFT and MP2 calculations. *RSC Adv.* **2019**, *9*, 24134–24145. [CrossRef]
36. Hankins, D.; Moskowitz, J.W.; Stillinger, F.H. Water molecule interactions. *J. Chem. Phys.* **1970**, *53*, 4544–4554. [CrossRef]
37. Xantheas, S.S. Ab-initio studies of cyclic water clusters $(H_2O)_n$, n=1-6. 2. Analysis of many-body interactions. *J. Chem. Phys.* **1994**, *100*, 7523–7534. [CrossRef]
38. Su, P.F.; Jiang, Z.; Chen, Z.C.; Wu, W. Energy Decomposition Scheme Based on the Generalized Kohn-Sham Scheme. *J. Phys. Chem. A* **2014**, *118*, 2531–2542. [CrossRef]
39. Schmidt, M.W.; Baldridge, K.K.; Boatz, J.A.; Elbert, S.T.; Gordon, M.S.; Jensen, J.H.; Koseki, S.; Matsunaga, N.; Nguyen, K.A.; Su, S.J.; et al. General atomic and molecular electronic-structure system. *J. Comp. Chem.* **1993**, *14*, 1347–1363. [CrossRef]
40. Bader, R.F.W. *Atoms in Molecules. A Quantum Theory*; Clarendon Press: Oxford, UK, 1990; pp. 1–456.
41. Keith, T.A. *AIMAll*, version 19.10.12; TK Gristmill Software: Overland Parks, KS, USA, 2019. Available online: aim.tkgristmill.com (accessed on 1 January 2023).
42. Reed, A.E.; Weinhold, F. Natural localized molecular-orbitals. *J. Chem. Phys.* **1985**, *83*, 1736–1740. [CrossRef]
43. Glendening, E.D.; Badenhoop, J.K.; Reed, A.E.; Carpenter, J.E.; Bohmann, J.A.; Morales, C.M.; Weinhold, F. *NBO 5.G*; Theoretical Chemistry Institute, University of Wisconsin: Madison, WI, USA, 2004.
44. Tkachenko, N.V.; Boldyrev, A.I. Chemical bonding analysis of excited states using the adaptive natural density partitioning method. *Phys. Chem. Chem. Phys.* **2019**, *21*, 9590–9596. [CrossRef] [PubMed]
45. Boto, R.A.; Peccati, F.; Laplaza, R.; Quan, C.; Carbone, A.; Piquemal, J.P.; Maday, Y.; Contreras-García, J. NCIPLOT4: Fast, robust and quantitative analysis of non-covalent interactions. *J. Chem. Theory Comp.* **2020**, *16*, 4150–4158. [CrossRef] [PubMed]
46. Savin, A.; Nesper, R.; Wengert, S.; Fassler, T.F. ELF: The electron localization function. *Angew. Chem. Int. Edit.* **1997**, *36*, 1809–1832. [CrossRef]
47. Santos, L.D.; Ramalho, T.C.; Hamlin, T.A.; Bickelhaupt, F.M. Intermolecular Covalent Interactions: Nature and Directionality. *Chem. Eur. J.* **2023**, *29*, e202203791. [CrossRef] [PubMed]

molecules

MDPI

Article

Computational Modeling Study of the Binding of Aging and Non-Aging Inhibitors with Neuropathy Target Esterase

Wenxiong Wu and Pan Wang *

Shenzhen Key Laboratory of Steroid Drug Discovery and Development, School of Medicine, The Chinese University of Hong Kong, Shenzhen 518172, China
* Correspondence: wangpan@cuhk.edu.cn

Abstract: Neuropathy target esterase (NTE) is a serine hydrolase with phospholipase B activity, which is involved in maintaining the homeostasis of phospholipids. It can be inhibited by aging inhibitors such as some organophosphorus (OP) compounds, which leads to delayed neurotoxicity with distal degeneration of axons. However, the detailed binding conformation of aging and non-aging inhibitors with NTE is not known. In this study, new computational models were constructed by using MODELLER 10.3 and AlphaFold2 to further investigate the inhibition mechanism of aging and non-aging compounds using molecular docking. The results show that the non-aging compounds bind the hydrophobic pocket much deeper than aging compounds and form the hydrophobic interaction with Phe1066. Therefore, the unique binding conformation of non-aging compounds may prevent the aging reaction. These important differences of the binding conformations of aging and non-aging inhibitors with NTE may help explain their different inhibition mechanism and the protection of non-aging NTE inhibitors against delayed neuropathy.

Keywords: computational modeling; neuropathy target esterase; molecular docking

1. Introduction

Neuropathy target esterase (NTE), also called patatin-like phospholipase domain-containing protein 6 (PNPLA6), is the sixth member of the protein family of patatin domain lipid hydrolase [1]. NTE is a serine hydrolase with phospholipase B activity [2], which has both phospholipase A1 and A2 activities [3] and could deacetylate both acyl bonds in phosphatidylcholine (PtdCho) [4–6]. NTE is important in maintaining the stability of membrane phospholipids [6–9]. NTE plays important roles in vasculogenesis [10] and axon maintenance [11].

In addition, NTE is involved in chronic neuropathies induced by organophosphorus compounds (OP) and other NTE inhibitors [12]. OP compounds are used widely in the chemical, agriculture, or pharmaceutical industries [6,13]. Aging inhibitors could cause delayed neurotoxicity, characterized by a distal degeneration of axons in the central nervous system (CNS) and peripheral nervous system (PNS) [6,12].

There are two types of NTE inhibitors. One is aging inhibitors, which include certain phosphates, phosphonates, and phosphoramidates [12,14–17]. The aging of NTE is the loss of reactivity of NTE [12,18–20]. The other type of NTE inhibitors is non-aging inhibitors, which usually include phosphinates, sulfonates, and carbamates. Non-aging inhibitors can inhibit NTE activity but would not cause aging on NTE due to the reactivation of NTE. The aged NTE could not be reactivated [12]. The non-aging inhibitors will not produce delayed neuropathy and can protect the animals from neurotoxicity induced by aging inhibitors [6,12]. For example, non-aging inhibitor phenylmethylsulfonyl fluoride (PMSF) can protect against the delayed neuropathy induced by tri-ortho-cresyl phosphate (TOCP) in hens [21] and protect against diisopropylfluorophosphate (DFP)-induced delayed neuropathy in cats [22,23] and hens [24].

A previous study reported the construction of a homology model of NTE esterase domain (NEST) using the template protein patatin isoform 17 (Pat17) to study the catalytic mechanism of NTE [25]. However, the identity between the sequences of NTE and Pat17 is rather low (18%) [25]. Moreover, little was known about the different binding of aging/non-aging inhibitors with NEST.

This study aims to construct a more precise model of NEST using templates with higher identity by using both AlphaFold2 and MODELLER 10.3. In addition, the possible mechanism of inhibition by various aging and non-aging inhibitors was investigated by using molecular docking. We revealed important differences of the binding conformations of aging and non-aging inhibitors with NTE, which may explain their different inhibition mechanism and the protection of non-aging NTE inhibitors against delayed neuropathy.

2. Results

2.1. Construction and Validation of the NEST Models

Homology modeling was used to construct the 3D models of the NTE catalytic domain NEST with homologous template protein. A BLAST search demonstrated that the native PlpD (PDB ID: 5FYA) shares 30.46% identity with the NEST sequence and the E value is 2×10^{-21} (Figure 1A). This sequence identity is much higher than that of the protein template used in the previous study, which used Pat17 with only 18% sequence identity with the NEST as the template protein to build NEST models [25]. The 3D structure analysis of NEST built by MODELLER10.3 indicates that the overall structure of NEST contains seven α-helices (named A1–A7) and five β-strands (named B1–B5) (Figure 1C). In addition, the model of NEST by AlphaFold2 contains eight α-helices (named A1–A8) and six β-strands (named B1–B6) with high model confidence (the predicted local-distance difference test (pLDDT) > 90) (Figure 1B).

The overlay between the template PlpD (PDB ID: 5FYA) and the NEST homology model built with MODELLER 10.3 indicates that they have high similarity with RMSD 0.141 Å (Figure 1E). The alpha-carbon RMSD value between the homology model built by MODELLER 10.3 and AlphaFold2 is 1.30 Å, which indicates that these two models are very similar to each other (Figure 1D,E).

The NEST models were evaluated by SAVES v6.0. The Ramachandran results for the models built with MODELLER 10.3 and AlphaFold2 are shown in Figure 2A,B, respectively, which shows that 98.6% of the amino acid residues are in the reasonable zone (86.4% of the residues in the "most favored" regions, 10.0% in the "additional allowed" regions, and 2.1% in the "generously allowed" regions). For the model built by the AlphaFold2, 100% of the amino acid residues are in the reasonable zone (91.4% of the residues in the "most favored" regions, 7.9% in the "additional allowed" regions, and 0.7% in the "generously allowed" regions). Therefore, both models were validated as acceptable models, and the model built by AlphaFold2 is relatively better and more reasonable compared to the model of MODELLER 10.3.

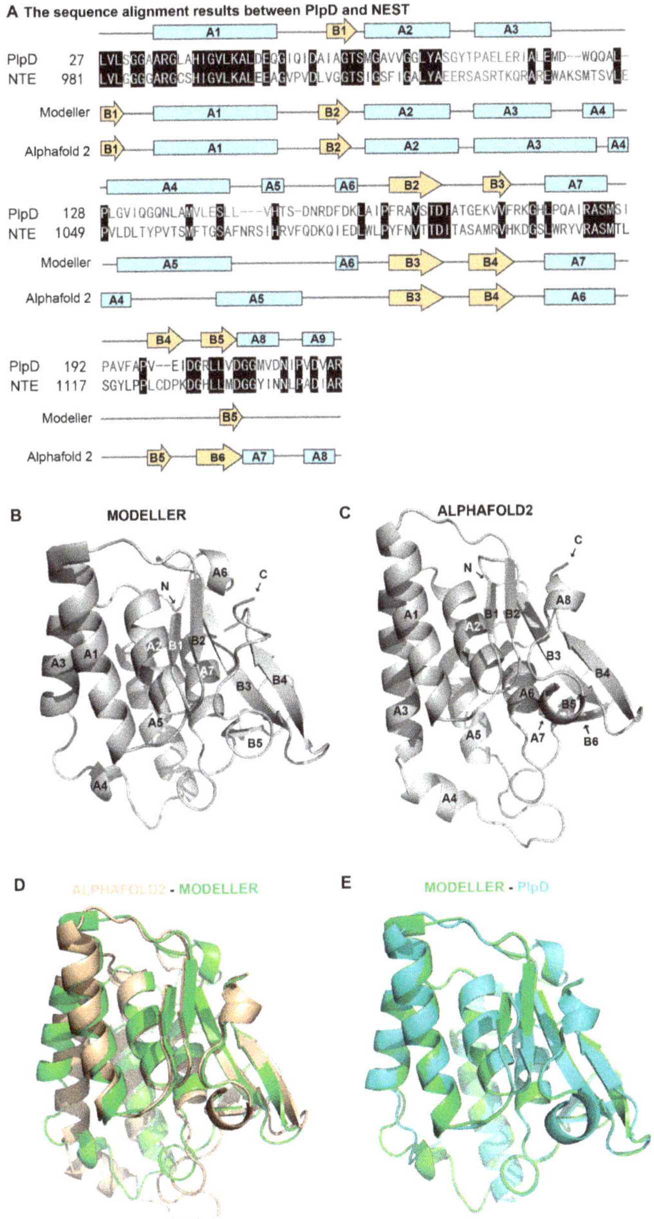

Figure 1. The constructed NEST models. (**A**) The sequence alignment results between PlpD and NEST. Identical residues have been boxed in the black color. Secondary structures of PlpD (shown on the top of the sequence) and two NEST models built by MODELLER 10.3 and the AlphaFold2 (shown below the sequence) are displayed. β-strands are shown as arrows colored in cream and α-helices are shown as rectangles colored in blue. (**B**,**C**) The overall structure models of NEST built by MODELLER 10.3 (**B**) and AlphaFold2 (**C**). (**D**,**E**) The overlay of AlphaFold model and MODELLER model (**D**) and the overlay of the MODELLER model with the PlpD structure (**E**). The PlpD structure, the MODELLER model, and the AlphaFold model are colored blue, green, and cream, respectively.

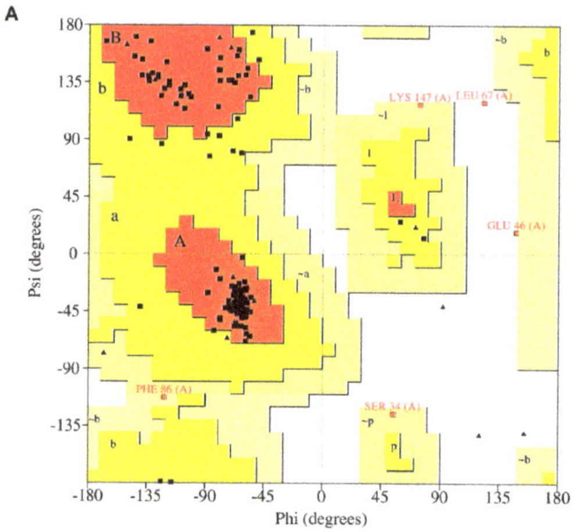

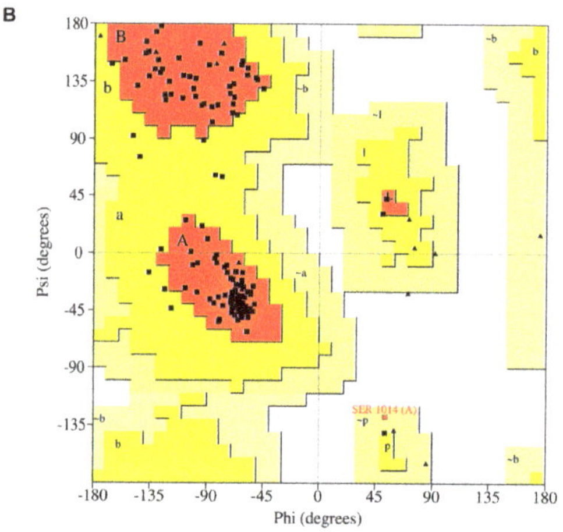

Figure 2. The assessment of the models of NEST. (**A**) Ramachandran Plot and Procheck result of the homology model built by MODELLER 10.3, in which 98.6% of the amino acid residues are in the reasonable zone. (**B**) Ramachandran Plot and Procheck result of the model built by the AlphaFold2, in which 100% of the amino acid residues are in the reasonable zone. In this Ramachandran Plot, the "most favored" regions are colored in red, "additional allowed" regions are colored in deep yellow, and the "generously allowed" regions are colored in light yellow while "disallowed" regions are colored in white. One black point represents one residue. Triangles indicate the glycine residues while rectangles indicate other residues. "A", "a" and "~a" indicate the "most favored", "additional allowed" and "generously allowed" regions of right-handed alpha helices, respectively. "B", "b" and "~b" indicate the "most favored", "additional allowed" and "generously allowed" regions of beta sheets, respectively. "L", "l" and "~l" indicate the "most favored", "additional allowed" and "generously allowed" regions of left-handed alpha helices, respectively. "p" and "~p" indicate the regions of "additional allowed" and "generously allowed" epsilon.

2.2. *The NEST Active Site Structures*

In the NEST model built with MODELLER 10.3 (Figure 3A), the overlay of the active sites of the two models including Ser1014, Asp1134, and oxyanion hole residues show that these key residues have very similar conformations (Figure 3C,D). Ser1014 and Asp1134 are located in the two conserved loops B2–A2 and B5–C terminal, respectively, and Ser1014 Oγ is located 3.9 Å from the Asp1134 Oδ_2. In the model built with AlphaFold2 (Figure 3B),

Ser1014 and Asp1134 are located in the conserved loops B2–A2 and B3–B4, and Ser1014 Oγ is located 2.7 Å from the Asp1134. Early studies have shown that Asp1008 is key to the activity of phenyl-valerate hydrolase in NEST with site-directed mutagenesis [26]. However, in our model, the distance between Asp1008 Oδ_2 and Ser1014 Oγ is 23.2 Å in the model built with MODELLER 10.3 and 25.4 Å in the model built with AlphaFold2. Therefore, this distance is too far for Asp1008 to play a direct role in catalysis.

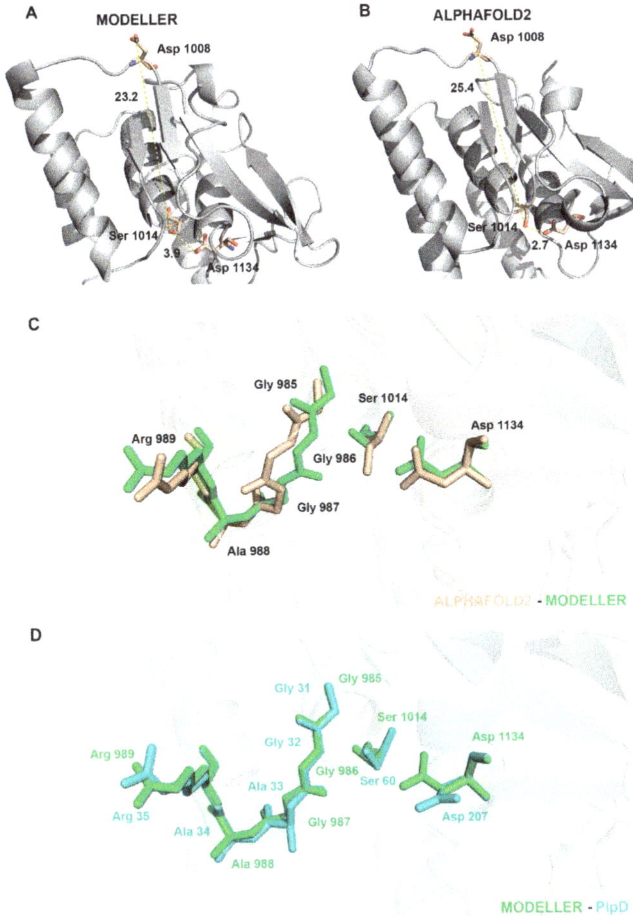

Figure 3. The overlay of the overall structures of NEST models and important active residues. (**A**,**B**) The enlarged view of the predicted active site of NEST. Important residues, such as Ser1014, Asp1008, and Asp1134 are shown as sticks. Distances between Ser1014 Oγ and Asp1008 O$_{\delta 2}$ and Ser966 Oγ and Asp1134 O$_{\delta 2}$ are shown as dotted lines. (**C**,**D**) The overlay of important residues (including Ser1014, Asp1134, and oxyanion hole residues between AlphaFold model and MODELLER model (**C**) and between the MODELLER model and the PlpD (**D**). The PlpD structure, the MODELLER model, and the AlphaFold model are colored blue, green, and cream, respectively.

Hydrophobicity analysis of the NEST ligand binding pocket built with both MODELLER 10.3 and AlphaFold2 shows that there are two vital areas in the NEST ligand binding site (Figure 4). One is the hydrophilic pocket consisting of active sites Ser1014-Asp1134 and oxyanion hole mainly consisting of Gly986-Gly987-Ala988-Arg989 (Figure 4C,F). The other is the hydrophobic pocket (Figure 4C,F) made up of several nonpolar amino acids and aromatic amino acids, for example Leu1116, Met1114, and Phe1066.

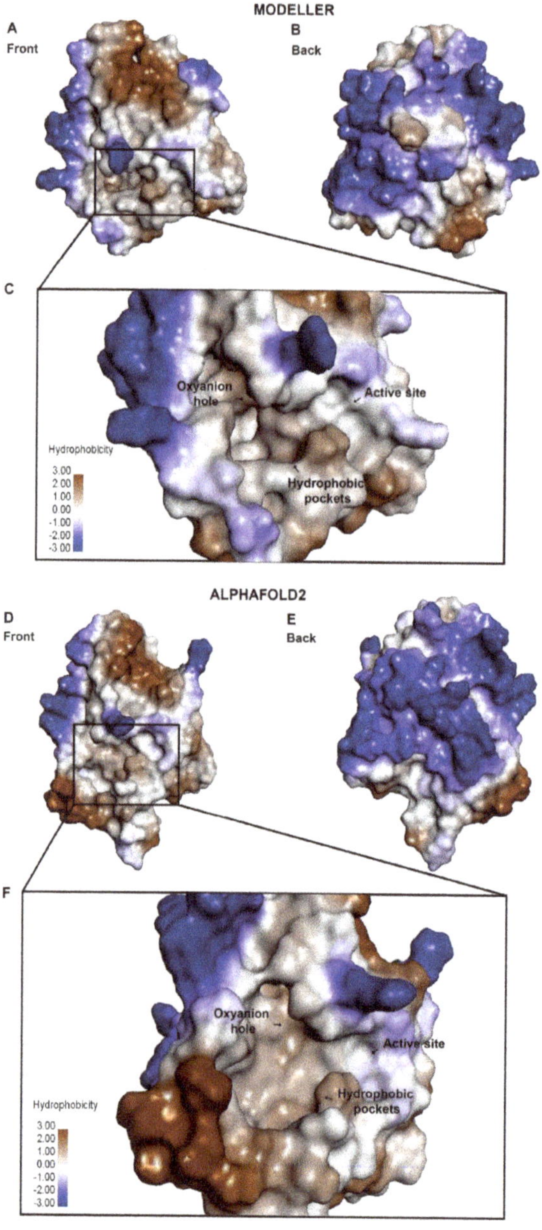

Figure 4. Hydrophobicity analysis of the NEST models built by MODELLER 10.3 and AlphaFold2. Hydrophobic regions are shown in brown and hydrophilic regions in blue. (**A**) The front view of the NEST homology model by MODELLER 10.3. (**B**) The back view of the NEST homology model by MODELLER 10.3. (**C**) The zoom-in view of the active site by MODELLER 10.3. (**D**) The front view of NEST model by AlphaFold2. (**E**) The back view of the NEST model by AlphaFold2. (**F**) The zoom-in view of the active sites by AlphaFold2.

The surface of NEST can be divided into two areas, i.e., the hydrophilic area and the hydrophobic area (Figures 4 and 5). In Figure 5, the red, blue, and white colors indicate that the area is negatively charged, positively charged, and neutral, respectively. It is possible that the hydrophilic area of NEST is exposed to the cytosol while the opposite side is located in the ER membrane. The ligand binding pocket is mostly located in the hydrophobic area surrounded by the hydrophilic area (Figures 4 and 5).

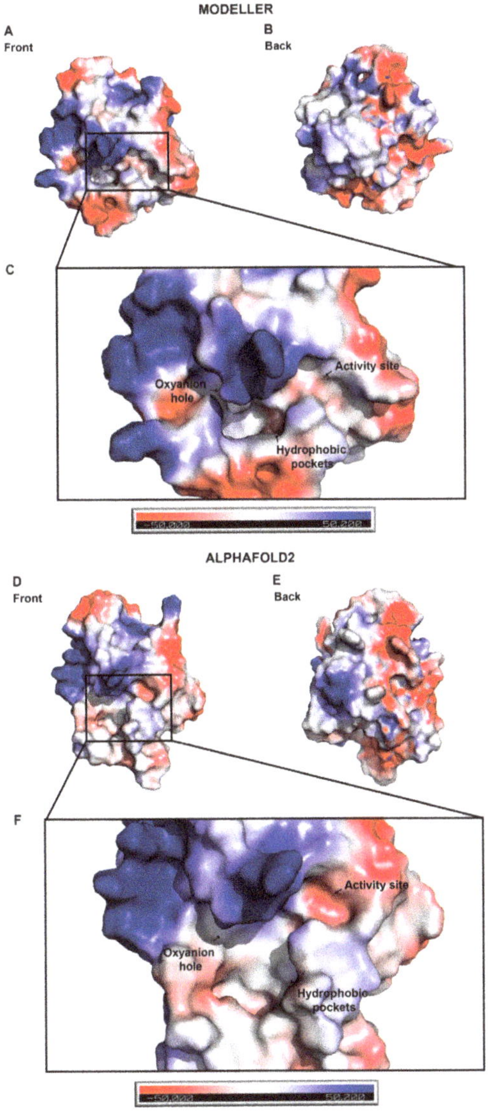

Figure 5. Electrostatic surface potential of the NEST models built by MODELLER 10.3 and AlphaFold2. Acidic regions are shown in red and basic regions in blue. (**A**) The front view of NEST homology model by MODELLER 10.3. (**B**) The back view of NEST homology model by MODELLER 10.3. (**C**) The zoom-in view of the active site by MODELLER 10.3. (**D**) The front view of NEST model by AlphaFold2. (**E**) The back view of NEST model by AlphaFold2. (**F**) The zoom-in view of the active site by AlphaFold2.

2.3. *Binding of Aging and Non-Aging Inhibitors with NEST*

Then, we studied the binding of five inhibitors and one artificial substrate phenyl valerate (PV) with NEST. The five inhibitors include three aging inhibitors, i.e., Mipafox, DFP, and 2-(ortho-cresyl)-4H-1,2,3-benzodioxaphosphoran-2-one (CBDP), as well as two non-aging inhibitors, i.e., phosphinic acid, PMSF. These ligands were selected and classified according to previous studies [6,12,27]. The CDOCKER interaction energy values of these six ligands with NEST are shown in Table 1. Two aging inhibitors CBDP and mipafox have similar binding energy as the substrate PV with NEST, indicating that their binding affinity with NEST is similar to the substrate PV. Non-aging inhibitors PMSF, phosphinic acid, and aging inhibitor DFP have relatively lower binding affinity with NEST compared to the substrate PV as indicated by their higher CDOCKER energy values.

Table 1. Predicted CDOCKER energy of 5 NTE inhibitors and 1 substrate with NEST AlphaFold model by molecular docking.

Ligand	Ligand Type	CDOCKER Energy (kcal/mol)
CBDP	Aging inhibitors	−30.8
PV	Substrate	−29.5
Mipafox	Aging inhibitors	−29.2
Phosphinic acid	Non-aging inhibitors	−25.6
DFP	Aging inhibitors	−21.7
PMSF	Non-aging inhibitors	−19.9

The binding interaction of the inhibitors and NEST predicted by molecular docking is shown in Figure 6. Most hydrogen bonds are formed with the oxyanion hole and Ser1014-Asp1134. Residue Ser1014 forms hydrogen bonds with all the ligands. Compared to Ser1014, Asp1134 is less involved in hydrogen bond formation with ligands in both aging and non-aging inhibitors (Table 2).

Table 2. Interactions of NTE inhibitors with important amino acid residues of NEST. "+" and "−" indicate whether there is a direct interaction (hydrophobic interaction or hydrogen bond) between NEST and ligands.

Regions in NEST	Residues	Aging NTE Inhibitors			Non-Aging NTE Inhibitors		Substrate
		Mipafox	DPF	CBDP	PMSF	Phosphinic Acid	PV
Catalytic dyad	Ser1014	+	+	+	+	+	+
	Asp1134	+	−	+	−	+	−
Oxyanion holes	Gly986	+	−	+	−	+	+
	Gly987	+	+	+	+	+	+
	Arg989	−	−	−	−	+	−
Hydrophobic pockets	Phe1066	−	−	−	+	+	+
	Met1114	−	+	+	+	+	−
	Leu1116	−	+	+	−	+	−
Others	Ile1015	−	+	+	+	−	+
	Gly985	−	−	−	−	+	+
	Trp1039	−	−	−	−	+	−

Moreover, our results show that the key oxyanion hole residues Gly986 and Gly987 are involved in the formation of hydrogen bonds with inhibitors and an artificial substrate (PV). It indicates that those two important residues may stabilize and fix the phosphate group of the ligands at the position to stabilize the negative charge of the phosphate group (Table 2 and Figure 6). Gly987 forms hydrogen bonds with all the ligands, which indicates that Gly987 is essential to all ligands which can bind to NEST.

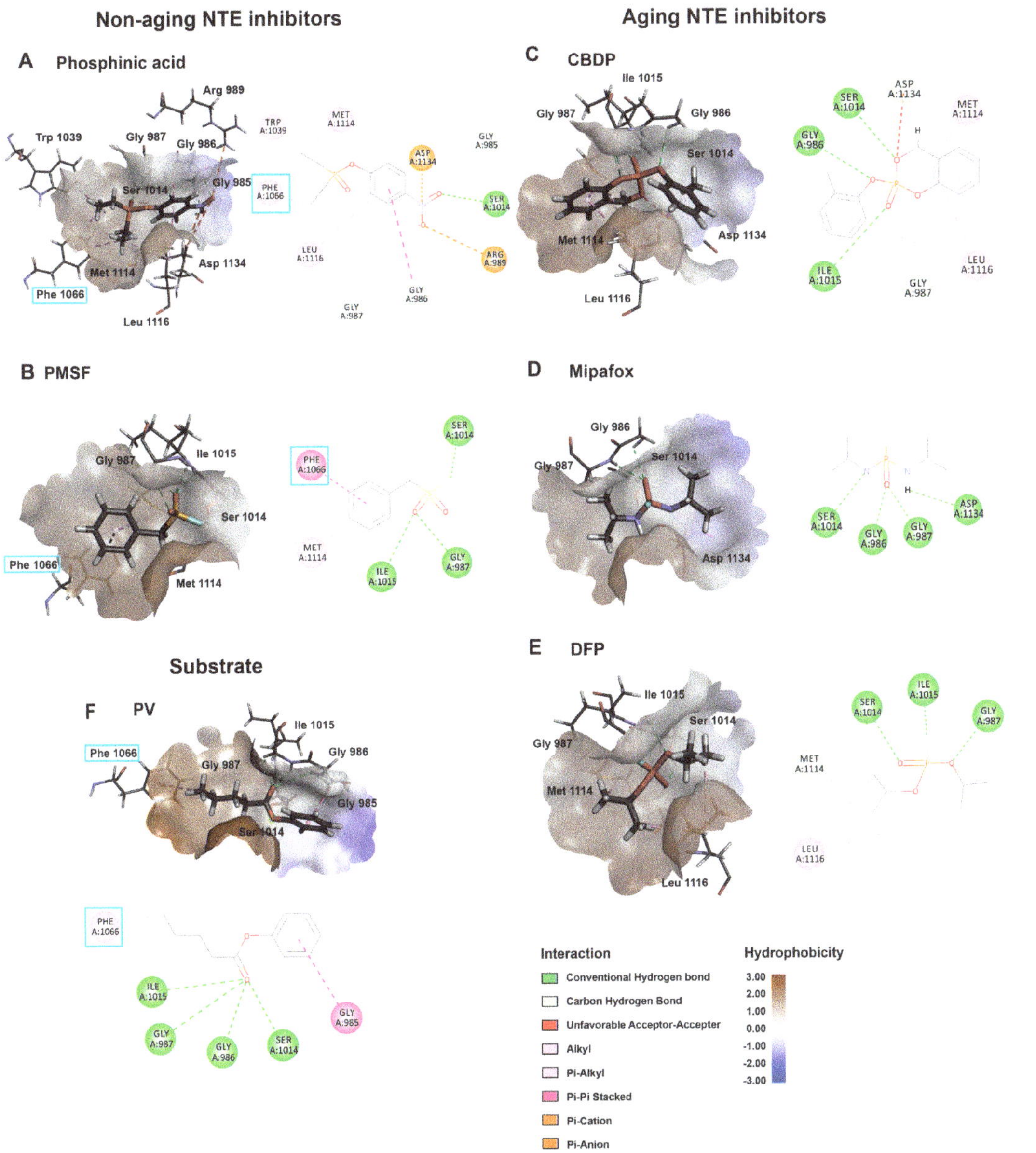

Figure 6. Docking results between NEST and different ligands. (**A**) Phosphinic acid. (**B**) PMSF. (**C**) CBDP. (**D**) Mipafox. (**E**) DFP. (**F**) PV. The ligands are shown in stick and surface of the protein is shown as surface and colored by hydrophobicity. Phosphinic acid and PMSF are non-aging inhibitors while CBDP, Mipafox, and DFP are aging compounds. PV is an artificial substrate of NTE. Dash lines with different colors in the 2D maps indicate different types of interactions, which are shown in the figure. Phe1066 is highlighted with a blue box.

Furthermore, the non-polar amino acids in the hydrophobic pocket such as Met 1114 and Leu1116 form the hydrophobic interactions with the alkyl group of the corresponding ligands, which indicates that these hydrophobic amino acids are also important for the binding of ligands to NEST. It is of note that Phe1066 in the NEST binding site has hydrophobic interaction with substrate PV and non-aging inhibitors, but not with aging inhibitors. This result indicates that Phe1066 in the hydrophobic pocket may play a unique role in the inhibition mechanism of non-aging inhibitors.

The docking results show that the non-aging inhibitors have a different binding interaction with NEST compared to aging inhibitors. Overall, the non-aging inhibitors bind the hydrophobic pocket much deeper than aging inhibitors and non-aging inhibitors and the substrate are closer to Phe1066 and form the hydrophobic interaction with Phe1066 (Figure 7, Table 3). The average values of the shortest distance to Phe1066 for aging inhibitors and non-aging inhibitors (plus the substrate) are 6.0 ± 0.2 Å and 4.7 ± 0.1 Å, respectively (Table 3). There is a statistical significance between these two groups ($p = 0.001 < 0.05$). These differences in binding interactions between non-aging and aging inhibitors may contribute to the differences in their inhibition mechanism for NEST.

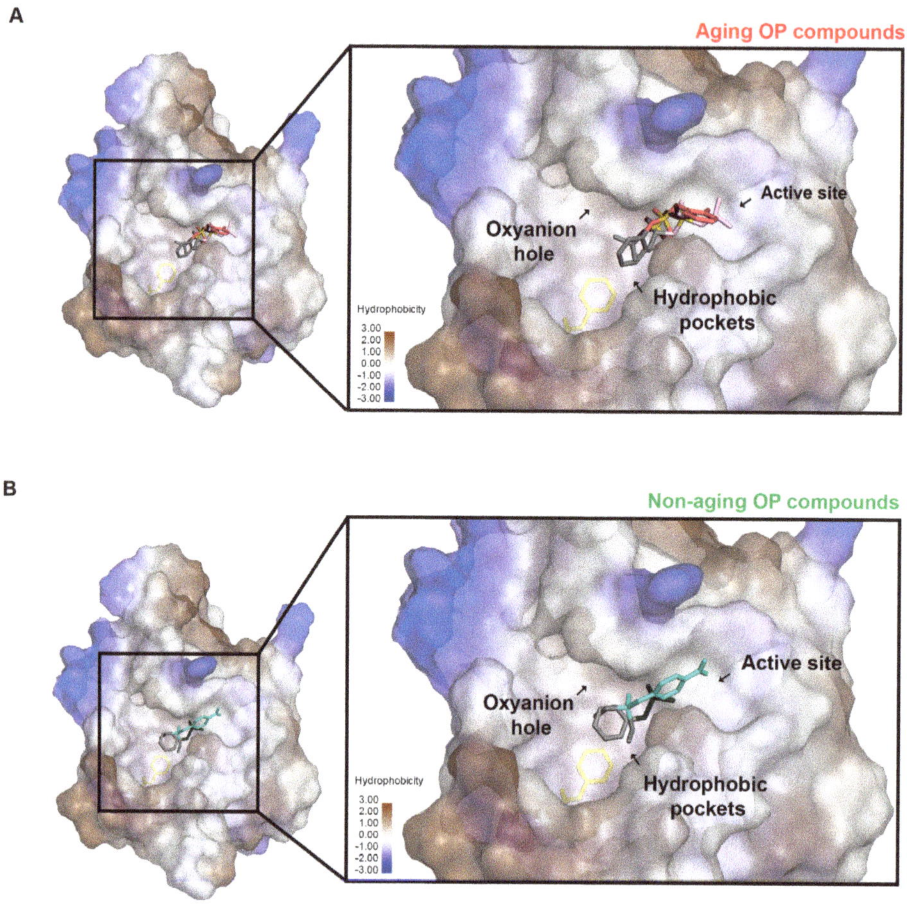

Figure 7. Overlay of the docking results of NEST with non-aging and aging inhibitors. (**A**) Binding of three aging inhibitors CBDP (light red), Mipafox (pink), and DFP (dark red) with the NEST model. (**B**) Binding of two non-aging inhibitors PMSF (dark green), and phosphinic acid (light green) with the NEST model. The protein is shown in surface and colored by hydrophobicity. The phosphate groups of the inhibitors are colored in yellow and the alkane groups of the inhibitors are colored in gray.

Table 3. The shortest distance between the carbon atoms of the inhibitors and the center point of the benzene ring of NEST Phe1066.

Ligands	Aging NTE Inhibitors			Non-Aging NTE Inhibitors		Substrate
	Mipafox	DPF	CBDP	PMSF	Phosphinic Acid	PV
Distance (Å)	6.0	6.2	5.7	4.8	4.7	4.62

3. Materials and Method

3.1. Modeling of NEST

The protein sequence of neuropathy target esterase (isoform 2, primary accession: Q8IY17) was obtained from Uniprot (available at https://www.uniprot.org/ (accessed on 24 October 2022)). NTE esterase domain (NEST) is the catalytic domain with amino acids from 727 to 1216. Then, the amino acid sequence of NEST was input into the National Center for Biotechnology Information (NCBI, available at https://www.ncbi.nlm.nih.gov/ (accessed on 24 October 2022)) database and the basic local alignment search tool (BLAST, available at https://blast.ncbi.nlm.nih.gov/Blast.cgi (accessed on 24 October 2022)) was used to search the similar protein sequences within sequence databases using blastp (protein-protein BLAST). In the Research Collaboratory for Structural Bioinformatics (RCSB) Protein Data Bank (PDB) database, there are in total three proteins that had similar primary structures with NEST, namely, type III effector protein (ExoU) from *Pseudomonas fluorescens A506* (PDB ID: 4QMK), vacuolar protein sorting inhibitor protein D (VipD) from *Legionella pneumophila* (PDB ID: 4AKF), and the bacterial patatin-like proteins (PlpD) from *Pseudomonas aeruginosa PAO1* (PDB ID: 5FYA). The PlpD (PDB ID: 5FYA) was selected as the modeling template because of the high identity (30.46%) with the NEST sequence.

MODELLER 10.3 (https://salilab.org/modeller/ (accessed on 24 October 2022)) was used for constructing the homology model of NEST. The alignment was automatically generated by MODELLER 10.3. In total, 100 models were built, and the conformation with the lowest discrete optimized protein energy (DOPE) value was selected as the model for verification and analysis [28].

Another model of NEST was predicted by using the AlphaFold Monomer v2.0 pipeline from AlphaFold Protein Structure Database (https://alphafold.ebi.ac.uk/ accessed on 24 October 2022) developed by DeepMind and EMBL-EBI [29]. The two models were evaluated with the residual percentage in a Ramachandran plot and Verify3D score via SAVES v6.0 (https://saves.mbi.ucla.edu/ (accessed on 24 October 2022)). Pymol (DeLano Scientific LLC, San Carlos, CA, USA) was utilized to visualize the tertiary structure.

3.2. Molecular Docking

The three-dimensional chemical structure of ligands (.sdf files) was obtained from PubChem (https://pubchem.ncbi.nlm.nih.gov/ (accessed on 24 October 2022)). DiscoveryStudio2019 v19.1.0.18287 (Dassault Systemes Biovia, San Diego, CA, USA) was used for molecular docking for NEST. The model generated by AlphaFold2 was used for the molecular docking because this model has better assessment results based on Ramachandran plots and Procheck analysis.

The NEST protein and the small ligands were cleaned and prepared using DiscoveryStudio2019 v19.1.0.18287. All ligands were processed by using the DiscoveryStudio2019 v19.1.0.18287 ligand preparation module. The coordinates for the center of the docking site were x: 29.839702, y: −4.038964, and z: −22.620865 with a radius of 10 Å. Flexible docking was used as the docking protocol. Five important residues (Gly986, Gly987, Ala988, Arg989, Ser1014, Asp1134) were allowed to change conformations according to the study of human cytosolic phospholipase A_2 ($cPLA_2$) [30].

The molecular docking conformations were evaluated according to their docking energy. For each conformation, we chose the smallest CDOCKER energy for each ligand as the best conformation, since a lower energy value indicates a more feasible binding [31]. CDOCKER energy calculates the intermolecular energy between the ligand and receptor.

Pymol and DiscoveryStudio2019 v19.1.0.18287 were used to perform electrostatic surface potential and hydrophobicity analysis, respectively.

4. Discussion

In this study, more precise 3D models built based on templates with higher sequence identity were constructed for the NTE catalytic domain compared to the previous study [25]. The models were built by using both MODELLER 10.3 and the AlphaFold2. Compared to the previous study which used the protein template Pat17 with a low sequence identity (18%) [25], our study used the template with 30% sequence identity.

In addition, we carried out molecular docking analysis of several representative NTE inhibitors. We find that the hydrophobic pocket residues below the oxyanion hole participates in the stabilization of NTE inhibitors, which has not been reported in the previous study. In addition, we find that the binding interaction with the non-aging inhibitors and aging inhibitors are different, and the hydrophobic pocket, especially Phe1066, plays an important role in the binding of non-aging inhibitors, but not in the binding of aging inhibitors.

The hydrophobicity analysis shows that the ligand binding pocket of NEST is generally hydrophilic and located in the hydrophobic side. The docking result shows that there are three important components of the binding site which recognize inhibitors (Figure 6 and Table 2). They are the catalytic dyad (Ser1014-Asp1134), hydrophobic pocket residues (Phe1066, Met1114, Leu1116), and oxyanion hole residues (Gly986, Gly987). The polar part of the ligands, i.e., the phosphate group tends to fall into the hydrophilic pocket of the NEST binding site, while the alkane group of the ligands tends to be trapped by the hydrophobic amino acids. The oxyanion hole plays an important role in stabilizing the phosphate group (Figure 6, Table 2).

Notably, we find that the binding of aging inhibitors to NEST is different compared to non-aging inhibitors. Firstly, compared to Ser1014, Asp1134 seems to play a less essential role in the binding of NTE inhibitors. In addition, Phe1066 directly interacts with the substrate (PV) and non-aging inhibitors but not with aging inhibitors (Figure 7). The overlay docking results among non-aging inhibitors and aging inhibitors indicate that the non-aging inhibitors bind deeper in the hydrophobic pocket than aging inhibitors and form hydrophobic interaction with Phe1066, which is the key feature of binding of non-aging inhibitors (Figure 7). Furthermore, the distance between the aging inhibitors and Phe106 is significantly longer than that of the non-aging inhibitors and the substrate (Table 3). Therefore, non-aging inhibitors may be trapped in the hydrophobic pocket, which may prevent the aging reaction.

In addition, the binding of non-aging inhibitors may prevent the subsequent binding of aging inhibitors. This may help explain the phenomena that the non-aging inhibitors can prevent delayed neuropathy when they are given before aging compounds, but cannot prevent delayed neuropathy when they are given after aging compounds [32]. When non-aging inhibitors are given before aging inhibitors, they may block the binding of aging inhibitors and NTE would not age, but when non-aging inhibitors are given after aging inhibitors, they cannot block the binding of aging inhibitors and cannot prevent NTE aging.

It is of note that the binding of the substrate PV to NEST is more similar to the non-aging inhibitors than to the aging inhibitors (Figure 6, Table 2). PV forms hydrophobic interaction with Phe1066 and there is no interaction with Asp1134. The distance of PV to Phe1066 is shorter compared to aging inhibitors (Table 3). This may explain why the catalytic reaction with the substrate usually does not result in the aging of NEST.

Early studies have shown that Asp1008 is key to the activity of phenyl-valerate hydrolase in NEST with site-directed mutagenesis [26]. However, our study shows that Asp1008 is at a far distance from the active site and is not likely to directly participate in the catalytic process. One possible explanation for the role of Asp1008 is that the mutation of this residue will change its nearby structure and folding of the anti-parallel β sheet.

Further computational studies such as molecular dynamics simulations may be used to further test the current hypothesis. Molecular dynamic simulation could evaluate the dynamics of the binding complexes of NEST with different inhibitors and provide more information on the interaction of inhibitors with NEST. Computation using quantum mechanics may also help understand the reaction mechanism of aging inhibitors or substrates with the amino acids in the active site of NEST. Moreover, structural and biochemical studies using Cryo-EM or mass spectrometry are needed to reveal the interaction of NEST with inhibitors experimentally.

In summary, our computational modeling and docking analysis provide insights into the different binding interactions of aging and non-aging NTE inhibitors, which may contribute to the different mechanisms of aging and non-aging NTE inhibitors. Therefore, computational modeling might be used to help predict the aging ability of chemicals on NTE in order to predict their potential of inducing delayed neuropathy.

Author Contributions: Conceptualization, P.W.; Methodology, W.W. and P.W.; Software, W.W.; Formal analysis, W.W.; Investigation, W.W.; Writing—original draft, W.W. and P.W.; Supervision, P.W.; Funding acquisition, P.W. All authors have read and agreed to the published version of the manuscript.

Funding: This study is supported by Ganghong Young Scholars Development Fund (No. 2021E0009, 2022E0032), Shenzhen Basic Science Project (No. JCYJ20190813170607348), Shenzhen Key Laboratory of Steroid Drug Discovery and Development (No. ZDSYS20190902093417963) and Shenzhen-Hong Kong Cooperation Zone for Technology and Innovation (No. HZQB-KCZYB-2020056).

Data Availability Statement: The data presented in this study are available upon request.

Conflicts of Interest: The authors declare no conflict of interest.

References

1. Kienesberger, P.C.; Oberer, M.; Lass, A.; Zechner, R. Mammalian patatin domain containing proteins: A family with diverse lipolytic activities involved in multiple biological functions. *J. Lipid Res.* **2009**, *50*, S63–S68. [CrossRef]
2. UniProt. UniProtKB—Q8IY17 (PLPL6_HUMAN). 2019. Available online: http://www.uniprot.org/uniprot/Q8IY17 (accessed on 26 November 2019).
3. Glynn, P. Neuronal phospholipid deacylation is essential for axonal and synaptic integrity. *Biochim. Biophys. Acta Mol. Cell Biol. Lipids* **2013**, *1831*, 633–641. [CrossRef]
4. Van Tienhoven, M.; Atkins, J.; Li, Y.; Glynn, P. Human Neuropathy Target Esterase Catalyzes Hydrolysis of Membrane Lipids. *J. Biol. Chem.* **2002**, *277*, 20942–20948. [CrossRef] [PubMed]
5. Zaccheo, O.; Dinsdale, D.; Meacock, P.A.; Glynn, P. Neuropathy Target Esterase and Its Yeast Homologue Degrade Phosphatidylcholine to Glycerophosphocholine in Living Cells. *J. Biol. Chem.* **2004**, *279*, 24024–24033. [CrossRef] [PubMed]
6. Richardson, R.J.; Fink, J.K.; Glynn, P.; Hufnagel, R.B.; Makhaeva, G.F.; Wijeyesakere, S.J. Neuropathy target esterase (NTE/PNPLA6) and organophosphorus compound-induced delayed neurotoxicity (OPIDN). In *Advances in Neurotoxicology*; Academic Press: Cambridge, MA, USA, 2020; Volume 4, pp. 1–78. [CrossRef]
7. Davis, C.S.; Richardson, R.J. Neurotoxic esterase: Characterization of the solubilized enzyme and the conditions for its solubilization from chicken brain microsomal membranes with ionic, zwitterionic, or nonionic detergents. *Biochem. Pharmacol.* **1987**, *36*, 1393–1399. [CrossRef] [PubMed]
8. Pope, C.N.; Padilla, S. Modulation of neurotoxic esterase activity in vitro by phospholipids. *Toxicol. Appl. Pharmacol.* **1989**, *97*, 272–278. [CrossRef]
9. Atkins, J.; Luthjens, L.H.; Hom, M.L.; Glynn, P. Monomers of the catalytic domain of human neuropathy target esterase are active in the presence of phospholipid. *Biochem. J.* **2002**, *361*, 119–123. [CrossRef]
10. Moser, M.; Li, Y.; Vaupel, K.; Kretzschmar, D.; Kluge, R.; Glynn, P.; Buettner, R. Placental failure and impaired vasculogenesis result in embryonic lethality for neuropathy target esterase-deficient mice. *Mol. Cell. Biol.* **2004**, *24*, 1667–1679. [CrossRef] [PubMed]
11. Read, D.J.; Li, Y.; Chao, M.V.; Cavanagh, J.B.; Glynn, P. Neuropathy Target Esterase Is Required for Adult Vertebrate Axon Maintenance. *J. Neurosci.* **2009**, *29*, 11594–11600. [CrossRef]
12. Richardson, R.J.; Hein, N.D.; Wijeyesakere, S.J.; Fink, J.K.; Makhaeva, G.F. Neuropathy target esterase (NTE): Overview and future. *Chem.-Biol. Interact.* **2013**, *203*, 238–244. [CrossRef]
13. Davis, C.S.; Richardson, R.J. Organophosphorus compounds. *Exp. Clin. Neurotoxicol.* **1980**, 527–544.
14. Johnson, M.K. Structure-activity relationships for substrates and inhibitors of hen brain neurotoxic esterase. *Biochem. Pharmacol.* **1975**, *24*, 797–805. [CrossRef] [PubMed]
15. Johnson, M.K. Organophosphorus esters causing delayed neurotoxic effects: Mechanism of action and structure activity studies. *Arch. Toxicol.* **1975**, *34*, 259–288. [CrossRef] [PubMed]

16. Johnson, M.K. Sensitivity and selectivity of compounds interacting with neuropathy target esterase. Further structure–activity studies. *Biochem. Pharmacol.* **1975**, *37*, 4095–4104. [CrossRef]
17. Johnson, M.K.; Henschler, D. The Delayed Neuropathy Caused by Some Organophosphorus Esters: Mechanism and Challenge. *CRC Crit. Rev. Toxicol.* **1975**, *3*, 289–316. [CrossRef] [PubMed]
18. Clothier, B.; Johnson, M.K. Rapid aging of neurotoxic esterase after inhibition by di-isopropyl phosphorofluoridate. *Biochem. J.* **1979**, *177*, 549–558. [CrossRef] [PubMed]
19. Clothier, B.; Johnson, M.K. Reactivation and aging of neurotoxic esterase inhibited by a variety of organophosphorus esters. *Biochem. J.* **1980**, *185*, 739–747. [CrossRef]
20. Kropp, T.J.; Glynn, P.; Richardson, R.J. The mipafox-inhibited catalytic domain of human neuropathy target esterase ages by reversible proton loss. *Biochemistry* **2004**, *43*, 3716–3722. [CrossRef]
21. Carrington, C.D.; Abou-Donia, M.B. The time course of protection from delayed neurotoxicity induced by tri-o-cresyl phosphate and O, O-diisopropyl phosphorofluoridate by phenyl methyl sulfonyl fluoride in chickens. *Toxicol. Lett.* **1983**, *18*, 251–256. [CrossRef]
22. Baker, T.; Lowndes, H.E.; Johnson, M.K.; Sandborg, I.C. The effects of phenylmethanesulfonyl fluoride on delayed organophosphorus neuropathy. *Arch. Toxicol.* **1980**, *46*, 305–311. [CrossRef]
23. Drakontides, A.B.; Baker, T. An electrophysiologic and ultrastructural study of the phenylmethanesulfonyl fluoride protection against a delayed organophosphorus neuropathy. *Toxicol. Appl. Pharmacol.* **1983**, *70*, 411–422. [CrossRef] [PubMed]
24. Caroldi, S.; Lotti, M.; Masutti, A. Intra-arterial injection of diisopropylfluorophosphate or phenylmethanesulphonyl fluoride produces unilateral neuropathy or protection, respectively, in hens. *Biochem. Pharmacol.* **1984**, *33*, 3213–3217. [CrossRef] [PubMed]
25. Wijeyesakere, S.J.; Richardson, R.J.; Stuckey, J.A. Modeling the tertiary structure of the patatin domain of neuropathy target esterase. *Protein J.* **2007**, *26*, 165–172. [CrossRef]
26. Atkins, J.; Glynn, P. Membrane Association of and Critical Residues in the Catalytic Domain of Human Neuropathy Target Esterase. *J. Biol. Chem.* **2000**, *275*, 24477–24483. [CrossRef] [PubMed]
27. Lotti, M.; Johnson, M.K. Neurotoxicity of organophosphorus pesticides: Predictions can be based on in vitro studies with hen and human enzymes. *Arch. Toxicol.* **1978**, *41*, 215–221. [CrossRef]
28. Shen, M.; Sali, A. Statistical potential for assessment and prediction of protein structures. *Protein Sci.* **2006**, *15*, 2507–2524. [CrossRef]
29. Jumper, J.; Evans, R.; Pritzel, A.; Green, T.; Figurnov, M.; Ronneberger, O.; Tunyasuvunakool, K.; Bates, R.; Žídek, A.; Potapenko, A.; et al. Highly accurate protein structure prediction with AlphaFold. *Nature* **2021**, *596*, 583–589. [CrossRef]
30. Dessen, A.; Tang, J.; Schmidt, H.; Stahl, M.; Clark, J.D.; Seehra, J.; Somers, W.S. Crystal structure of human cytosolic phospholipase A2 reveals a novel topology and catalytic mechanism. *Cell* **1999**, *97*, 349–360. [CrossRef] [PubMed]
31. Wu, G.; Robertson, D.H.; Brooks, C.L., III; Vieth, M. Detailed analysis of grid-based molecular docking: A case study of CDOCKER-A CHARMm-based MD docking algorithm. *J. Comput. Chem.* **2003**, *24*, 1549–1562. [CrossRef] [PubMed]
32. Pope, C.N.; Padilla, S. Potentiation of organophosphorus-induced delayed neurotoxicity by phenylmethylsulfonyl fluoride. *J. Toxicol. Environ. Health* **1990**, *31*, 261–273. [CrossRef]

molecules

MDPI

Article

Fusing Sequence and Structural Knowledge by Heterogeneous Models to Accurately and Interpretively Predict Drug–Target Affinity

Xin Zeng [1], Kai-Yang Zhong [1], Bei Jiang [2] and Yi Li [1,*]

1 College of Mathematics and Computer Science, Dali University, Dali 671003, China; xinzeng@dali.edu.cn (X.Z.); zkaiyang2022@163.com (K.-Y.Z.)
2 Yunnan Key Laboratory of Screening and Research on Anti-Pathogenic Plant Resources from Western Yunnan, Dali University, Dali 671000, China; jiangbei@dali.edu.cn
* Correspondence: yili@dali.edu.cn

Abstract: Drug–target affinity (DTA) prediction is crucial for understanding molecular interactions and aiding drug discovery and development. While various computational methods have been proposed for DTA prediction, their predictive accuracy remains limited, failing to delve into the structural nuances of interactions. With increasingly accurate and accessible structure prediction of targets, we developed a novel deep learning model, named S2DTA, to accurately predict DTA by fusing sequence features of drug SMILES, targets, and pockets and their corresponding graph structural features using heterogeneous models based on graph and semantic networks. Experimental findings underscored that complex feature representations imparted negligible enhancements to the model's performance. However, the integration of heterogeneous models demonstrably bolstered predictive accuracy. In comparison to three state-of-the-art methodologies, such as DeepDTA, GraphDTA, and DeepDTAF, S2DTA's performance became more evident. It exhibited a 25.2% reduction in mean absolute error (MAE) and a 20.1% decrease in root mean square error (RMSE). Additionally, S2DTA showed some improvements in other crucial metrics, including Pearson Correlation Coefficient (PCC), Spearman, Concordance Index (CI), and R^2, with these metrics experiencing increases of 19.6%, 17.5%, 8.1%, and 49.4%, respectively. Finally, we conducted an interpretability analysis on the effectiveness of S2DTA by bidirectional self-attention mechanism. The analysis results supported that S2DTA was an effective and accurate tool for predicting DTA.

Keywords: graph neural network; convolutional neural network; drug–target affinity; sequence and structural knowledge; heterogeneous models

1. Introduction

Target proteins are crucial biological macromolecules that play diverse roles in cells. Identifying drug–target interactions is an important process in drug discovery [1]. Calculating a drug molecule (small molecule compound) with high binding affinity against the target is a crucial step in early-stage drug research [2]. The successful identification of drug–target affinity (DTA) has received long-term attention from academia and industry due to its significant role in drug discovery and design [3]. Typically, drug molecules interact with specific targets through direct physicochemical interactions in shallow hydrophobic depressions on the surface of targets or buried deep inside the targets known as pockets [4,5]. These pockets contain charged amino acid residues and hydrophobic residues. Biological experimental observations suggest that the amino acid composition and properties of these pockets are vital determinants for drug–target interactions [4].

Some traditional biological methods to determine DTA are measuring the half maximal inhibitory concentration (IC50), inhibition constant (K_i), and dissociation constant (K_d)

between a pair of drugs and targets using bioassays [6]. However, traditional biological experiments to measure DTA are time-consuming, labor-intensive, and resource-intensive [7]. With the increasing number of drug molecules and targets in databases, experimental characterization of all possible drug–target pairs becomes nearly impossible. To address this, numerous computational methods have been developed to predict DTA [8–11], with some deep learning models based on sequence or/and structural knowledge showing promising results [12,13].

Most DTA prediction models based on deep learning, including DeepDTA [9], DeepDTAF [10], DeepAffinity [14], FusionDTA [15], WideDTA [16], AttentionDTA [17], DeepCDA [18], and MRBDTA [19], primarily rely on drugs' SMILES (Simplified Molecular Input Line Entry System) [20] and target sequences. These sequence-based methods employed Convolutional Neural Networks (CNNs) [21], Recurrent Neural Networks (RNNs) [22], or the attention mechanism [23] to extract high-level features from the drug SMILES and target sequences. For instance, DeepDTA used three consecutive 1D-convolutional layers to extract local features from drug SMILES and the sequences of targets, which were then passed through three fully connected (FC) layers for DTA prediction. On the other hand, DeepDTAF employed three independent CNN blocks to learn high-level features from the drug SMILES and the sequences of targets and pockets. These CNN blocks consisted of a 1D-convolutional layer for extracting local features from pockets and two 1D dilated convolutional networks [24] for deciphering global features of drugs and targets, respectively. MRBDTA utilized the Trans block with a multi-head attention mechanism to extract molecule features from drug SMILES and the sequences of targets. Then, interaction learning integrated these extracted features for DTA prediction. While sequence-based approaches are effective in predicting DTA, it is evident that their performance can be further enhanced if structural knowledge is incorporated. For example, in DeepDTAF, the pocket knowledge which is essentially obtained from both its sequences and structures may achieve higher performance. Furthermore, these methods are limited in providing a reasonable interpretation of drug–target interactions due to the lack of structural knowledge.

While the structures of targets are challenging to obtain experimentally, some methods, such as GraphDTA [11], MgraphDTA [25], and DGraphDTA [26], have successfully extracted high-level features from a limited number of the structures of drugs and targets to predict DTA. For instance, GraphDTA recognized the structures of small molecules and employed graph convolutional networks (GCNs) [27] to extract graph-based features. In MgraphDTA, multiscale graph neural networks (MGNN) were utilized to extract multiscale features from the input graph of the drug. In DGraphDTA, the structures of targets were represented as graphs based on the contact maps and then combined with the graphs of drugs to predict DTA by two GCNs. Although structure-based methods offer better interpretation, they do have some limitations: (1) the full potential of structural knowledge of drugs and targets remains untapped; (2) most models lack the utilization of semantic knowledge in sequence format, relying solely on structural data for DTA prediction; (3) the 3D spatial positions of the atoms in targets are not fully characterized, as some methods only use contact maps.

In recent years, with the accumulation of experimentally resolved structures of targets and the emergence of algorithms like AlphaFold [28], which can predict the structures of monomeric proteins and proteins without extensive disordered regions, there is a certain degree of relief in obtaining protein structure. To address the problem of insufficient feature extraction from the sequences and structures of drugs and targets in existing methods, a novel deep learning model called S2DTA was proposed. S2DTA predicted DTA by integrating sequence features of drug SMILES, targets, and pockets, and their corresponding graph structural features using a heterogeneous model that combined GCNs and CNNs. Experimental results demonstrated that S2DTA outperformed three other state-of-the-art models by using simple feature representations. Additionally, we conducted an interpretability analysis on the effectiveness of S2DTA using a bidirectional self-attention mechanism, supporting that S2DTA is an effective and accurate tool for predicting DTA.

2. Results and Discussion

2.1. Limited Performance Improvement by Using Only Sequence Knowledge

Numerous predictive models for DTA primarily rely on the sequences of targets due to the challenge of obtaining the experimental structures of targets. In our study, we conducted comprehensive tests involving both ML and DL models, comparing models that used solely sequence data against a fundamental GCN model that incorporated the structural insights of targets (Table 1). Within the realm of ML models, RF exhibited the most promising performance across all evaluation metrics. Specifically, MAE, RMSE, PCC, Spearman, CI, and R^2 achieved values of 1.221, 1.527, 0.726, 0.713, 0.760, and 0.504, respectively. Turning to DL models, sequence-oriented DL architectures like 1D-CNN and RNN demonstrated relatively weaker performance compared to RF with label encoding. This outcome highlighted that complex DL models did not necessarily lead to performance enhancements. However, 1D-CNN, relying on one-hot encoding, delivered superior performance across all evaluation metrics. This underscored that one-hot encoding provided robust feature representation, enhancing the efficacy of sequence-based DL models. To further enrich the understanding of targets, we incorporated the structural knowledge of targets into the GCN model by characterizing the sequence knowledge of targets as vertex features within the graphs of targets. Empirical findings demonstrated that even a straightforward GCN model with label encoding surpassed RF and 1D-CNN. Notably, when the sequence feature representations of targets employed one-hot encoding, the GCN performed better than RF and 1D-CNN. Specifically, the GCN outperformed RF with a reduction of 0.063 in both MAE and RMSE, and exhibited increments of 0.012, 0.016, 0.009, and 0.04 in PCC, Spearman, CI, and R^2, respectively. All experimental results clearly showed that the simple GCN model with a baseline feature representation outperformed all sequence-based models. Only using the sequence knowledge of targets to predict DTA limited performance improvement.

Table 1. Performance of machine learning and deep learning models using only sequence knowledge compared to graph-based models fusing structure knowledge.

Category	Model	Feature Representation	MAE	RMSE	PCC	Spearman	CI	R^2
Machine learning models	RR	label encoding	1.471	1.864	0.515	0.521	0.685	0.261
	KNN	label encoding	1.377	1.732	0.604	0.565	0.703	0.362
	DT	label encoding	1.409	1.882	0.590	0.577	0.707	0.247
	SVR	label encoding	1.749	2.150	0.156	0.231	0.527	0.017
	RF	label encoding	1.221	1.527	0.726	0.713	0.760	0.504
Sequence-based deep learning models	1D-CNN	label encoding	1.569	1.938	0.535	0.531	0.683	0.201
		one-hot encoding	1.159	1.473	0.752	0.741	0.776	0.539
	RNN	label encoding	1.815	2.227	0.175	0.329	0.611	−0.054
		one-hot encoding	1.669	2.045	0.459	0.460	0.657	0.111
Structure-based deep learning models	GraphSAGE	label encoding	1.288	1.599	0.675	0.664	0.739	0.449
		one-hot encoding	1.127	1.417	0.758	0.745	0.778	0.573

2.2. Performance Comparison of Different Graph-Based Models

Although GraphSAGE has demonstrated superior performance as depicted in Table 1, it is important to note that graph neural networks (GNNs) offer an array of models employing diverse approaches to extract features from graphs. Presently, GNNs are broadly categorized into two types: spectral-based GNNs and spatial-based GNNs. Spectral-based GNNs like GCN and ChebyNet operate on Laplacian matrices, focusing on single graphs. However, for the context of this study, ChebyNet was excluded due to interpretability concerns. Diverging from spectral-based models, spatial-based GNNs such as GraphSAGE, GAT, and GIN gather knowledge from neighboring vertices directly in the spatial domain.

The methodology entailed initial one-hot encoding of drug SMILES and the sequences of targets, utilizing them as the features of vertices in the respective graphs. Subsequently, four distinct graph-based models were harnessed for DTA prediction. The ensuing comparison of experimental results (Table 2) highlighted that GraphSAGE consistently outperformed the other three models across all evaluation metrics. Noteworthiness was its achievement of MAE and RMSE reductions of no less than 0.266 and 0.326 compared to the other models. Moreover, GraphSAGE accomplished improvements of no less than 0.157, 0.156, 0.071, and 0.219 in PCC, Spearman, CI, and R^2, respectively, all of which demonstrated values exceeding 0.5. The better performance of GraphSAGE underscored its efficacy in capturing interactions within the tertiary structures of drugs and targets, reinforcing the significance of structural knowledge in DTA prediction.

Table 2. Performance comparison of different graph-based models.

Model	Feature Representation	MAE	RMSE	PCC	Spearman	CI	R^2
GCN	one-hot encoding	1.393	1.743	0.601	0.589	0.707	0.354
GIN	one-hot encoding	1.524	1.842	0.532	0.549	0.692	0.279
GAT	one-hot encoding	1.708	2.091	0.296	0.313	0.606	0.070
GraphSAGE	one-hot encoding	1.127	1.417	0.758	0.745	0.778	0.573

2.3. *Complex Feature Representations Do Not Improve Performance*

In addition to employing one-hot encoding, the utilization of complex feature representations for the drug SMILES and target sequences could potentially exert a substantial impact on the performance of DTA prediction. Within the scope of this investigation, a variety of complex feature representations were explored for the sequences of targets, encompassing the physical–chemical properties of amino acids, position-specific scoring matrix (PSSM) [29], Hidden Markov Models matrix (HMM) [30], Protbert [31], CNN, and Word2Vec [32]. For drugs, only the physical–chemical properties of atoms were considered. It is noteworthy that when alternative complex feature representations were applied to the sequences of targets, the drug SMILES representation was employed using one-hot encoding.

Experimental results (Table 3) demonstrated that GraphSAGE attained comparable performance across various representations, including one-hot encoding, Protbert, CNN, and Word2Vec. For instance, the key evaluation metrics like RMSE and R^2 hovered around 1.4 and 0.5, respectively. However, it is pertinent to acknowledge that generating semantic features for the sequences of targets through complex representations like Protbert, CNN, and Word2Vec can be time-intensive. Despite this investment of time, the performance improvements garnered from these complex representations have not been notably substantial.

Table 3. Performance comparison of GraphSAGE with different feature representations.

Feature Representation	MAE	RMSE	PCC	Spearman	CI	R^2
one-hot encoding	1.127	1.417	0.758	0.745	0.778	0.573
physical–chemical properties	1.444	1.770	0.583	0.585	0.706	0.334
PSSM	1.236	1.526	0.722	0.713	0.761	0.505
HMM	1.342	1.664	0.644	0.639	0.729	0.411
semantic feature (Protbert)	1.139	1.485	0.734	0.720	0.768	0.531
semantic feature (CNN)	1.158	1.477	0.733	0.723	0.769	0.536
semantic feature (Word2Vec)	1.136	1.411	0.760	0.746	0.777	0.577

2.4. *Impact of Pocket Knowledge on Drug–Target Affinity Prediction*

As demonstrated by DeepDTAF, pockets played a crucial role in predicting DTA, and their characteristics directly influenced the binding affinity of drugs with their respective targets. The maximum sequence length for pockets was 125, serving as the fixed sequence length. In cases where the pocket sequences were shorter than 125, padding was applied using zeros. The representation of the sequences of pockets employed one-hot encoding. The structures of pockets were utilized for constructing the graphs, following a methodology consistent with the graph construction of the target.

Three distinct experiments were conducted based on the sequences and structures of targets, pockets, and their combination. Experimental results (Table 4) showed that the performance of GraphSAGE utilizing the sequences and structures of targets was comparable to that using the sequences and structures of pockets across all evaluation metrics. Notably, the MAE, RMSE, PCC, Spearman, CI, and R^2 exhibited values of approximately 1.1, 1.4, 0.7, 0.7, 0.7, and 0.5, respectively. The distribution plots depicting the real and predictive affinity based on targets and pockets were illustrated in Figure 1. However, noteworthy enhancements were observed in the performance of GraphSAGE when incorporating a combination of the sequences and structures of targets and pockets. As compared to using only the knowledge of targets, the MAE and RMSE exhibited reductions of 0.047 (4.2%) and 0.04 (2.8%), respectively. Moreover, improvements of 0.018 (2.4%) in PCC, 0.028 (3.8%) in Spearman, 0.013 (1.7%) in CI, and 0.024 (4.2%) in R^2 were realized. Similarly, in comparison to using only the knowledge of pockets, the MAE and RMSE experienced decreases of 0.021 (1.9%) and 0.43 (3%), while enhancements of 0.017 (2.2%) in PCC, 0.018 (2.4%) in Spearman, 0.009 (1.2%) in CI, and 0.026 (4.6%) in R^2 were achieved. The distribution of real and predictive affinity, incorporating both the sequence and structural knowledge of targets and pockets, is visualized in Figure 1.

Table 4. Performance comparison of GraphSAGE with targets or combining the knowledge of pockets.

Name	MAE	RMSE	PCC	Spearman	CI	R^2
Targets	1.127	1.417	0.758	0.745	0.778	0.573
Pockets	1.101	1.420	0.759	0.755	0.782	0.571
Targets + Pockets	1.080	1.377	0.776	0.773	0.791	0.597

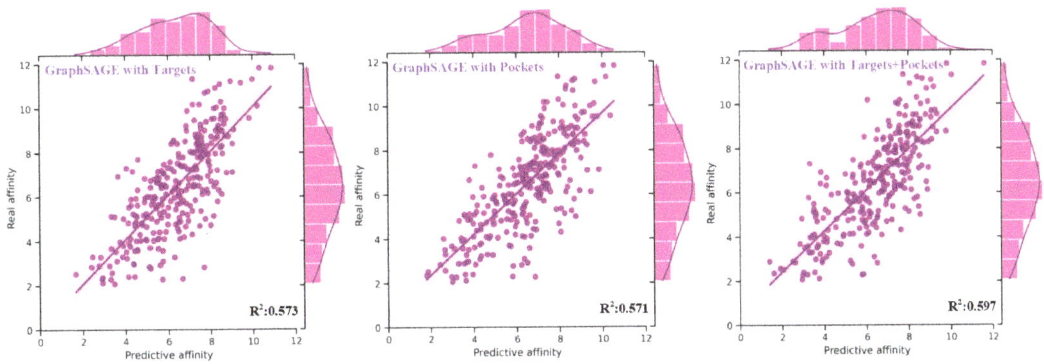

Figure 1. The distribution of real and predictive affinity of GraphSAGE with targets, pockets, and their combination based on the test set. The performance of GraphSAGE, when harnessed in conjunction with both the sequence and structural knowledge of targets and pockets, surpassed that achieved by solely relying on the sequence and structural features of targets or pockets individually. Specifically, as compared to using only the features of targets, the MAE and RMSE exhibited reductions of 0.047 and 0.04, respectively. Moreover, improvements of 0.018 in PCC, 0.028 in Spearman, 0.013 in CI, and 0.024 in R^2 were realized. Similarly, in comparison to using only the features of pockets, the MAE and RMSE experienced decreases of 0.021 and 0.43, while enhancements of 0.017 in PCC, 0.018 in Spearman, 0.009 in CI, and 0.026 in R^2 were achieved.

2.5. Fusing Sequence and Structure Knowledge with Heterogeneous Models

Although the semantic features extracted from the sequences of targets using Protbert, CNN, and Word2Vec did not lead to an improvement in the performance of GraphSAGE, fusing sequence knowledge with CNN to create an end-to-end model could enhance the performance. The process involved employing one-hot encoding for the sequences of targets, which was then fed into a CNN to extract sequence knowledge. This extracted sequence knowledge was utilized as vertex features within the graphs of targets, subsequently combined with the graphs of targets and input into a three-layer GCN network to extract the fusion knowledge. This approach was also used to extract the sequence features of pockets, but not to drug SMILES. Finally, the knowledge extracted from drugs, targets, and pockets by three independent three-layer GCN networks was concatenated and fed into the FC network to predict DTA.

Experimental results (Table 5) demonstrated some enhancements in performance by integrating CNN into the model alongside GCN and FC. In comparison to the model without CNN (Table 4), the MAE and RMSE were reduced by 0.090 (8.3%) and 0.078 (5.7%), respectively. Moreover, the PCC, Spearman, CI, and R^2 exhibited improvements of 0.025 (3.2%), 0.027 (3.5%), 0.013 (1.6%) and 0.044 (7.4%), respectively. In addition to CNN, several other sequence-processing models, such as Long Short-Term Memory (LSTM) [33] and Bidirectional LSTM (BiLSTM) [34], are also capable of extracting sequence knowledge. Extensive experiments were conducted on these commonly utilized sequence-processing models, with the findings (Table 5) indicating that integrating sequence knowledge with CNN consistently yielded superior performance. The distribution of predictive affinity resulting from the fusion of sequence knowledge with sequence-processing models is depicted in Figure 2.

Table 5. Performance comparison of heterogeneous models with fused sequence and structural knowledge.

Sequence Processing Model	Structural Model	MAE	RMSE	PCC	Spearman	CI	R^2
LSTM	GraphSAGE	1.084	1.397	0.766	0.749	0.781	0.585
BiLSTM		1.100	1.399	0.764	0.765	0.785	0.584
CNN + LSTM		1.075	1.370	0.776	0.771	0.789	0.601
CNN		0.990	1.299	0.801	0.800	0.804	0.641

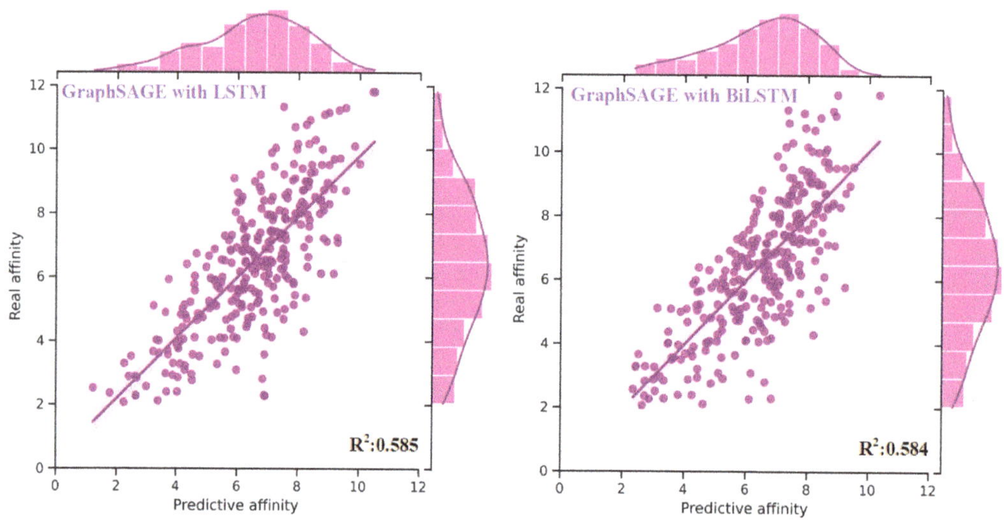

Figure 2. *Cont.*

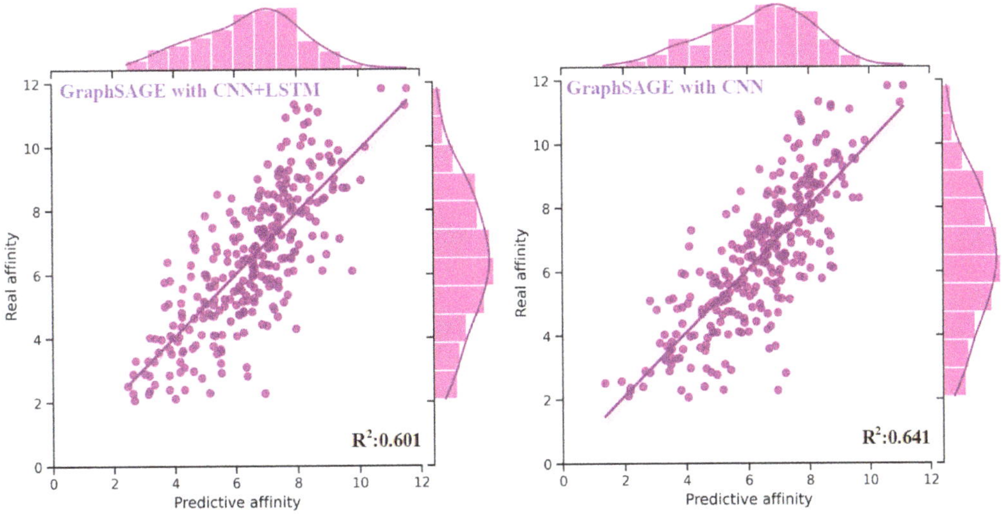

Figure 2. The distribution of real and predictive affinity of heterogeneous models to fuse sequence and structural knowledge from drugs, targets, and pockets based on the test set. Among multiple sequence-processing models, including CNN, LSTM, BiLSTM, and their combinations, the end-to-end heterogeneous model composed of CNN and GCN exhibited the best performance. In comparison to the model without CNN, the MAE and RMSE were reduced by 8.3% and 5.7%, respectively. Moreover, the PCC, Spearman, CI, and R^2 exhibited improvements of 3.2%, 3.5%, 1.6% and 7.4%, respectively.

In summary, the proposed method, denoted as S2DTA, which encompassed CNN, GCN, and FC, exhibited the highest performance by amalgamating sequence and structural knowledge from drugs, targets, and pockets. The process involved extracting the sequence knowledge from the sequences of targets and pockets, and incorporating structures and the extracted sequence knowledge from targets and pockets into GCN to attain the high-level fused features. For drugs, high-level features were directly extracted from drug SMILES and their structures through GCN. Ultimately, all of these features were concatenated and fed into a FC network to predict DTA.

2.6. Comparison with State-of-the-Art Methods

To comprehensively assess the predictive capability of S2DTA for DTA prediction, we conducted a comparative analysis against three deep learning methods: DeepDTA, DeepDTAF, and GraphDTA. Firstly, we trained the three compared methods on the training set provided in this study. Then employing the test dataset sourced from the PDBbind database (2016 version), S2DTA emerged as the superior performer across all evaluation metrics (Table 6). Experimental findings showed that S2DTA exhibited a 25.2% reduction in MAE and a 20.1% decrease in RMSE when compared to the three state-of-the-art methodologies. Additionally, S2DTA displayed some improvements in other vital metrics: PCC, Spearman, CI, and R^2, with increases of 19.6%, 17.5%, 8.1%, and 49.4%, respectively. Finally, these experiments underscored the robustness and reliability of S2DTA as a predictive model for DTA.

Table 6. Performance comparison of S2DTA with three state-of-the-art methods.

Model	MAE	RMSE	PCC	Spearman	CI	R^2
DeepDTA	1.323	1.641	0.667	0.681	0.744	0.428
GraphDTA	1.367	1.715	0.618	0.612	0.717	0.374
DeepDTAF	1.327	1.625	0.670	0.654	0.734	0.429
S2DTA	0.990	1.299	0.801	0.800	0.804	0.641

2.7. Performance Comparison between the Real Structures and Predictive Structures of Targets

Currently, a lot of targets possess known sequences but lack clear structures. This limitation hampers the application of S2DTA. However, the emergence of AlphaFold offers the potential to predict the structures of monomeric proteins and proteins without extensive disordered regions. To evaluate the adaptability of S2DTA on these predictive structures of targets, we selected 225 sequences from the test dataset, each with a length of approximately 500 amino acid residues. Employing AlphaFold, we obtained predictive structures for these 225 sequences. The outcomes of S2DTA using both real and predictive structures of targets are presented in Table 7. For S2DTA utilizing the predictive structures of targets, the MAE and RMSE stood at 1.133 (compared to 1.046 with real structures) and 1.432 (1.316), respectively. The PCC, Spearman, CI, and R^2 for S2DTA employing the predictive structures of targets were 0.748 (compared to 0.792 with real structures), 0.749 (0.787), 0.78 (0.799), and 0.558 (0.627), respectively. The evaluation metrics collectively indicated that the performance of S2DTA was similar to that when using real structures of targets. S2DTA demonstrated good generalization proficiency when using the predictive structures of targets.

Table 7. Performance comparison of S2DTA based on the real and predictive structures of targets.

Model	Structures of Targets	MAE	RMSE	PCC	Spearman	CI	R^2
S2DTA	Real structures	1.046	1.316	0.792	0.787	0.799	0.627
	Predictive structures	1.133	1.432	0.748	0.749	0.780	0.558

2.8. Structure-Based Interpretable Analysis of S2DTA

To explain the effectiveness of S2DTA, we employed a bidirectional self-attention mechanism to extract crucial information regarding drug–target binding affinity. This was performed after the pooling operation in S2DTA, resulting in attention score matrices that capture the interactions between drugs and targets. By making use of the structural knowledge of drugs and targets, we aimed to provide interpretability for the effectiveness of S2DTA. The pockets are the primary sites where drug–target interactions take place. Hence, we conducted a structural explanatory analysis of S2DTA by explaining the atoms constituting the drugs and the amino acids present in the targets, using different attention score matrices. For drugs, we computed the binding data between each atom (excluding hydrogen atoms) and all amino acids in the targets. Subsequently, we performed data statistics (Table 8) to determine whether the top 5, top 10, and top 20 amino acids in the attention score matrices were located within the pockets. The statistical findings revealed that the top 5, top 10, and top 20 amino acids within the pockets accounted for 13.7%, 12.8%, and 12.4%, respectively. In the upper row of Figure 3, we displayed four targets for drug binding. In the middle row of Figure 3, we showcased the top 10 amino acids (depicted in purple) identified by four drugs (PubChem ID: 43158872, 444128, 44629533, and 53308636) within the pockets of their respective targets (PDB ID: 3AO4, 1A30, 2QNQ, and 3O9I). These proportions were relatively low due to the length of target sequences, which made it challenging to identify any interactions between each atom and the amino acids in the pockets. However, S2DTA demonstrated its ability to effectively identify the amino acids within the pockets that interacted with the atoms. Conversely, we also examined the interaction between each amino acid and all atoms (excluding hydrogen atoms). Statistical analysis (Table 8) was performed on the top 3, top 5, and top 8 atoms in the attention score matrices. Interestingly, these atoms accounted for 48.6%, 56%, and 64.8% of the total number of atoms in the drugs, respectively. The lower row of Figure 3 presented the visualization results of the top 5 atoms (represented by the purple logo) with the highest scores in the interactions between the amino acids constituting the targets and the corresponding drugs. These above statistical results indicated that S2DTA could effectively identify the specific atoms that interacted with the amino acids. Hence, the attention score matrices derived from bidirectional self-attention served as a key explanation for the effectiveness of S2DTA in accurately predicting drug–target binding affinity. This underscores the reliability and interpretability of S2DTA as a practical tool for drug–target affinity prediction.

Table 8. Data statistics for interpretability analysis of S2DTA.

Atoms of Drugs			Amino Acids of Targets		
Top 5	Top 10	Top 20	Top 3	Top 5	Top 8
13.7%	12.8%	12.4%	48.6%	56%	64.8%

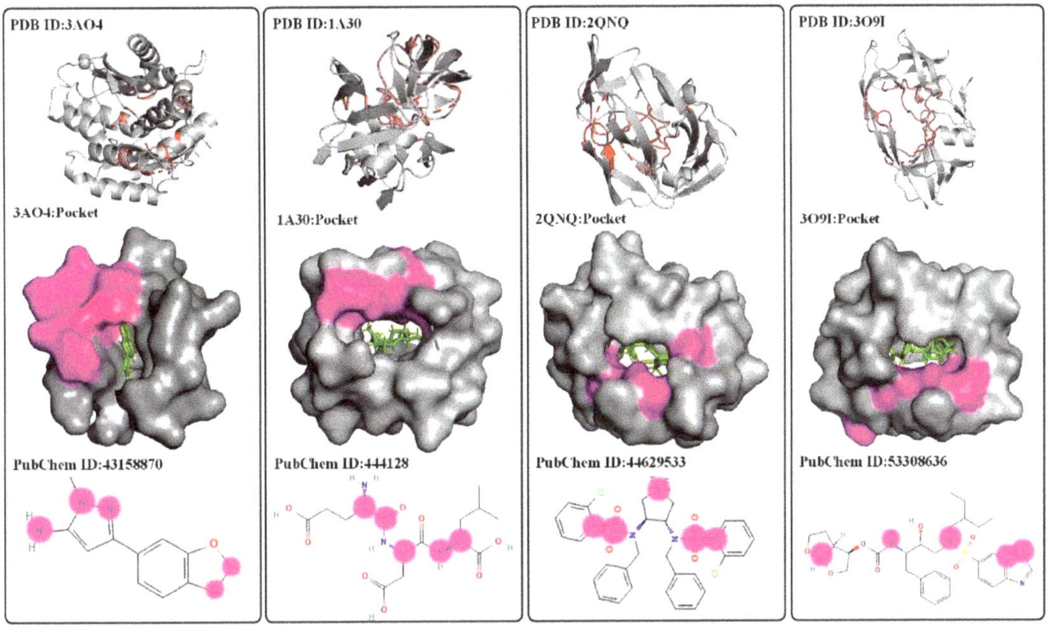

Figure 3. The visualization results of drug–target binding affinity by a bidirectional self-attention mechanism for explaining the effectiveness of S2DTA. In the **upper row**, we present the targets of amino acids with higher drug interaction scores. In the **middle row**, we show the top 10 amino acids (depicted in purple) identified by four drugs (PubChem ID: 43158872, 444128, 44629533, and 53308636) within the pockets of their respective targets (PDBID: 3AO4, 1A30, 2QNQ, and 3O9I). The statistical findings reveal that the top 5, top 10, and top 20 amino acids within the pockets account for 13.7%, 12.8%, and 12.4%, respectively. The **lower row** presents the visualization results of the top 5 atoms (represented by the purple logo) with the highest scores in the interactions between the amino acids constituting the targets and the corresponding drugs. Statistical analysis was performed on the top 3, top 5, and top 8 atoms in the attention score matrices. These atoms accounted for 48.6%, 56%, and 64.8% of the total number of atoms in the drugs, respectively.

2.9. Case Study

To further validate the effectiveness of the model, we compared S2DTA with DeepDTAF and MM-GBSA/PBSA [35] in specific samples. Firstly, we randomly selected 50 drug–target binding samples from the test set of the PDBbind database (2016 version) and calculated the affinity values of all samples using S2DTA, the predictive model provided by the advanced method DeepDTAF, and MM-GBSA/PBSA to calculate the binding free energy of each sample. We then normalized the absolute values of the real affinity values, predictive affinity values, and binding free energy (most of which are negative). Finally, we evaluated the correlation between the normalized real affinity values and the normalized predictive values using PCC and R^2. Our experimental results (Figure 4) showed that the binding energy calculated by MM-GBSA/PBSA has a poor linear fit with the real affinity values and two evaluation metrics (PCC = 0.597, R^2 = 0.356) reflected their weak correlation. However, both S2DTA (PCC = 0.969, R^2 = 0.941) and DeepDTAF (PCC = 0.921, R^2 = 0.847) exhibited strong correlations with the real affinity values in terms of linear fit and evaluation metrics. Furthermore, we found that S2DTA demonstrated slightly better performance than DeepDTAF, as confirmed by the linear fitting degree and the two evaluation metrics PCC and R^2 in Figure 4. Therefore, our experimental results suggest that S2DTA is an effective and reliable tool for predicting drug–target binding affinity.

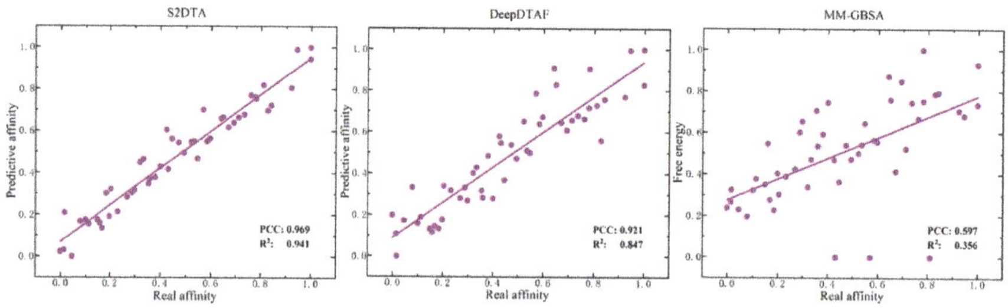

Figure 4. The correlation analysis between the real affinity values and predictive values of S2DTA, DeepDTAF, and MM-GBSA/PBSA. The results show that the real affinity values of the selected samples exhibited a strong linear relationship with the predictive affinity values of both S2DTA (PCC = 0.969, R^2 = 0.941) and DeepDTAF (PCC = 0.921, R^2 = 0.847). Notably, S2DTA showed slightly superior performance compared to DeepDTAF. In contrast, we observed that the real affinity values of the selected samples did not align well with the free energy predicted by MM-GBSA/PBSA (PCC = 0.597, R^2 = 0.356) in terms of a linear relationship, and its predictive values exhibited a weak correlation with the real affinity values.

3. Materials and Methods

3.1. Datasets

In line with current competing methods [9–11], we utilized a benchmark dataset from the PDBbind database (2016 version) [36] to train, validate, and test our model. This dataset comprises 13,283 high-resolution structures of drug–target complexes sourced from the Protein Data Bank (PDB, https://www.rcsb.org (accessed on 16 September 2023)), along with experimental values of DTA, typically represented by pK_d. The training and testing datasets were constructed from the PDBbind database (2016 version), resulting in 12,993 and 290 samples, respectively. To ensure the training efficiency of the model, the sequence length of targets was fixed at 2100 amino acids to cover 98.74% of samples (Supplementary Figure S1) of the training dataset. A total of four samples were excluded due to processing issues with the Biopython package [37]. Consequently, the training dataset comprises 12,823 samples, while the testing dataset has 289 samples. The maximum length of drug SMILES was chosen as the fixed length. Sequences of targets or drug SMILES shorter than their fixed lengths were zero-padded.

3.2. Sequence Representation

We utilized label encoding and one-hot encoding as baseline representations to depict the 20 commonly used amino acids in targets and the 27 different atoms in drugs. Label encoding employs a single number to differentiate objects, while one-hot encoding maps encoded objects to status registers, generating encoding vectors. To characterize the diverse amino acids of targets and atoms of drugs, we selected 24 (Supplementary Table S1) [38] and 9 physical–chemical properties (Supplementary Table S2), respectively. For the sequence features of targets, we extracted the PSSM and utilized HHBlits for aligning homologous sequences to achieve the HMM. PSSM and HMM matrices, with dimensions of 20, encapsulate the features of the 20 commonly used amino acids in the sequences of targets. Furthermore, we harnessed multiple sequence-processing models, including Protbert, CNN, and Word2Vec, to extract semantic features from the sequences of targets.

3.3. Structure Representation

To exploit structural insight, we devised a graph-based data format for drugs, targets, and pockets. In the case of drugs, the individual atoms composing the drug formed the vertices of its molecular graph. Covalent bonds from the .sdf or .mol2 files of drugs were employed to establish edges. Regarding the graphs of targets, the carbon α atoms of amino acid residues in the structures of targets were designated as vertices, and edges were established between any two carbon α atoms within a Euclidean distance of 8 Å. The structure representation of the pocket was consistent with the target.

3.4. Proposed S2DTA Approach

In this study, a deep learning method consisting of heterogeneous models, called S2DTA (see Figure 5), was proposed to accurately and interpretively predict DTA by fusing the sequence and structural knowledge of drugs and targets. S2DTA is an end-to-end model that contained four modules: data input module, sequence learning module, structure learning module, and feature fusion module.

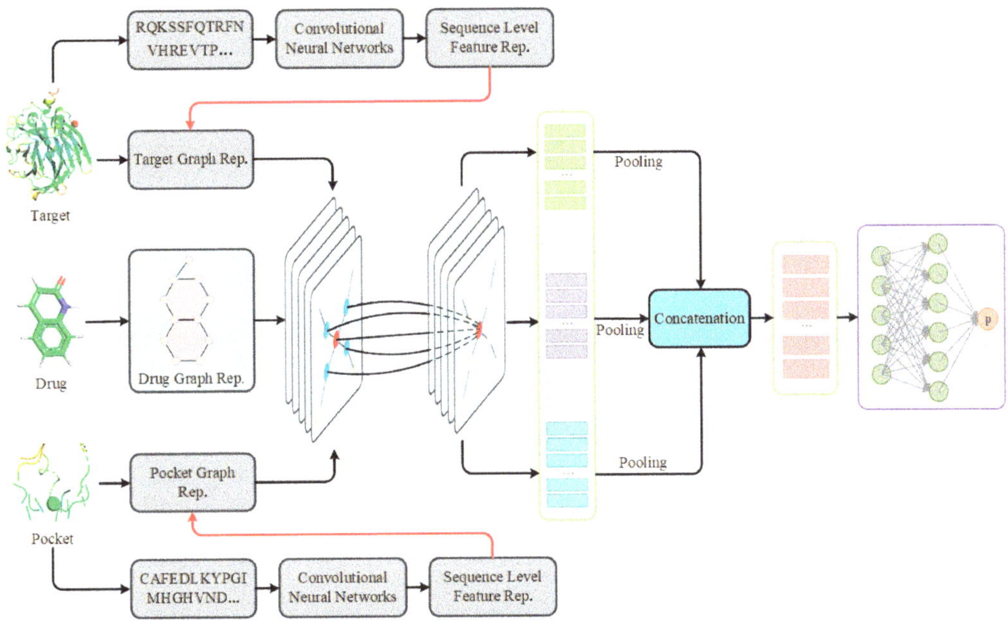

Figure 5. The architecture of S2DTA. S2DTA is an advanced model designed to predict drug–target affinity (DTA) by leveraging sequence features of drug SMILES, targets, and pockets, and their corresponding structural features. This model comprises four essential modules: (1) The data input module was responsible for representing the sequence and structural data of drugs, targets, and pockets. (2) Within the sequence learning module, a 1D-CNN layer was employed to extract semantic features from both targets and pockets, based on their sequence data. (3) In the structure learning module, three independent Graph Convolutional Networks (GCNs) were employed to extract high-level features from the vertices present in the graph-based representation of drugs, targets, and pockets. (4) The feature fusion module encompasses a two-layer fully connected (FC) network, which played a pivotal role in predicting DTA by integrating the extracted features.

(1) Data input module: S2DTA processed the drug SMILES and their structure data, targets, and pockets, where input sequences were represented as a one-hot encoded feature vector, and structure data included the molecular graphs for drugs and the graphs of targets and pockets.

(2) Sequence learning module: A 1D-CNN layer was employed to extract semantic features from both targets and pockets based on their sequence data. The 20-dimensional sequence data of targets underwent convolution with a kernel size of 5, and a stride of 1, resulting in an output dimension of 128. Meanwhile, the 20-dimensional sequence data of pockets utilized a convolution kernel of size 3, with the remaining parameters identical to those used for targets.

(3) Structure learning module: Three independent GCN networks, utilizing the message-passing mechanism, were employed to extract high-level features from vertices within the graph-based representation of drugs, targets, and pockets. Each GCN network consisted of three layers of graph convolutions. In the case of targets and pockets, the feature dimension of each vertex was 128, with the three graph convolutional layers having dimensions of 32, 64, and 128, sequentially. A 128-dimensional feature vector obtained from the final graph convolutional layer for targets and pockets was linearly transformed into a 256-dimensional feature vector. For drugs, the 27-dimensional sequence data consisted in the feature dimension of their vertices, and the dimensions of the three graph convolutional layers were 32, 64, and 128. Similarly, a 128-dimensional feature vector of the drug from the third graph convolutional layer was linearly transformed into a 256-dimensional feature vector.

(4) Feature fusion module: The features extracted from drugs, targets, and pockets through GCN networks were concatenated, resulting in a comprehensive 256-dimensional feature vector. This feature vector was then fed into a FC network to predict DTA. The FC network comprised two linear layers. The initial linear layer, acting as a hidden layer, comprised 256 neurons. The subsequent linear layer, serving as the output layer, consisted of a single neuron responsible for producing the affinity.

3.5. Training Process

S2DTA underwent a training process spanning 300 iterations, with optimization carried out using the Adam optimizer [39]. The LeakyReLU activation function [40] was employed, and mean square error (MSE) served as the chosen loss function. The hyperparameters adopted for S2DTA were as follows: a learning rate of 0.0001, a batch size of 128, and a dropout rate of 0.4.

3.6. Comparison Strategy

A comprehensive set of machine learning (ML) and deep learning (DL) models, and MM-GBSA/PBSA were employed in this study to establish a robust baseline for evaluation. The ML models encompassed ridge regression (RR), K nearest neighbor (KNN), support vector machine regression (SVR), decision tree (DT), and random forest (RF). On the other hand, the DL models comprised 1D-CNN and RNN. Additionally, a variety of graph-based models, including GCN, ChebyNet [41], GraphSAGE [42], GAT [43], and GIN [44], were tested.

The implementation of ML models was facilitated through the scikit-learn library (https://scikit-learn.org (accessed on 16 September 2023)), while the DL models were created using the Pytorch package (https://pytorch.org/ (accessed on 16 September 2023)). Furthermore, the graph-based models applied the capabilities of the Deep Graph Library (DGL, https://www.dgl.ai/ (accessed on 16 September 2023)) to carry out their computations and operations.

3.7. Evaluation Metrics

Six evaluation metrics including mean absolute error (MAE), root mean square error (RMSE), Pearson Correlation Coefficient (PCC), Spearman, Concordance Index (CI), and

R-squared (R^2) were used to comprehensively evaluate the performance of the model. MAE is used to measure the mean absolute error between predictive and real values. It reflects the size of the real prediction error; RMSE is often used to measure the deviation between predictive and real values; PCC is used to measure the mutual relationship between two variables; Spearman is a nonparametric measure of the dependence of two variables; CI is an index used to evaluate the prediction accuracy of the model; R^2 is mainly used to measure how well the predictive value fits the real value.

4. Conclusions

In this study, we proposed S2DTA, a graph-based deep learning model designed to predict DTA with accuracy and interpretability. By leveraging heterogeneous models and fusing sequence and structural knowledge, S2DTA demonstrated novel advancements in DTA prediction. The S2DTA model introduces two main innovations. Firstly, it integrates sequence features of drug SMILES, targets, and pockets and their corresponding structural features. This builds upon previous research which primarily focused on drug SMILES and the sequences of targets, thereby enhancing the overall performance of the model. Secondly, the incorporation of structural knowledge not only boosts predictive accuracy but also offers interpretability for DTA. However, it is important to recognize three limitations of S2DTA.

AlphaFold can only effectively predict the structures of monomeric proteins and proteins without extensive disordered regions.

In cases where pocket knowledge for drug–target binding is missing from datasets, precise pocket identification tools or biological experiments become necessary to provide this crucial knowledge.

S2DTA is a deep-learning model specifically designed to predict drug–target binding affinity. It should be noted that there are certain limitations when it comes to predicting the binding affinity between drugs and other entities, such as drugs and nucleic acids or drugs and other drugs.

While S2DTA has demonstrated itself as an effective and interpretable predictive model for DTA, its refinement is still underway with a focus on two key aspects: (1) Although S2DTA utilizes CNN effectively to extract local sequence knowledge, it will be advanced to incorporate the extraction of long-range dependence knowledge. This expansion will broaden the understanding of the intricate relationship of the model within sequences and contribute to its predictive performance. (2) Recognizing the substantial impact that surface pocket features have on DTA, S2DTA will undergo further development to extract and integrate these features into its predictive framework.

In summary, the accessibility of drug molecular maps and target tertiary structures has paved the way for exciting prospects in drug–target affinity research. Applying CNN, GNN, and 3D-CNN models to extract high-level features from the drug SMILES, target sequences, and their structures, along with complementary fusion of the extracted heterogeneous knowledge, enables us to fully harness the potential of both sequence and structural data. Undoubtedly, the integration of deep learning with the drug SMILES, target sequences, and their structures will emerge as a prominent and promising research direction in the future.

Supplementary Materials: The following supporting information can be downloaded at: https://www.mdpi.com/article/10.3390/molecules28248005/s1, Figure S1: A sequence length distribution statistic chart of targets in 2016 version dataset of PDBbind database; Table S1: 24 physical-chemical properties of amino acids in proteins; Table S2: 9 physical-chemical properties of atoms in drugs

Author Contributions: X.Z.: Ideas, writing—original manuscript, writing—review and editing, writing—revised manuscript. K.-Y.Z.: Creation of models, data curation, implementation of the computer code and supporting algorithms. B.J.: Supervision. Y.L.: Development of methodology, formal analysis, writing—review and editing S1, modification guidance. All authors have read and agreed to the published version of the manuscript.

Funding: This work was supported by the National Natural Sciences Foundation of China (No. 62241602), Yunnan Fundamental Research Projects (No. 202101BA070001-227), and the Yunnan Key Laboratory of Screening and Research on Anti-pathogenic Plant Resources from Western Yunnan (No. APR202110).

Institutional Review Board Statement: Not applicable.

Informed Consent Statement: Not applicable.

Data Availability Statement: The source data and code repository can be accessed at https://github.com/dldxzx/S2DTA (accessed on 16 September 2023).

Acknowledgments: We wish to thank Shu-Qun Liu and Yuan-Ling Xia from Yunnan University for their advice on the revision.

Conflicts of Interest: The authors declare no conflict of interest.

References

1. Chen, X.; Yan, C.C.; Zhang, X.; Zhang, X.; Dai, F.; Yin, J.; Zhang, Y. Drug–Target Interaction Prediction: Databases, Web Servers and Computational Models. *Brief. Bioinform.* **2016**, *17*, 696–712. [CrossRef]
2. Santos, R.; Ursu, O.; Gaulton, A.; Bento, A.P.; Donadi, R.S.; Bologa, C.G.; Karlsson, A.; Al-Lazikani, B.; Hersey, A.; Oprea, T.I.; et al. A Comprehensive Map of Molecular Drug Targets. *Nat. Rev. Drug Discov.* **2017**, *16*, 19–34. [CrossRef]
3. Mofidifar, S.; Sohraby, F.; Bagheri, M.; Aryapour, H. Repurposing Existing Drugs for New AMPK Activators as a Strategy to Extend Lifespan: A Computer-Aided Drug Discovery Study. *Biogerontology* **2018**, *19*, 133–143. [CrossRef]
4. Du, X.; Li, Y.; Xia, Y.-L.; Ai, S.-M.; Liang, J.; Sang, P.; Ji, X.-L.; Liu, S.-Q. Insights into Protein–Ligand Interactions: Mechanisms, Models, and Methods. *Int. J. Mol. Sci.* **2016**, *17*, 144. [CrossRef]
5. Sehnal, D.; Svobodová Vařeková, R.; Berka, K.; Pravda, L.; Navrátilová, V.; Banáš, P.; Ionescu, C.-M.; Otyepka, M.; Koča, J. MOLE 2.0: Advanced Approach for Analysis of Biomacromolecular Channels. *J. Cheminform.* **2013**, *5*, 39. [CrossRef]
6. Hu, G.; Li, X.; Zhang, X.; Li, Y.; Ma, L.; Yang, L.-M.; Liu, G.; Li, W.; Huang, J.; Shen, X.; et al. Discovery of Inhibitors To Block Interactions of HIV-1 Integrase with Human LEDGF/P75 via Structure-Based Virtual Screening and Bioassays. *J. Med. Chem.* **2012**, *55*, 10108–10117. [CrossRef]
7. Zhang, H.; Liao, L.; Cai, Y.; Hu, Y.; Wang, H. IVS2vec: A Tool of Inverse Virtual Screening Based on Word2vec and Deep Learning Techniques. *Methods* **2019**, *166*, 57–65. [CrossRef] [PubMed]
8. Abbasi, K.; Razzaghi, P.; Poso, A.; Ghanbari-Ara, S.; Masoudi-Nejad, A. Deep Learning in Drug Target Interaction Prediction: Current and Future Perspectives. *Curr. Med. Chem.* **2021**, *28*, 2100–2113. [CrossRef] [PubMed]
9. Öztürk, H.; Özgür, A.; Ozkirimli, E. DeepDTA: Deep Drug–Target Binding Affinity Prediction. *Bioinformatics* **2018**, *34*, i821–i829. [CrossRef]
10. Wang, K.; Zhou, R.; Li, Y.; Li, M. DeepDTAF: A Deep Learning Method to Predict Protein–Ligand Binding Affinity. *Brief. Bioinform.* **2021**, *22*, bbab072. [CrossRef] [PubMed]
11. Nguyen, T.; Le, H.; Quinn, T.P.; Nguyen, T.; Le, T.D.; Venkatesh, S. GraphDTA: Predicting Drug–Target Binding Affinity with Graph Neural Networks. *Bioinformatics* **2021**, *37*, 1140–1147. [CrossRef]
12. Ain, Q.U.; Aleksandrova, A.; Roessler, F.D.; Ballester, P.J. Machine-Learning Scoring Functions to Improve Structure-Based Binding Affinity Prediction and Virtual Screening: Machine-Learning SFs to Improve Structure-Based Binding Affinity Prediction and Virtual Screening. *Wiley Interdiscip. Rev. Comput. Mol. Sci.* **2015**, *5*, 405–424. [CrossRef]
13. Kimber, T.B.; Chen, Y.; Volkamer, A. Deep Learning in Virtual Screening: Recent Applications and Developments. *Int. J. Mol. Sci.* **2021**, *22*, 4435. [CrossRef]
14. Karimi, M.; Wu, D.; Wang, Z.; Shen, Y. DeepAffinity: Interpretable Deep Learning of Compound–Protein Affinity through Unified Recurrent and Convolutional Neural Networks. *Bioinformatics* **2019**, *35*, 3329–3338. [CrossRef] [PubMed]
15. Yuan, W.; Chen, G.; Chen, C.Y.-C. FusionDTA: Attention-Based Feature Polymerizer and Knowledge Distillation for Drug-Target Binding Affinity Prediction. *Brief. Bioinform.* **2022**, *23*, bbab506. [CrossRef] [PubMed]
16. Öztürk, H.; Ozkirimli, E.; Özgür, A. WideDTA: Prediction of Drug-Target Binding Affinity. *arXiv* **2019**, arXiv:1902.04166.
17. Zhao, Q.; Duan, G.; Yang, M.; Cheng, Z.; Li, Y.; Wang, J. AttentionDTA: Drug-Target Binding Affinity Prediction by Sequence-Based Deep Learning with Attention Mechanism. *IEEE/ACM Trans. Comput. Biol. Bioinform.* **2022**, *20*, 852–863. [CrossRef]
18. Abbasi, K.; Razzaghi, P.; Poso, A.; Amanlou, M.; Ghasemi, J.B.; Masoudi-Nejad, A. DeepCDA: Deep Cross-Domain Compound–Protein Affinity Prediction through LSTM and Convolutional Neural Networks. *Bioinformatics* **2020**, *36*, 4633–4642. [CrossRef]
19. Zhang, L.; Wang, C.-C.; Chen, X. Predicting Drug–Target Binding Affinity through Molecule Representation Block Based on Multi-Head Attention and Skip Connection. *Brief. Bioinform.* **2022**, *23*, bbac468. [CrossRef]
20. Weininger, D. SMILES, a Chemical Language and Information System. 1. Introduction to Methodology and Encoding Rules. *J. Chem. Inf. Model.* **1988**, *28*, 31–36. [CrossRef]
21. Kang, L.; Ye, P.; Li, Y.; Doermann, D. Convolutional Neural Networks for No-Reference Image Quality Assessment. In Proceedings of the 2014 IEEE Conference on Computer Vision and Pattern Recognition, Columbus, OH, USA, 23–28 June 2014; pp. 1733–1740.

22. Cho, K.; van Merrienboer, B.; Gulcehre, C.; Bahdanau, D.; Bougares, F.; Schwenk, H.; Bengio, Y. Learning Phrase Representations Using RNN Encoder–Decoder for Statistical Machine Translation. In Proceedings of the 2014 Conference on Empirical Methods in Natural Language Processing (EMNLP), Doha, Qatar, 25–29 October 2014; pp. 1724–1734.
23. Vaswani, A.; Shazeer, N.; Parmar, N.; Uszkoreit, J.; Jones, L.; Gomez, A.N.; Kaiser, L.; Polosukhin, I. Attention Is All You Need 2017. Available online: https://arxiv.org/abs/1706.03762 (accessed on 5 December 2023).
24. Yu, F.; Koltun, V. Multi-Scale Context Aggregation by Dilated Convolutions. In Proceedings of the ICLR, San Juan, Puerto Rico, 2–4 May 2016. Available online: https://arxiv.org/abs/1511.07122 (accessed on 5 December 2023).
25. Yang, Z.; Zhong, W.; Zhao, L.; Yu-Chian Chen, C. MGraphDTA: Deep Multiscale Graph Neural Network for Explainable Drug–Target Binding Affinity Prediction. *Chem. Sci.* **2022**, *13*, 816–833. [CrossRef]
26. Jiang, M.; Li, Z.; Zhang, S.; Wang, S.; Wang, X.; Yuan, Q.; Wei, Z. Drug–Target Affinity Prediction Using Graph Neural Network and Contact Maps. *RSC Adv.* **2020**, *10*, 20701–20712. [CrossRef] [PubMed]
27. Bronstein, M.M.; Bruna, J.; LeCun, Y.; Szlam, A.; Vandergheynst, P. Geometric Deep Learning: Going beyond Euclidean Data. *IEEE Signal Process. Mag.* **2017**, *34*, 18–42. [CrossRef]
28. Jumper, J.; Evans, R.; Pritzel, A.; Green, T.; Figurnov, M.; Ronneberger, O.; Tunyasuvunakool, K.; Bates, R.; Žídek, A.; Potapenko, A.; et al. Highly Accurate Protein Structure Prediction with AlphaFold. *Nature* **2021**, *596*, 583–589. [CrossRef] [PubMed]
29. Jeong, J.C.; Lin, X.; Chen, X.-W. On Position-Specific Scoring Matrix for Protein Function Prediction. *IEEE/ACM Trans. Comput. Biol. Bioinform.* **2011**, *8*, 308–315. [CrossRef]
30. Remmert, M.; Biegert, A.; Hauser, A.; Söding, J. HHblits: Lightning-Fast Iterative Protein Sequence Searching by HMM-HMM Alignment. *Nat. Methods* **2012**, *9*, 173–175. [CrossRef]
31. Zhang, Y.; Zhu, G.; Li, K.; Li, F.; Huang, L.; Duan, M.; Zhou, F. HLAB: Learning the BiLSTM Features from the ProtBert-Encoded Proteins for the Class I HLA-Peptide Binding Prediction. *Brief. Bioinform.* **2022**, *23*, bbac173. [CrossRef] [PubMed]
32. Adjuik, T.A.; Ananey-Obiri, D. Word2vec Neural Model-Based Technique to Generate Protein Vectors for Combating COVID-19: A Machine Learning Approach. *Int. J. Inf. Technol.* **2022**, *14*, 3291–3299. [CrossRef]
33. Zhang, J.; Zhu, M.; Qian, Y. Protein2vec: Predicting Protein-Protein Interactions Based on LSTM. *IEEE/ACM Trans. Comput. Biol. Bioinform.* **2020**, *19*, 1257–1266. [CrossRef]
34. Wei, B.; Zhang, Y.; Gong, X. DeepLPI: A Novel Deep Learning-Based Model for Protein–Ligand Interaction Prediction for Drug Repurposing. *Sci. Rep.* **2022**, *12*, 18200. [CrossRef]
35. Yang, M.; Bo, Z.; Xu, T.; Xu, B.; Wang, D.; Zheng, H. Uni-GBSA: An Open-Source and Web-Based Automatic Workflow to Perform MM/GB(PB)SA Calculations for Virtual Screening. *Brief. Bioinform.* **2023**, *24*, bbad218. [CrossRef] [PubMed]
36. Wang, R.; Fang, X.; Lu, Y.; Yang, C.-Y.; Wang, S. The PDBbind Database: Methodologies and Updates. *J. Med. Chem.* **2005**, *48*, 4111–4119. [CrossRef] [PubMed]
37. Cock, P.J.A.; Antao, T.; Chang, J.T.; Chapman, B.A.; Cox, C.J.; Dalke, A.; Friedberg, I.; Hamelryck, T.; Kauff, F.; Wilczynski, B.; et al. Biopython: Freely Available Python Tools for Computational Molecular Biology and Bioinformatics. *Bioinformatics* **2009**, *25*, 1422–1423. [CrossRef] [PubMed]
38. Elhanati, Y.; Sethna, Z.; Marcou, Q.; Callan, C.G., Jr.; Mora, T.; Walczak, A.M. Inferring Processes Underlying B-Cell Repertoire Diversity. *Philos. Trans. R. Soc. B Biol. Sci.* **2015**, *370*, 20140243. [CrossRef]
39. Isola, P.; Zhu, J.-Y.; Zhou, T.; Efros, A.A. Image-to-Image Translation with Conditional Adversarial Networks. In Proceedings of the 2017 IEEE Conference on Computer Vision and Pattern Recognition (CVPR), Honolulu, HI, USA, 21–26 July 2017; pp. 5967–5976.
40. Tang, Z.; Luo, L.; Xie, B.; Zhu, Y.; Zhao, R.; Bi, L.; Lu, C. Automatic Sparse Connectivity Learning for Neural Networks. *IEEE Trans. Neural Netw. Learn. Syst.* **2022**, *34*, 7350–7364. [CrossRef]
41. Defferrard, M.; Bresson, X.; Vandergheynst, P. Convolutional Neural Networks on Graphs with Fast Localized Spectral Filtering. In Proceedings of the 30th Conference on Neural Information Processing Systems (NIPS 2016), Barcelona, Spain, 9 December 2016; pp. 3844–3852.
42. Hamilton, W.; Ying, Z.; Leskovec, J. Inductive Representation Learning on Large Graphs. In Proceedings of the Advances in Neural Information Processing Systems; Guyon, I., Luxburg, U.V., Bengio, S., Wallach, H., Fergus, R., Vishwanathan, S., Garnett, R., Eds.; Curran Associates, Inc.: Red Hook, NY, USA, 2017; Volume 30, pp. 1024–1034.
43. Xie, Y.; Zhang, Y.; Gong, M.; Tang, Z.; Han, C. MGAT: Multi-View Graph Attention Networks. *Neural Netw.* **2020**, *132*, 180–189. [CrossRef]
44. Xu, K.; Hu, W.; Leskovec, J.; Jegelka, S. How Powerful Are Graph Neural Networks? In Proceedings of the ICLR 2019, New Orleans, LA, USA, 6–9 May 2019.

molecules

MDPI

Article

Core-Hole Excitation Spectra of the Oxides and Hydrates of Fullerene C_{60} and Azafullerene $C_{59}N$

Xiong Li [1,2], Shuyi Wang [1,3], Jingdong Guo [3], Ziye Wu [4], Changrui Guo [4], Shaohong Cai [4,5,*] and Mingsen Deng [4,*]

1 College of Big Data and Information Engineering, Guizhou University, Guiyang 550025, China; lix@ecut.edu.cn (X.L.); shuyiwang@gznc.edu.cn (S.W.)
2 School of Science, East China University of Technology, Nanchang 330013, China
3 Guizhou Provincial Key Laboratory of Computational Nano-Material Science, Guizhou Education University, Guiyang 550018, China; jingdongguo@126.com
4 School of Information, Guizhou University of Finance and Economics, Guiyang 550025, China; zywu@mail.gufe.edu.cn (Z.W.); guochangrui@mail.gufe.edu.cn (C.G.)
5 Department of Resources and Environment, Moutai Institute, Renhuai 564507, China
* Correspondence: caish@mail.gufe.edu.cn (S.C.); msdeng@mail.gufe.edu.cn (M.D.)

Abstract: The interaction of fullerenes and their derivatives with environmental molecules such as oxygen or water was crucial for the rational design of low-dimensional materials and devices. In this paper, the near-edge X-ray absorption fine structure (NEXAFS), X-ray emission spectroscopy (XES) and X-ray photoelectron spectroscopy (XPS) shake-up satellites were employed to distinguish the oxides and hydrates of the fullerene C_{60} and azafullerene $C_{59}N$ families. The study includes various isomers, such as the open [5,6] and closed [6,6] isomers of $C_{60}O$, $C_{60}H(OH)$, C_{60}-O-C_{60}, $C_{60}H$-O-$C_{60}H$, $C_{59}N(OH)$ and $C_{59}N$-O-$C_{59}N$, based on density functional theory. These soft X-ray spectra offered comprehensive insights into the molecular orbitals of these azafullerene molecular groups. The oxygen *K*-edge NEXAFS, carbon and oxygen *K*-edge XPS shake-up satellite spectra provided valuable tools for distinguishing oxides or hydrates of fullerene C_{60} and azafullerene $C_{59}N$. Our findings could significantly benefit the development of fullerene functional molecular materials and expand the application scope of soft X-ray spectroscopy as a molecular fingerprinting tool for the fullerene family.

Keywords: azafullerene; oxide; hydrate; NEXAFS; XES; XPS shake-up

1. Introduction

The carbon allotrope fullerene C_{60} and its molecular crystals are considered fundamental units in probing the mechanisms of transport and photoelectric processes. Their fast carrier mobility and strong electron affinity position fullerenes are at the forefront of electronic device applications, including field-effect transistors [1] or photovoltaics [2], nanomedicine [3], nanoelectronics [4], organic and inorganic nanocomposites [5,6]. The azafullerene $C_{59}N$ monomer, notable for its unpaired electrons and resultant free radical behavior [7], remains a vibrant research area despite its instability. For example, the report of Martín-Gomis et al. showed promising implications in the development of organic solar cells built with covalent donor-acceptor systems using $C_{59}N$ as the acceptor [8]. Further, the $C_{59}N$ system has been shown to achieve a more persistent charge-separated state [9], underscoring its potential in advanced applications.

However, the insertion of extrinsic species, such as oxygen and water, could substantially alter the physical properties of these electronic materials [10]. Numerous studies have reported evidence for the sensitivity of fullerene or azafullerene to the incorporation of such impurities [11–14]. To further understand these impacts, Lin et al. conducted a detailed study on the oxidation process of azafullerene $C_{59}N$ [15]. Research has established that oxides and hydrates of fullerene and azafullerene are key materials in constructing a variety of fullerene-based nanostructures [16,17]. As a straightforward derivative of fullerene,

$C_{60}O$ has emerged as a foundational material for producing more complex fullerene oxides, playing a crucial role in the synthesis of diverse compounds [18–20]. Fedurco et al. and Jones et al. demonstrated that N@$C_{60}O$ represented a valuable intermediary for researchers engaged in quantum information processing at the molecular level [21]. The work of Mondal et al. has also highlighted the complex interplay between thermal and mechanical transformations within the $C_{60}O$ cluster, which could influence its thermoelectric properties through the modulation of intermolecular interactions [22]. Moreover, the $C_{59}N$ oxides have been identified as robust recyclable catalysts, featuring a rate-determining energy barrier of 0.80 eV [23].

The presence of oxides and water-related reactions in fullerene and azafullerene compounds often leads to substantial modifications in the physical properties of organic semiconductors. These modifications are instrumental in the fabrication of numerous fullerene nanostructures. However, the properties of the oxides and hydrates exhibit significant variations [18]. Consequently, understanding their electronic and chemical characteristics is essential for further exploration and utilization. In this regard, Palotás et al. have made a noteworthy contribution by providing the initial experimental infrared multiple-photon dissociation vibrational spectra of gaseous $C_{60}O^+$ and $C_{60}OH^+$. However, challenges persist, as the experimental band centered at 1333 cm^{-1} was not accurately reproduced by the theoretical [6,6]$C_{60}OH^+$ spectrum [24]. This discrepancy demands considerable efforts dedicated to acquiring a deeper understanding of these materials, as evident in the studies mentioned earlier [25–28], where X-ray spectra have proven to be valuable experimental methods for investigating the electronic structures of aza[60]fullerene [29–31].

Near-edge X-ray absorption fine structure (NEXAFS) is a well-established spectroscopic approach employed for examining the local molecular structure and acquiring insights into unoccupied molecular orbitals. X-ray emission spectroscopy (XES) was frequently utilized to investigate the occupancy density of molecular electronic states. Both NEXAFS and XES offered element and site-specific capabilities, making them potent tools for elucidating intricate electronic properties of materials. Additionally, X-ray photoelectron spectroscopy (XPS) shake-up satellites, engendered by excitations of valence electrons concomitant with the ionization of core electrons, provided information regarding the valence band of materials in the vicinity of the core-hole screening [32–39]. For instance, Klein et al. utilized the delta-self-consistent-field (ΔSCF) and transition potential (TP) methods for calculating XPS and NEXAFS transitions, allowing the prediction of *K*-shell information [40]. Meanwhile, Armillotta et al. investigated the mechanisms of metal intermolecular coordination using NEXAFS technology [41].

The open [5,6] and closed [6,6] isomers of $C_{60}O$, as well as C_{60}-O-C_{60}, are examples of molecular crystals that had been synthesized [42–44]. Creegan et al. identified the epoxide isomer of $C_{60}O$ through ^{13}C NMR and FTIR examinations, revealing that the oxygen atom bridges to carbon atoms of two hexagons with C_{2v} symmetry [42]. Weisman et al. further demonstrated that $C_{60}O$ can be prepared in both closed [6,6] epoxide and open [5,6] annulene isomers, depending on the synthesis method [44]. The C_{60}-O-C_{60} molecular crystal consists of a C_{60} dimer, and an oxygen atom bridged in a furan-like ring (Figure 1e) [43], while the two isomers of $C_{60}O$ are composed of a C_{60} monomer and an oxygen atom in the form of epoxide and annulene, respectively (Figure 1b,c) [42]. Erbahar et al. discovered that the most stable structure for a single oxygen atom within $(C_{59}N)_2$ involves the oxygen atom in an ether configuration, forming a cross-linking bond between two ($C_{59}N$) units [45], namely $C_{59}N$-O-$C_{59}N$ (Figure 1j). In the reaction between H_2O and C_{60}, the configuration of the cross-linking bond could also be considered, specifically $C_{60}H$-O-$C_{60}H$ (Figure 1f). In this study, the NEXAFS, XES and XPS shake-up satellites for seven exemplary members of the hydrates and oxides of fullerene C_{60} and aza[60]fullerene $C_{59}N$, namely open [5,6] and closed [6,6] isomers of $C_{60}O$, $C_{60}H(OH)$, C_{60}-O-C_{60}, $C_{60}H$-O-$C_{60}H$, $C_{59}N(OH)$ and $C_{59}N$-O-$C_{59}N$, were studied by employing the full core-hole (FCH) potential method as well as equivalent core-hole Kohn–Sham (ECH-KS) approach.

The study provides valuable insights that can support and guide further experimental investigations related to these oxide materials and water-involved reactions.

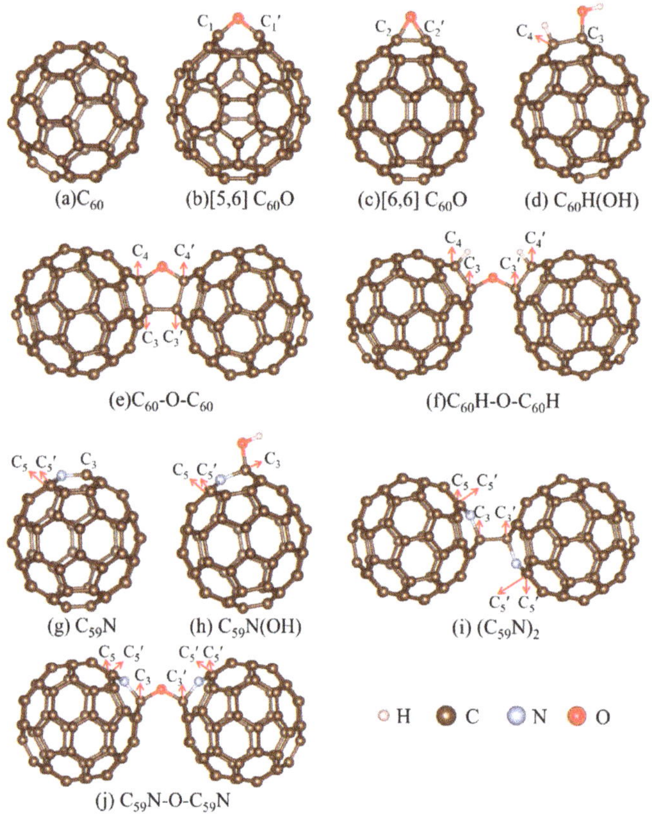

Figure 1. The optimized molecular models of the fullerene C_{60}, aza[60]fullerene $C_{59}N$, their oxides, and hydrates, the carbon atoms C_1 and C_1', C_2 and C_2', C_3 and C_3', C_4 and C_4' and C_5 and C_5' are symmetrically equivalent.

2. Results and Discussion

The optimized molecular models of the fullerene C_{60}, aza[60]fullerene $C_{59}N$, their oxides and hydrates are shown in Figure 1. The carbon atoms C_1 and C_1', C_2 and C_2', C_3 and C_3', C_4 and C_4' and C_5 and C_5' are symmetrically equivalent. The binding energies (BE) of the H atom, O atom, OH, $C_{59}N$, and C_{60} to the $C_{59}N$ and C_{60} monomers were calculated to obtain the stability of oxides and hydrates of the fullerenes and azafullerenes by

$$E_{BE} = E_{tot} - \sum_{i=1}^{n} E_i \quad (1)$$

where E_{tot} is the electronic energy of C_{60}, open [5,6] and closed [6,6] isomers of $C_{60}O$, $C_{60}H(OH)$, C_{60}-O-C_{60}, $C_{60}H$-O-$C_{60}H$, $C_{59}N$, $C_{59}N(OH)$, $(C_{59}N)_2$ and $C_{59}N$-O-$C_{59}N$ molecules, E_i is the electronic energy of the isolated H atom, O atom, OH, $C_{59}N$ and C_{60} molecule, which are the monomers that constitute the ten molecules mentioned above. HOMO-LUMO energy gap (E_{gap}) is calculated by

$$E_{gap} = E_{LUMO} - E_{HOMO} \quad (2)$$

where E_{LUMO} are LUMO energies and E_{HOMO} are HOMO energies. The binding energies, HOMO and LUMO energies and E_{gap} are collected in Table 1, which are verified well with the theoretical investigations based on the B3LYP [46] and LDA-PW91 [45] level. The symmetry point groups and the number of symmetrically inequivalent carbon atoms for the ten molecule groups are also shown in Table 1. The energy gaps decrease upon oxidation and hydration of fullerene C_{60}, whereas the gaps increase for azafullerene $C_{59}N$. The energy gap change trends indicated that the fullerene C_{60} becomes more reactive after oxidation and hydration, while azafullerene $C_{59}N$ becomes less reactive after oxidation and hydration. The binding energies confirm that these oxides and hydrates are more stable than the $(C_{59}N)_2$ molecule, with the $C_{60}H$-O-$C_{60}H$ molecule identified as the most stable oxidation configuration.

Table 1. Binding energies calculated using the B3LYP/6-31G(d,p) level, HOMO and LUMO energies, E_{gap}, symmetry point groups and the number of symmetrically inequivalent carbon atoms for the fullerene C_{60}, aza[60]fullerene $C_{59}N$, their oxides and hydrates.

Molecule	HOMO Energy (eV)	LUMO Energy (eV)	E_{gap} (eV)	BE (kcal/mol)	Sym.	Number of Symmetrically Inequivalent Carbon
C_{60}	−5.99	−3.23	2.76	-	I_h	1
[5,6]$C_{60}O$	−5.90	−3.30	2.60	−78.92	C_s	32
[6,6]$C_{60}O$	−5.94	−3.33	2.61	−76.81	C_{2v}	16
$C_{60}H(OH)$	−5.75	−3.19	2.56	−122.22	C_s	32
C_{60}-O-C_{60}	−5.77	−3.31	2.46	−94.02	C_{2v}	32
$C_{60}H$-O-$C_{60}H$	−5.71	−3.27	2.44	−220.17	C_{2v}	32
$C_{59}N$	−4.58 (α spin) −5.94 (β spin)	−3.29 (α spin) −3.46 (β spin)	1.29 (α spin) 2.48 (β spin)	-	C_s	31
$C_{59}N(OH)$	−5.62	−3.24	2.38	−72.53	C_s	31
$(C_{59}N)_2$	−5.52	−3.29	2.23	−31.81	C_{2h}	31
$C_{59}N$-O-$C_{59}N$	−5.55	−3.33	2.22	−123.60	C_{2v}	31

The HOMOs and LUMOs of the ten molecular groups, calculated with B3LYP/6-311+G*, are presented in Figure S1, which are plotted with VESTA [47]. The isovalues for C_{60}-O-C_{60}, $C_{60}H$-O-$C_{60}H$, $(C_{59}N)_2$ and $C_{59}N$-O-$C_{59}N$ molecules were set to 0.02, while that of the other six molecular groups was set to 0.04. The calculated energy of HOMO and LUMO of C_{60}, $C_{59}N$ and $(C_{59}N)_2$ are in good agreement with the works of Mohammadi and colleagues [48–50]. It is notable that the two isomers of $C_{60}O$ had nearly the same energy gap and only a 2.11 kcal/mol difference in binding energy. Yet, they exhibit distinct electronic features, such as in O_3 adsorption [18].

2.1. Carbon K-edge NEXAFS and XES

The NEXAFS and XES spectra of carbon *K*-edges of the selected molecules, including the open [5,6] and closed [6,6] isomers of $C_{60}O$, $C_{60}H(OH)$, C_{60}-O-C_{60}, $C_{60}H$-O-$C_{60}H$, $C_{59}N(OH)$ and $C_{59}N$-O-$C_{59}N$, were plotted in Figure 2. The spectra of the C_{60}, $C_{59}N$, and $(C_{59}N)_2$ molecules were also plotted for comparison. Also, the experimental NEXAFS spectra of C_{60}, [6,6]$C_{60}O$ and $(C_{59}N)_2$ molecules we could find are reproduced from the work of Käämbre et al. [51] and Schulte et al. [52] for comparison, which are plotted in dashed lines. It is clear that simulated NEXAFS spectra of C_{60} and [6,6]$C_{60}O$ molecules are in good agreement with the experimental results. Our simulated spectra reproduced all the important characteristics of the experiment results.

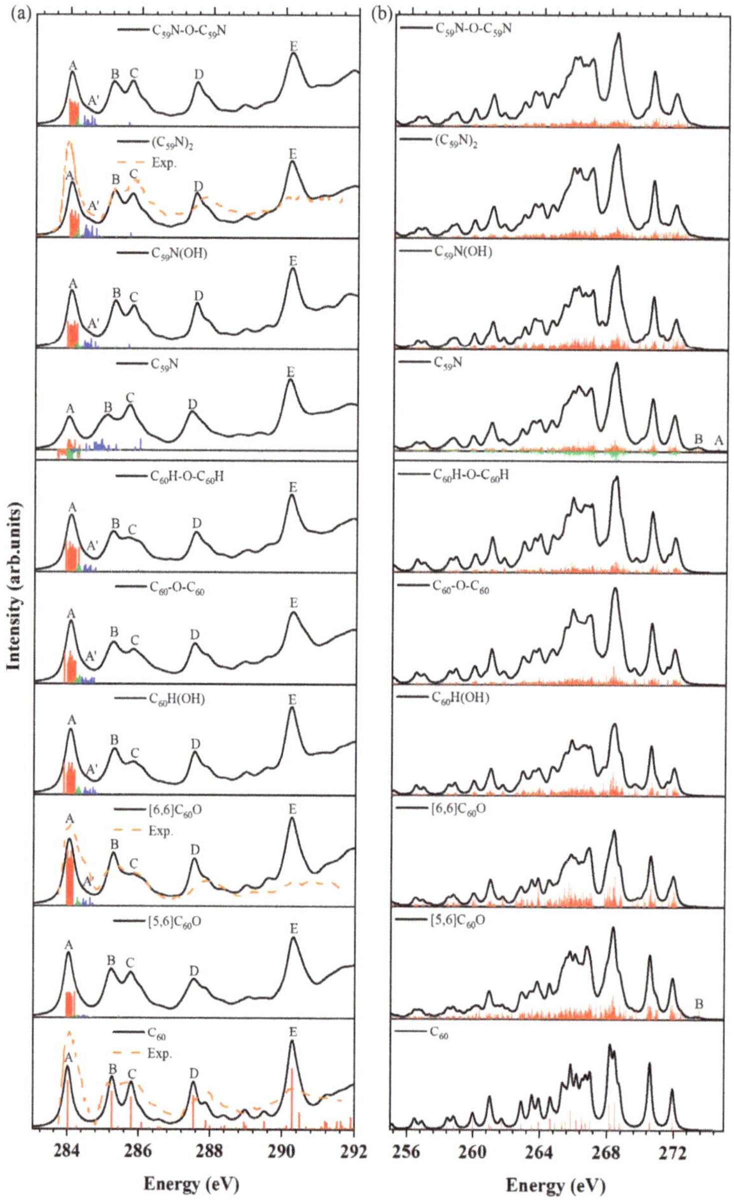

Figure 2. (**a**) NEXAFS spectra at the carbon *K*-edge of the fullerene C_{60}, aza[60]fullerene $C_{59}N$, their oxides and hydrates and $(C_{59}N)_2$. The orbital transitions from the 1*s* to the LUMO (shown as red bars), LUMO + 1 (shown as green bars) and LUMO + 2 (shown as blue bars) in the NEXAFS spectra of the nine molecules, along with the discrete intensities (shown as red bars) of the NEXAFS spectrum of the fullerene C_{60} molecule were plotted below each convoluted spectrum. For the $C_{59}N$ molecule, the contributions from β-spin electrons were marked as negative numbers. Experimental NEXAFS spectra of C_{60}, [6,6]$C_{60}O$ [51] and $(C_{59}N)_2$ molecules [52] have been shifted −0.5 eV (shown as dashed lines). (**b**) Simulated XES spectra at the carbon *K*-edge of the fullerene C_{60}, aza[60]fullerene $C_{59}N$, their oxides and hydrates and $(C_{59}N)_2$. The discrete XES intensities of all molecules were depicted in the form of red bars (α-spin) and green bars (β-spin, represented as negative numbers) within the convoluted XES spectra.

Carbon *K*-edge NEXAFS. The NEXAFS spectra in Figure 2a exhibit five distinct peaks (labeled A–E) at nearly identical energy positions for each molecule, as illustrated in Figure 1. In addition, the orbital transitions from the 1*s* to the LUMO (shown as red bars), LUMO + 1 (shown as green bars) and LUMO + 2 (shown as blue bars) in the NEXAFS spectra of the nine molecules, along with the discrete intensities (shown as red bars) of the NEXAFS spectrum of the fullerene C_{60} molecule were plotted below each convoluted spectrum. For the $C_{59}N$ molecule, the contributions from β-spin electrons are represented as negative numbers. Although the NEXAFS spectra of these ten molecules exhibit similar peak positions, some notable differences can be observed. Specifically, depending on their lower symmetry—C_{2h} for $(C_{59}N)_2$, C_{2v} for closed [6,6] isomer of $C_{60}O$, C_{60}-O-C_{60}, C_{60}H-O-C_{60}H, C_{59}N-O-C_{59}N and *Cs* for open [5,6] isomer of $C_{60}O$, $C_{60}H(OH)$ and $C_{59}N(OH)$, the NEXAFS spectra of these molecules exhibit a larger number of molecular orbital (MO) transitions from C 1*s* orbitals to unoccupied orbitals within each peak in comparison to that of C_{60}. From the feature peak E to their low-energy region, which corresponds to the C *K*-shell 1*s* absorption edges, signals that the C *K*-shell 1*s* absorption edges of all ten molecules are nearly identical. This indicated that both fullerene C_{60} and aza[60]fullerene $C_{59}N$ maintain the same valence state of carbon even after oxidation or exposure to water.

In the oxides and hydrates of fullerene C_{60}, the NEXAFS spectrum of the closed [6,6] isomer of $C_{60}O$ resembles that of the $C_{60}H(OH)$ and C_{60}H-O-C_{60}H, as they all have a relatively weak and broad peak C and a strong peak B, and a very weak feature peak A' between the peaks A and B can be observed. While peak B of the NEXAFS spectra of the other two molecules open the [5,6] isomer of $C_{60}O$, and C_{60}-O-C_{60} exhibit a slightly higher intensity than peak C, the feature peak A' only traced in the C_{60}-O-C_{60} molecule. The feature peak A' primarily arose from the carbon 1*s* orbital to the LUMO + 2. Notably, the transition 1*s* $\rightarrow$ LUMO + 2 is notably weaker in the open [5,6] isomer of $C_{60}O$. The feature peak A' is linked to the change of electronic structures of the C_{60} molecule upon its interaction with another atom or molecule.

It is evident that the NEXAFS spectra for $C_{59}N(OH)$, $(C_{59}N)_2$ and C_{59}N-O-C_{59}N molecules all have a weak feature peak A′, which is the marker of the oxides or reactions involving H_2O of aza[60]fullerene $C_{59}N$. However, they exhibit significant differences in comparison to that of the $C_{59}N$ molecule, which lacks peak A′. Remarkably, nearly all carbon atoms exhibit similar NEXAFS spectral feature peaks A, B, C, D and E, regardless of any specific molecule. However, there is one notable exception, where feature peak A lacks contribution from the nearest carbon atoms (C_1, C_2, C_3, C_4, C_5 in Figure 1) neighboring the nitrogen and oxygen atoms. This distinction in feature peak A among the spectra of the neighbor of the nitrogen, oxygen and other carbon atoms could be attributed to their distinct local (chemical) environments.

Carbon *K*-edge XES. The XES spectra at the carbon *K*-edge of the ten molecules were plotted in Figure 2b. The discrete XES intensities of [5,6]$C_{60}O$ and $C_{59}N$ are shown in red bars (α-spin) and green bars (β-spin). These bars are superimposed on their convoluted XES spectra. The XES spectra of these ten molecules exhibit limited discernible differences in contrast to the NEXAFS spectra, except for specific peaks. The peak A in the $C_{59}N$ molecule can be attributed to the MO transition from the HOMO to the 1*s* orbital of carbon atom C_3 [46], and the peak B can be attributed to the transitions from the HOMO to the 1*s* orbitals of other carbon atoms. Similarly, peak B in the [5,6]$C_{60}O$ isomer primarily arises from the MO transitions between the HOMO and the 1*s* orbital of the carbon atoms. Consequently, it might be beneficial to utilize the feature peak B in the XES spectrum of the [5,6]$C_{60}O$ isomer to examine its presence within C_{60}. Although the experimental aspects might pose certain challenges, employing these distinctive features can provide valuable insights into their existence within the C_{60} crystalline and provide effective evidence for distinguishing the open [5,6] isomers of $C_{60}O$ from oxides and hydrates of C_{60}.

In summary, the carbon *K*-edge NEXAFS and XES spectra provide electronic structure information about the oxides and hydrates of C_{60} and $C_{59}N$, which can be used to distinguish the open [5,6] isomer of $C_{60}O$. However, distinguishing other oxides and hydrates using these spectra remains challenging.

2.2. Nitrogen K-edge NEXAFS and XES

To detect the presence of $C_{59}N(OH)$ and $C_{59}N$-O-$C_{59}N$, which are the oxides or hydrates of aza[60]fullerene $C_{59}N$, the nitrogen *K*-edge NEXAFS and XES spectra were simulated, as shown in Figure 3. The spectra of the $C_{59}N$ and $(C_{59}N)_2$ were also plotted for comparison. The experimental NEXAFS spectrum of $(C_{59}N)_2$ molecule is shown in dashed lines reproduced from the work of Schulte and colleagues [52]. It can be found that the simulated and experimental spectra are in good agreement.

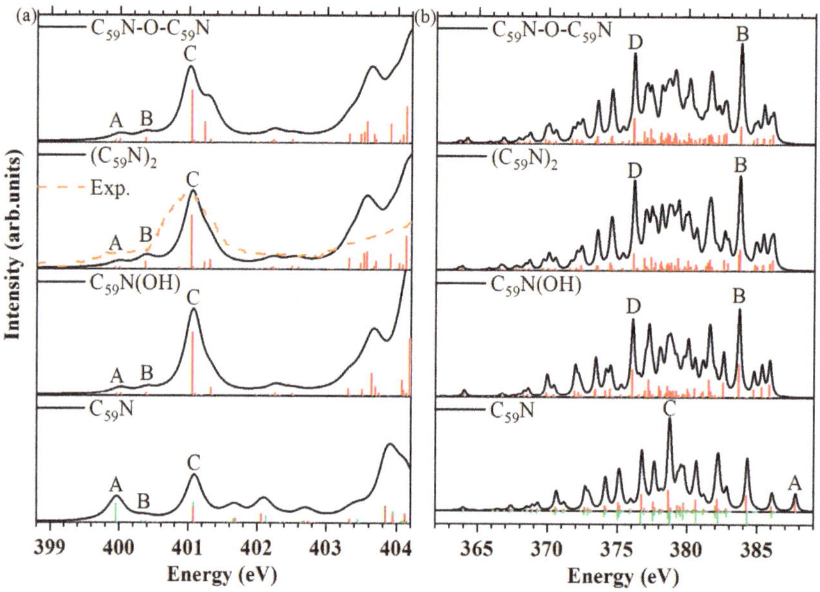

Figure 3. (**a**) Simulated NEXAFS spectra at the nitrogen *K*-edge of $C_{59}N$, $C_{59}N(OH)$, $(C_{59}N)_2$ and $C_{59}N$-O-$C_{59}N$ molecules, discrete intensities before convolution were represented by red bars for α-spin and green bars for β-spin. The experimental NEXAFS spectra of $(C_{59}N)_2$ molecule [52] are shown in dashed lines. (**b**) Simulated XES spectra at the nitrogen *K*-edge of these four molecules, discrete intensities of XES of all molecules were in red (α-spin) and green bars (β-spin, represented as negative numbers) within the convoluted XES spectra.

Nitrogen *K*-edge NEXAFS. Both the $C_{59}N(OH)$ and $C_{59}N$-O-$C_{59}N$ exhibit a faint shoulder A in the NEXAFS spectra (Figure 3a), which is attributed to the nitrogen 1*s* to LUMO transition. In contrast, the $C_{59}N$ molecule presents a relatively more pronounced first peak (peak A) at lower excitation energy, resulting from the nitrogen 1*s* → LUMO transition of β-spin. The MO transitions and corresponding energies of peak or shoulder A are available in Table S1. Despite the significance of the core-hole effect in comprehending NEXAFS spectra, the disparity observed in the nitrogen 1*s* → LUMO MO transition among the four molecules could be elucidated by considering the distribution of the LUMOs in the Ground state (GS). The LUMOs of $C_{59}N$(β-spin), $C_{59}N(OH)$, $(C_{59}N)_2$ and $C_{59}N$-O-$C_{59}N$ were plotted in Figure S1g–j, respectively. The LUMO of $C_{59}N$ exhibits a noticeable distribution around the nitrogen atom, whereas the LUMO of $C_{59}N(OH)$ exhibits a pronounced carbon character, similar to the $(C_{59}N)_2$, which fits well with prior results [53,54]. The LUMO in the $C_{59}N$-O-$C_{59}N$ molecule is predominantly concentrated on the intermolecular bond of carbon, which contributes to the existence of a weak shoulder A in their NEXAFS spectra.

Notably, the weak shoulder A observed in $C_{59}N(OH)$ and $C_{59}N$-O-$C_{59}N$ are valuable for distinguishing them from $C_{59}N$.

Nitrogen *K*-edge XES. The XES spectra at the nitrogen *K*-edge provide valuable insights, particularly when combining information obtained from GS calculations. In the case of the $C_{59}N(OH)$, $(C_{59}N)_2$ and $C_{59}N$-O-$C_{59}N$, their HOMOs exhibit some nitrogen characteristics, as illustrated in Figure S1h–j. For those compounds, the transitions from the HOMO to the nitrogen 1*s* orbital occur at relatively lower energies compared with the energy corresponding to the distinct peak A in the $C_{59}N$ molecule. The peak A in $C_{59}N$ originates from the transition from the HOMO (α-spin) to nitrogen 1s orbital transition [46]. Interestingly, the XES spectra of both $C_{59}N(OH)$ and $C_{59}N$-O-$C_{59}N$ exhibit strong and sharp feature peaks B and D, resembling those observed in $(C_{59}N)_2$. This similarity is in stark contrast to the sharp and pronounced feature, peak C, observed in the $C_{59}N$. These peaks are a result of transitions originating from deeper valence orbitals to the nitrogen 1*s* orbital(s), allowing for a clear distinction between $C_{59}N(OH)$ and $C_{59}N$-O-$C_{59}N$ when compared with $C_{59}N$.

2.3. Oxygen K-edge NEXAFS and XES

Oxygen *K*-edge NEXAFS. The simulated NEXAFS spectra at the oxygen *K*-edge of the seven oxygen-containing molecules are shown in Figure 4a. The experimental NEXAFS spectrum of [6,6]$C_{60}O$ molecule is shown in the dashed line, which is reproduced from the work of Wohlers et al. at 370K [55]. The energy positions of peak (shoulder) A (corresponds to 1s $\rightarrow$ LUMO molecular orbital transition) and peaks B and C are listed in Table 2. Both the [6,6]$C_{60}O$ and the $C_{60}H$-O-$C_{60}H$ produce the first shoulder A or peak A with moderate intensity, which arises from the transition of oxygen 1*s* to the LUMO. The peak A of the NEXAFS spectra of the closed [6,6] isomer of $C_{60}O$ is observed at a higher excitation energy (532.35 eV), whereas it appears at a lower excitation energy in the $C_{60}H$-O-$C_{60}H$ (531.97 eV). Conversely, the oxygen 1*s* $\rightarrow$ LUMO transition only produces a weak peak A in the $C_{60}H(OH)$ and a weak shoulder A in the [5,6]$C_{60}O$, C_{60}-O-C_{60}, $C_{59}N(OH)$ and $C_{59}N$-O-$C_{59}N$ molecules. The disparity observed in the oxygen 1*s* $\rightarrow$ LUMO transition among these seven molecules could also be elucidated by examining the distribution of their LUMOs in the GS. Unlike the [5,6]$C_{60}O$, $C_{60}H(OH)$, C_{60}-O-C_{60}, $C_{59}N(OH)$ and $C_{59}N$-O-$C_{59}N$, whose LUMOs have strong carbon character (Figure S1b,d,e,h,j), the LUMOs of the [6,6]$C_{60}O$ and the $C_{60}H$-O-$C_{60}H$ molecules present here also exhibit some distribution around the oxygen atom (Figure S1c,f). Therefore, their NEXAFS spectra exhibit a moderate shoulder A or peak A. In particular, the moderately strong shoulder A with its relatively high excitation energy in the [6,6]$C_{60}O$ and the moderately strong peak A with its relatively low excitation energy in the $C_{60}H$-O-$C_{60}H$ can be considered as the fingerprint of their existence.

Table 2. Energy of peak (shoulder) A, peaks B and C in the O *K*-edge NEXAFS for the [5,6] and [6,6] isomers of $C_{60}O$, $C_{60}H(OH)$, C_{60}-O-C_{60}, $C_{60}H$-O-$C_{60}H$, $C_{59}N(OH)$ and $C_{59}N$-O-$C_{59}N$.

Molecule	Peak (Shoulder) A (eV) (1s $\rightarrow$ LUMO)	Peak B (eV)	Peak C (eV)
[5,6]$C_{60}O$	532.80	–	–
[6,6]$C_{60}O$	532.35	532.98	–
$C_{60}H(OH)$	532.64	533.93	–
C_{60}-O-C_{60}	532.03	532.64	–
$C_{60}H$-O-$C_{60}H$	531.97	533.00	–
$C_{59}N(OH)$	532.78	533.82	532.97
$C_{59}N$-O-$C_{59}N$	532.04	532.51	533.45

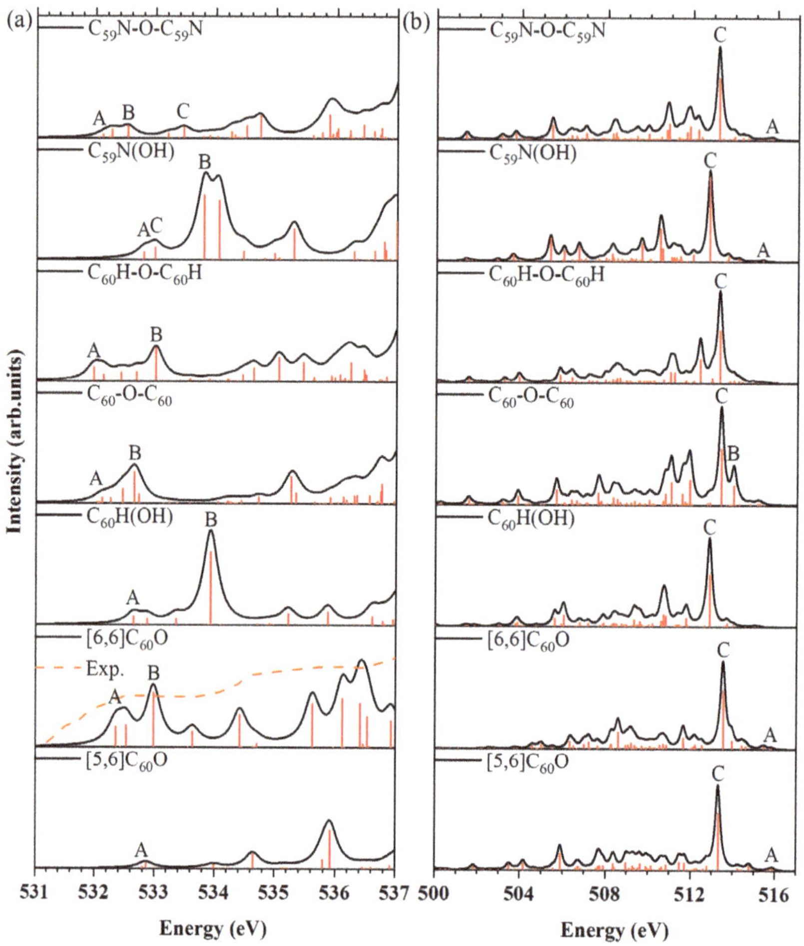

Figure 4. (**a**) Simulated NEXAFS spectra at the oxygen *K*-edge of the open [5,6] and closed [6,6] isomers of $C_{60}O$, $C_{60}H(OH)$, C_{60}-O-C_{60}, $C_{60}H$-O-$C_{60}H$, $C_{59}N(OH)$ and $C_{59}N$-O-$C_{59}N$, discrete intensities before convolution were in red bars. The experimental NEXAFS spectrum of [6,6]$C_{60}O$ molecule [55] is shown in dashed lines. (**b**) Simulated XES spectra at the oxygen *K*-edge of these seven oxygen-containing molecules, discrete intensities of XES of all molecules were in red bars within the convoluted XES spectra.

Furthermore, in the case of the [5,6]$C_{60}O$, $C_{60}H(OH)$ and $C_{59}N(OH)$, the oxygen 1*s* $\rightarrow$ LUMO transition is shown in the initial peak A or shoulder A with a weak intensity at nearly identical higher excitation energies (532.80 eV, 532.64 eV and 532.78 eV), among the five molecules with pronounced carbon characteristics in their LUMOs. The molecular groups, including the open [5,6] isomer of $C_{60}O$, $C_{60}H(OH)$ and $C_{59}N(OH)$, show that the absence of a strong peak B in comparison to the other two molecules, which can serve as the fingerprint of [5,6]$C_{60}O$. Additionally, $C_{60}H(OH)$ can be distinguished by the presence of a distinct peak B, while the presence of $C_{59}N(OH)$ can be recognized by the strong peak B and weak peak C. The spectra of the other two molecules, C_{60}-O-C_{60} and $C_{59}N$-O-$C_{59}N$, fall into another category, in which the oxygen 1*s* $\rightarrow$ LUMO transition generates the initial shoulder A with a weak intensity at nearly identical lower excitation energies (532.03 eV and 532.04 eV). These could be attributed to the influence of another carbon cage. At the same time, the energy positions of their peak B are also quite similar (532.64 eV and 532.51 eV). Nevertheless, the distinct feature peak C at 533.45 eV in the $C_{59}N$-O-$C_{59}N$ molecule serves as its distinguishing characteristic compared with C_{60}-O-C_{60}.

Oxygen *K*-edge XES. The XES spectra at the oxygen *K*-edge are shown in Figure 4b. The MO transitions and corresponding energies of peak A are compared in Table S2. The XES spectrum of the [5,6] and [6,6] isomers of $C_{60}O$, as well as the $C_{59}N$-O-$C_{59}N$, exhibit a discernible weak peak A at high energy (515.88 eV, 515.82 eV and 515.74 eV). These peaks arise from the HOMO to oxygen 1*s* orbital transition. The HOMOs of the [5,6] and [6,6] isomers of $C_{60}O$ and $C_{59}N$-O-$C_{59}N$, as depicted in Figure S1b,c,j, exhibit some characteristics of oxygen. This results in the manifestation of the moderate peak A in their XES spectra. Conversely, the HOMOs of $C_{60}H(OH)$, C_{60}-O-C_{60} and $C_{60}H$-O-$C_{60}H$ possess strong carbon attributes, as shown in Figure S1d–f. The transition from the HOMO to the oxygen 1*s* orbital does not result in the observation of any discernible peaks in their XES spectra. For the $C_{59}N(OH)$, although its HOMOs also exhibit little oxygen characteristics, the transition from the HOMO to the oxygen 1*s* orbital occurs at relatively lower energy (515.33 eV) compared with the [5,6] and [6,6] isomers of $C_{60}O$, as well as $C_{59}N$-O-$C_{59}N$. Significantly, the XES spectra reveal distinct characteristics, including the moderately strong peak B in the C_{60}-O-C_{60}, as well as the uniformly strong and sharp peak C in all seven molecules. These peaks arise from transitions originating in the deeper valence orbitals to the oxygen 1*s* orbital, allowing for the distinction of oxides or hydrates of different fullerene and azafullerene.

2.4. XPS Shake-Up Satellites

Carbon *K*-edge XPS shake-up satellites. The calculated XPS shake-up satellites at the carbon *K*-edge for the fullerene C_{60}, aza[60]fullerene $C_{59}N$, their oxides and reactions involving H_2O and $(C_{59}N)_2$ were plotted in Figure 5a. Additionally, spectra of C_{60}, $C_{59}N$ and $(C_{59}N)_2$ were also provided for comparison. The XPS main lines of all symmetrically inequivalent carbon atoms are represented as red bars underneath each XPS shake-up spectrum of the other nine members. Contributions from the β-spin electrons in the $C_{59}N$ molecule were plotted as green bars. The experimental XPS shake-up satellite spectrum of $(C_{59}N)_2$ molecule is reproduced from the work of Schulte et al. [52], which is in agreement very well with our simulated spectrum. It is evident that the XPS shake-up satellites of the oxides or hydrates of fullerene C_{60} show significantly different profiles compared with those of C_{60}, except for that of [6,6] $C_{60}O$. This observation signifies an alteration in the electronic structures of the valence orbitals following the oxidation or reaction with water of the C_{60} molecule. Similarly, the spectra of the $C_{59}N(OH)$ and $C_{59}N$-O-$C_{59}N$ exhibit distinct profiles compared with those of the $C_{59}N$ and aza[60]fullerene dimer $(C_{59}N)_2$.

The spectra of these ten molecules exhibit very similar profiles, featuring prominent peaks A and C located proximate to the XPS main line, except for C_{60} and [6,6]$C_{60}O$. The presence of shoulder B can be attributed to contributions from several carbon atoms in $C_{60}H(OH)$. The primary contributions to the peaks A and C in the nine molecules have been elucidated and are collected in Table 3. It is interesting that except for the peak C in the open [5,6] isomer of $C_{60}O$, $C_{59}N$, $C_{59}N(OH)$ and $C_{59}N$-O-$C_{59}N$, resulting from the shake-up mechanisms, the primary contributions to the peaks A and C in the other five molecules are actually the XPS main lines of the atoms C_1, C_2, C_3, C_4 and C_5. It is evident that the carbon atom C_1 (sp^2-like) is connected to the oxygen atom, C_3 (in the $C_{59}N$) and C_5 (in the $C_{59}N$, $C_{59}N(OH)$, $(C_{59}N)_2$ and $C_{59}N$-O-$C_{59}N$) are bonded to the nitrogen atom with sp^2-like, while the carbon atoms C_2 (connected to the oxygen atom), C_3 (in the $C_{60}H(OH)$, C_{60}-O-C_{60}, $C_{60}H$-O-$C_{60}H$, $C_{59}N(OH)$, $(C_{59}N)_2$, and $C_{59}N$-O-$C_{59}N$) and C_4 (in the $C_{60}H(OH)$, C_{60}-O-C_{60}, and $C_{60}H$-O-$C_{60}H$) are sp^3-like atoms in the closed [6,6] isomer of $C_{60}O$, $C_{60}H(OH)$, C_{60}-O-C_{60}, $C_{60}H$-O-$C_{60}H$, $C_{59}N(OH)$, $(C_{59}N)_2$ and $C_{59}N$-O-$C_{59}N$ (Figure 1b–j). Hence, their binding energies exhibit a slight increase compared with those of other carbon atoms. The hybridization of carbon atoms and the blue-shifted binding energy of these hybridized carbon atoms relative to the XPS main line for the oxides or hydrates of fullerene C_{60} and aza[60]fullerene $C_{59}N$ are collected in Table 4.

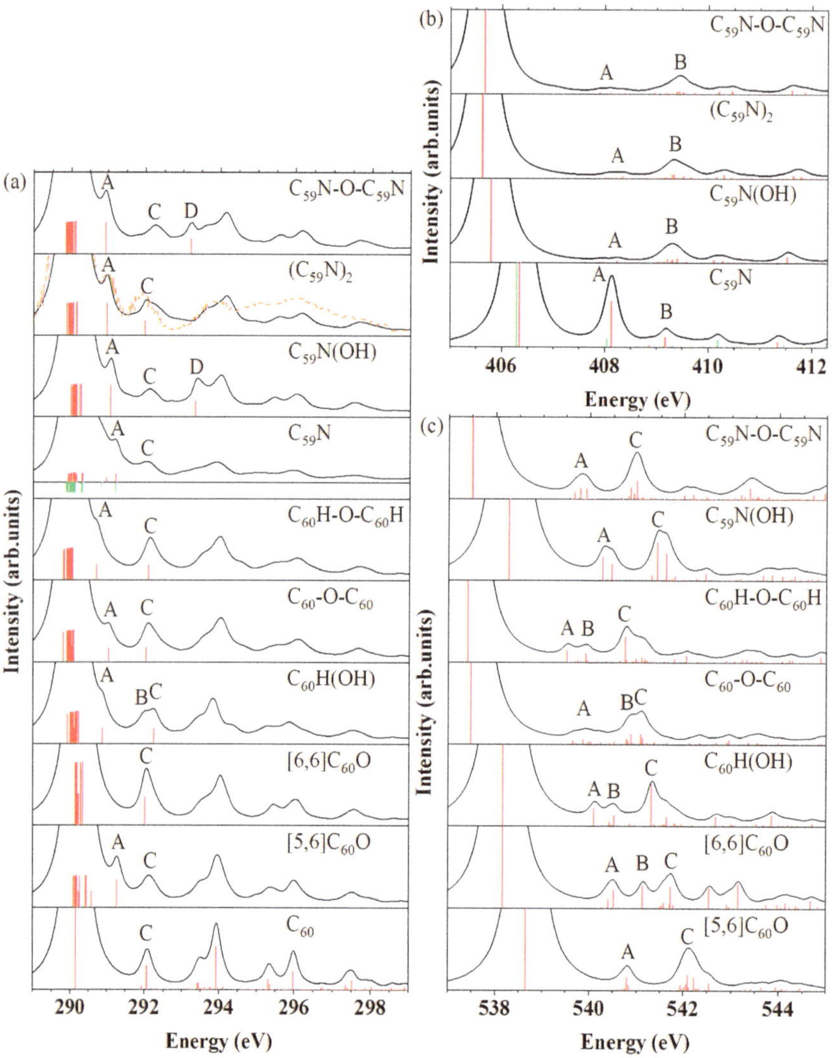

Figure 5. (**a**) XPS shake-up satellites at the carbon *K*-edge for the fullerene C_{60}, aza[60]fullerene $C_{59}N$, their oxides and hydrates and $(C_{59}N)_2$, along with the discrete intensities of the C_{60} molecule before convolution were depicted as red bars. The XPS main lines of all symmetrically inequivalent carbon atoms are represented as red bars underneath each XPS shake-up spectrum of the other nine members. Contributions from the β-spin electrons in the $C_{59}N$ molecule are shown as green bars. The experimental XPS shake-up satellite spectrum of $(C_{59}N)_2$ molecule [52] is shown as dashed lines. (**b**) XPS shake-up satellites at the nitrogen *K*-edge for $C_{59}N$, $C_{59}N(OH)$, $(C_{59}N)_2$ and $C_{59}N$-O-$C_{59}N$ molecules, along with the discrete intensities before convolution, are depicted as red bars. Contributions from the β-spin electrons are also shown as green bars under the spectrum in the $C_{59}N$ molecule. (**c**) XPS shake-up satellites at the oxygen *K*-edge for the open [5,6] and closed [6,6] isomers of $C_{60}O$, $C_{60}H(OH)$, C_{60}-O-C_{60}, $C_{60}H$-O-$C_{60}H$, $C_{59}N(OH)$ and $C_{59}N$-O-$C_{59}N$, along with the discrete intensities before convolution are shown as red bars.

Table 3. The C *K*-edge XPS shake-up satellites of the fullerene C_{60}, aza[60]fullerene $C_{59}N$, their oxides and hydrates.

Molecule	A	C
[5,6]$C_{60}O$	C_1: 291.284 eV	... [a]
[6,6]$C_{60}O$	-	C_2: 292.035 eV
$C_{60}H(OH)$	C_4: 290.888 eV	C_3: 292.257 eV
C_{60}-O-C_{60}	C_3: 291.055 eV	C_4: 292.064 eV
C_{60}H-O-C_{60}H	C_4: 290.735 eV	C_3: 292.114 eV
$C_{59}N$	C_5: 291.245 eV (α spin) C_5: 291.228 eV (β spin)	... [a]
$C_{59}N(OH)$	C_5: 291.096 eV	... [a]
$(C_{59}N)_2$	C_5: 290.989 eV	C_3: 292.003 eV
C_{59}N-O-C_{59}N	C_5: 290.952 eV	... [a]

[a] XPS shake-up satellites originating from multiple carbon atoms.

Table 4. The hybridization of carbon atoms and the blue-shifted binding energy of these hybridized carbon atoms relative to the XPS main line in the oxides or hydrates of fullerene C_{60} and aza[60]fullerene $C_{59}N$.

Molecule	sp^2-like Carbon Atom (Blue-Shift)	sp^3-like Carbon Atom (Blue-Shift)
[5,6]$C_{60}O$	C_1 (~1 eV)	-
[6,6]$C_{60}O$	-	C_2 (~2 eV)
$C_{60}H(OH)$	-	C_3 (~2 eV)
	-	C_4 (~1 eV)
C_{60}-O-C_{60}	-	C_3 (~1 eV)
	-	C_4 (~2 eV)
C_{60}H-O-C_{60}H	-	C_3 (~2 eV)
	-	C_4 (~1 eV)
$C_{59}N$	C_3 (-)	-
	C_5 (~1 eV)	-
$C_{59}N(OH)$	C_5 (~1 eV)	C_3 (~3 eV)
$(C_{59}N)_2$	C_5 (~1 eV)	C_3 (~2 eV)
C_{59}N-O-C_{59}N	C_5 (~1 eV)	C_3 (~3 eV)

The local electronic structures of the C_3 and C_4 atoms in the oxides or hydrates of fullerene C_{60} show minimal changes after bonding with other atoms (like H atom), such as (C_4 in the $C_{60}H(OH)$ and C_4 in the C_{60}H-O-C_{60}H), or a large molecule (like C_3 in the C_{60}-O-C_{60}), although they become an sp^3-like carbon, and consequently lead to blue-shifted binding energy—around 1 eV in the $C_{60}H(OH)$ (C_4), C_{60}-O-C_{60} (C_3) and C_{60}H-O-C_{60}H (C_4) in comparison to their XPS main line. In contrast, sp^3-like carbon atoms bonded to oxygen atoms are shown a pronounced blue-shifted in their binding energy—approximately 2 eV in the closed [6,6] isomer of $C_{60}O(C_2)$, $C_{60}H(OH)$ (C_3), C_{60}-O-C_{60} (C_4) and C_{60}H-O-C_{60}H (C_3) when compared with their XPS main line. The larger blue shifts demonstrated the local electronic structures of the sp^3-like C_2, C_3 and C_4 atoms changing a lot after being connected to an oxygen atom (such as C_2 in the [6,6]$C_{60}O$, C_3 in the $C_{60}H(OH)$ and C_{60}H-O-C_{60}H, C_4 in the C_{60}-O-C_{60}). The C *K*-edge XPS shake-up satellite spectra reveal the distinctive features of the [6,6]$C_{60}O$ molecule, characterized by a strong peak C and the absence of peak A, while $C_{60}H(OH)$ exhibits a unique shoulder B in its spectrum. Therefore, C *K*-edge XPS shake-up satellites serve as a valuable technique for detecting the presence of [6,6]$C_{60}O$ and $C_{60}H(OH)$ molecules, enabling their differentiation from other fullerene C_{60} oxides and hydrates.

In the oxides or hydrates of azafullerene $C_{59}N$, the binding energy of carbon atom C_5 is blue-shifted by approximately 1 eV relative to the XPS main line, while the binding energy of carbon atom C_3 remains unchanged in the $C_{59}N$ molecule [46]. However, upon oxidation (e.g., C_{59}N-O-C_{59}N) or reaction with water (e.g., $C_{59}N(OH)$), the local electronic structure of the carbon atom C_3 undergoes significant alteration, transforming it into a sp^3-like carbon atom in the $C_{59}N(OH)$ and $C_{59}N$-O-$C_{59}N$. Consequently, a substantial blue-shift binding energy of around 3 eV is observed in the $C_{59}N(OH)$ and $C_{59}N$-O-$C_{59}N$ compared to their XPS main line, while the blue-shift is approximately 2 eV in the $(C_{59}N)_2$[46]. The unique peak D in the spectra of $C_{59}N(OH)$ and $C_{59}N$-O-$C_{59}N$ molecules distinguish them from $C_{59}N$ and $(C_{59}N)_2$ molecules. However, the C *K*-edge XPS shake-up satellite spectra could not differentiate between $C_{59}N(OH)$ and $C_{59}N$-O-$C_{59}N$ molecules.

It is noteworthy that only C_1 exhibits a sp^2-like hybridization, leading to a subtle blue shift in binding energy relative to the XPS main line among all the carbon atoms bonded to an oxygen atom. The uniqueness of the C_1 atom demonstrates a remarkable change in the localized electronic structure of C_1 in the [5,6]$C_{60}O$ isomer after breaking the five- and six-membered ring structures of the fullerene cage.

Nitrogen *K*-edge XPS shake-up satellites. The alterations in electronic structures can be elucidated through the analysis of nitrogen *K*-edge XPS shake-up satellites (Figure 5b). The energy of peak B in the spectra for the $C_{59}N$, $C_{59}N(OH)$, $(C_{59}N)_2$ and $C_{59}N$-O-$C_{59}N$ are shown in Table 5. In both $C_{59}N(OH)$ and $C_{59}N$-O-$C_{59}N$, peak A is relatively weaker compared with that of the $C_{59}N$. Furthermore, peak B in $C_{59}N$-O-$C_{59}N$ exhibits a blue shift of approximately 0.15 eV compared with peak B of the $C_{59}N(OH)$. An extensive assignment of the principal contributions to peaks A and B, which are observed in the nitrogen *K*-edge XPS shake-up satellites of these four members within the $C_{59}N$ family, is provided in Table S3. The information can serve as a valuable reference for future experimental investigations. It is evident that, apart from $C_{59}N$ and $C_{59}N(OH)$, feature B in the other two molecules is attributed to excitations originating from deeper valence orbitals.

Table 5. Energy of peak B in the N *K*-edge XPS shake-up satellites for the $C_{59}N$, $C_{59}N(OH)$, $(C_{59}N)_2$ and $C_{59}N$-O-$C_{59}N$ molecules.

Molecule	Energy of Peak B (eV)
$C_{59}N$	409.19
$C_{59}N(OH)$	409.31
$(C_{59}N)_2$	409.34
$C_{59}N$-O-$C_{59}N$	409.46

Oxygen *K*-edge XPS shake-up satellites. The alterations in electronic structures can also be discerned from the oxygen *K*-edge XPS shake-up satellites, as depicted in Figure 5c. The energies of peaks A, B and C in the oxygen *K*-edge XPS shake-up satellites are shown in Table 6 for the seven oxygen-containing molecules. The [6,6]$C_{60}O$, $C_{60}H(OH)$, C_{60}-O-C_{60} and $C_{60}H$-O-$C_{60}H$ exhibit three feature peaks A, B and C. The other three molecules, open [5,6] isomer of $C_{60}O$, $C_{59}N(OH)$ and $C_{59}N$-O-$C_{59}N$, do not show a discernible peak B, highlighting a noticeable difference between the [6,6]$C_{60}O$, $C_{60}H(OH)$, C_{60}-O-C_{60} and $C_{60}H$-O-$C_{60}H$. In comparison to the [5,6]$C_{60}O$ molecule, the featured peak A of $C_{59}N(OH)$ demonstrates a red shift of 0.52 eV, whereas the characteristic peak A of the $C_{59}N$-O-$C_{59}N$ exhibits a red shift of 1.00 eV. In contrast, in comparison with [6,6]$C_{60}O$, the characteristic peak A of $C_{60}H(OH)$ shows a red shift of 0.42 eV, the feature peak A of the C_{60}-O-C_{60} exhibits a red shift of 0.66 eV, and the characteristic peak A of the $C_{60}H$-O-$C_{60}H$ shows a red shift of 1.00 eV. Hence, the oxygen *K*-edge XPS shake-up satellites offer a valuable technique to detect the presence of these seven molecules, which can also be used to distinguish them from other oxides or hydrates family members of fullerene C_{60} and aza[60]fullerene $C_{59}N$.

The detailed assignments of the primary contributions to peaks A, B and C within the oxygen *K*-edge XPS shake-up satellites for these seven oxygen-containing molecules are collected in Table S4, which can provide a valuable reference for future experiments. It is notable that peak B includes excitations from deeper valence orbitals in C_{60}-O-C_{60}. With the exception of $C_{60}H(OH)$ and $C_{59}N(OH)$ molecules, peak C also encompasses excitations from deeper valence orbitals in the remaining five molecules.

Table 6. Energy of peaks A, B and C in the O *K*-edge XPS shake-up satellites for the [5,6] and [6,6] isomers of $C_{60}O$, $C_{60}H(OH)$, C_{60}-O-C_{60}, $C_{60}H$-O-$C_{60}H$, $C_{59}N(OH)$ and $C_{59}N$-O-$C_{59}N$.

Molecule	Energy of Peak A (eV)	Energy of Peak (Shoulder) B (eV)	Energy of Peak C (eV)
[5,6]$C_{60}O$	540.81	-	542.11
[6,6]$C_{60}O$	540.53	541.14	541.75
$C_{60}H(OH)$	540.11	540.54	541.33
C_{60}-O-C_{60}	539.87	540.89	541.10
$C_{60}H$-O-$C_{60}H$	539.53	539.94	540.76
$C_{59}N(OH)$	540.29	-	541.44
$C_{59}N$-O-$C_{59}N$	539.81	-	541.00

3. Computational Details

The ten groups of molecular models are presented in Figure 1, including fullerene(C_{60}), $C_{60}O$ with oxygen in the [5,6] open annulene and [6,6] closed epoxide isomers, $C_{60}H(OH)$, ether-oxygen-bridged fullerene dimer C_{60}-O-C_{60}, $C_{60}H$-O-$C_{60}H$, aza[60]fullerene $C_{59}N$, $C_{59}N(OH)$, $(C_{59}N)_2$ and $C_{59}N$-O-$C_{59}N$, in which C_{60}, $C_{59}N$ and $(C_{59}N)_2$ molecules are used for comparison. All the molecular configurations were optimized, and the single point energy calculations were performed with the Gaussian16 program at the B3LYP/6-31G(d,p) level [56], and the optimized configurations were subsequently validated as energy minima through frequency calculations. All structures were optimized without initial symmetry constraints. The geometries with Cartesian coordinates of all optimized structures are listed in Table S7 The lowest frequencies of these structures are collected in Table S5, while the energies of the ten molecules with different spin multiplicities are collected in Table S6. The electronic state with the lowest energy was applied for the soft X-ray spectra simulations. Nitrogen, oxygen and their attached carbon atoms are shown in Figure 1b–j, where atoms C_1 and C_1', C_2 and C_2', C_3 and C_3', C_4 and C_4' and C_5 and C_5' are symmetrically equivalent atoms.

The Z + 1 approximation and TP method have been employed for the efficient computation of large systems [57–59], which are characterized by a high number of MOs in the NEXAFS and XES spectroscopy calculations. To ensure computing precision, the FCH potential method was adopted for the accurate characterization of NEXAFS spectra [60–64]. The triple-ζ quality individual gauge for localized orbital (IGLO-III) basis set was selected for the excited carbon, nitrogen or oxygen atoms, while other non-excited atoms were calculated by employing model core potentials (MCP), which was defined as the Huzinaga's procedure [65] implemented in the StoBe program [66–68]. This computing strategy ensured the accuracy and speed for large-scale computing in our soft X-ray spectroscopy calculations [60,61]. Additionally, to enhance the accuracy, an augmented diffuse basis set (19s, 19p and 19d) for the excited carbon, nitrogen or oxygen atoms was utilized in the NEXAFS spectra calculations.

The NEXAFS spectra and ΔKohn–Sham (ΔKS) [60] calculations were carried out using StoBe [66–68]. The relativistic effects were accounted for by introducing energy shifts of +0.2 eV for the carbon *K*-edge, +0.3 eV for the nitrogen *K*-edge and +0.5 eV for the oxygen *K*-edge [69]. The NEXAFS spectra were generated by convoluting discrete intensities with a Lorentzian function, which incorporated spectral broadening. For energies below the ionization potential (IP), the full width at half maximum (FWHM) was set to 0.3 eV. In the range from the IP to 5 eV above, the FWHM increased linearly up to 1.0 eV. For energies beyond that range, a constant FWHM of 1.0 eV was employed. The XES spectra were simulated employing the BioNano-LEGO developed by Yi Luo's group [70]. GS wave functions were derived from the Gaussian16 program [56] at the B3LYP/6-311+G* level. The computed XES spectra underwent additional convolution with a Lorentzian function, employing an FWHM of 0.3 eV.

The ECH-KS method was employed for XPS shake-up satellite spectra calculations [71–73], which has been introduced for the calculation of the numerous excited states necessary for the XPS shake-up satellites in large systems [74]. The 6-31G basis set was applied to the non-ionized atoms, while the IGLO-III basis set was utilized for the core-ionized carbon, nitrogen or oxygen atom [75]. The shake-up satellites were calculated by the ECH-KS method, which was consistent with ECH-TDDFT calculations and experimental results, as previously validated [75–77]. In our calculations, the XPS shake-up satellites were also performed by BioNano-LEGO [70]. The shake-up intensities were convolved with a Lorentzian function with an FWHM of 0.3 eV.

4. Conclusions

The core-hole excitation spectroscopic techniques, specifically NEXAFS, XES and XPS shake-up satellites, were employed to differentiate between various members of the oxides or hydrates family members of fullerene C_{60} and aza[60]fullerene $C_{59}N$. These members include the open [5,6] and closed [6,6] isomers of $C_{60}O$, $C_{60}H(OH)$, C_{60}-O-C_{60}, $C_{60}H$-O-$C_{60}H$, $C_{59}N(OH)$ and $C_{59}N$-O-$C_{59}N$. These spectra provided us with comprehensive insights into both the unoccupied and occupied molecular orbitals of the molecular groups under study. Meanwhile, the oxygen *K*-edge NEXAFS, carbon and oxygen *K*-edge XPS shake-up satellite spectra were employed to distinguish oxides or hydrates of fullerene C_{60} and azafullerene $C_{59}N$. The approach allowed us to delve deeply insights into the relationships between individual spectral peaks and the electronic and geometric structures of the molecules within the fullerene C_{60} and aza[60]fullerene $C_{59}N$ families. These findings not only enhance our understanding of these complex molecular systems but also pave the way for future research and applications in the field of nanomaterial science.

Supplementary Materials: The following supporting information can be downloaded at: https://www.mdpi.com/article/10.3390/molecules29030609/s1, Figure S1: The HOMOs and LUMOs of the fullerene C_{60}, aza[60]fullerene, their oxides and reactions involving H2O calculated at B3LYP/6-311+G* level. The carbon atoms C_1 and C_1', C_2 and C_2', C_3 and C_3' and C_4 and C_4', which were connected to the oxygen atom, were symmetrically equivalent; Table S1: MO transitions and the corresponding energies for peak or shoulder A in the nitrogen K-edge NEXAFS of the $C_{59}N$, $C_{59}N(OH)$, $(C_{59}N)2$ and $C_{59}N$-O-$C_{59}N$ molecules; Table S2: MO transitions and the corresponding energies for peak A in the oxygen K-edge XES of the open [5,6] and closed [6,6] isomers of $C_{60}O$, $C_{59}N(OH)$ and $C_{59}N$-O-$C_{59}N$ molecules; Table S3: Assignments of the nitrogen K-edge XPS shake-up satellites of the $C_{59}N$, $C_{59}N(OH)$, $(C_{59}N)2$ and $C_{59}N$-O-$C_{59}N$ molecules; Table S4: Assignments of the oxygen K-edge XPS shake-up satellites of the open [5,6] and closed [6,6] isomers of $C_{60}O$, $C_{60}H(OH)$, C_{60}-O-C_{60}, $C_{60}H$-O-$C_{60}H$, $C_{59}N(OH)$ and $C_{59}N$-O-$C_{59}N$ molecules; Table S5: The lowest frequencies of all optimized structures; Table S6: The energies of the fullerene C_{60}, aza[60]fullerene $C_{59}N$, their oxides and hydrates molecules at different spin multiplicities; Table S7: Molecule coordinate of the open [5,6] and closed [6,6] isomers of $C_{60}O$, $C_{60}H(OH)$, C_{60}-O-C_{60}, $C_{60}H$-O-$C_{60}H$, $C_{59}N(OH)$ and $C_{59}N$-O-$C_{59}N$ molecules.

Author Contributions: Conceptualization, X.L.; methodology, M.D.; software, X.L., M.D. and J.G.; validation, S.W., Z.W. and C.G.; writing—original draft preparation, X.L.; writing—review and editing, M.D.; visualization, X.L.; project administration, S.C. and M.D. All authors have read and agreed to the published version of the manuscript.

Funding: This research was funded by the National Natural Science Foundation of China (No. 11964004, 22103018), the Program for Innovative Research Team of Guizhou Province (Grant QKHP-TRC(2020)5023) and the Natural Science and Technology Foundation of Guizhou Province (grant QKHBasic[2019]1044).

Institutional Review Board Statement: Not applicable.

Informed Consent Statement: Not applicable.

Data Availability Statement: Data are contained within the article and Supplementary Materials.

Acknowledgments: The authors acknowledge Hujun Shen in GZEU for useful discuss.

Conflicts of Interest: The authors declare no conflicts of interest.

References

1. Nam, S.; Khim, D.; Martinez, G.T.; Varambhia, A.; Nellist, P.D.; Kim, Y.; Anthopoulos, T.D.; Bradley, D.D. Significant Performance Improvement in n-Channel Organic Field-Effect Transistors with C_{60}:C_{70} Co-Crystals Induced by Poly (2-ethyl-2-oxazoline) Nanodots. *Adv. Mater.* **2021**, *33*, 2100421. [CrossRef] [PubMed]
2. Benatto, L.; Marchiori, C.; Talka, T.; Aramini, M.; Yamamoto, N.; Huotari, S.; Roman, L.; Koehler, M. Comparing C_{60} and C_{70} as acceptor in organic solar cells: Influence of the electronic structure and aggregation size on the photovoltaic characteristics. *Thin Solid Films* **2020**, *697*, 137827. [CrossRef]
3. Alipour, E.; Alimohammady, F.; Yumashev, A.; Maseleno, A. Fullerene C_{60} containing porphyrin-like metal center as drug delivery system for ibuprofen drug. *J. Mol. Model.* **2020**, *26*, 7. [CrossRef]
4. Hong, I.-H.; Gao, C.-J.; Lin, K.-B.; Kaun, C.-C. Self-organized C_{70}/C_{60} heterojunction nanowire arrays on Si(110) for Si-based molecular negative differential resistance nanodevices. *Appl. Surf. Sci.* **2020**, *531*, 147338. [CrossRef]
5. Chen, L.; Wang, X.; Kumar, S. Thermal transport in fullerene derivatives using molecular dynamics simulations. *Sci. Rep.* **2015**, *5*, 12763. [CrossRef] [PubMed]
6. Socol, M.; Preda, N.; Costas, A.; Borca, B.; Popescu-Pelin, G.; Mihailescu, A.; Socol, G.; Stanculescu, A. Thin films based on cobalt phthalocyanine: C_{60} fullerene: ZnO hybrid nanocomposite obtained by laser evaporation. *Nanomaterials* **2020**, *10*, 468. [CrossRef]
7. Tanuma, Y.; Knaflič, T.; Anézo, B.; Stangel, C.; Volkmann, J.; Tagmatarchis, N.; Wegner, H.A.; Arčon, D.; Ewels, C.P. Long Spin Coherence Times on $C_{59}N$-C_{60} Heterodimer Radicals Entrapped in Cycloparaphenylene Rings. *J. Phys. Chem. C* **2023**, *127*, 6552–6561. [CrossRef]
8. Martín-Gomis, L.; Rotas, G.; Ohkubo, K.; Fernández-Lázaro, F.; Fukuzumi, S.; Tagmatarchis, N.; Sastre-Santos, Á. Does a nitrogen matter? Synthesis and photoinduced electron transfer of perylenediimide donors covalently linked to $C_{59}N$ and C_{60} acceptors. *Nanoscale* **2015**, *7*, 7437–7444. [CrossRef]
9. Rotas, G.; Stranius, K.; Tkachenko, N.; Tagmatarchis, N. Ultralong 20 Milliseconds Charge Separation Lifetime for Photoilluminated Oligophenylenevinylene-Azafullerene Systems. *Adv. Funct. Mater.* **2018**, *28*, 1702278. [CrossRef]
10. Yin, H.; Lin, H.; Zong, Y.; Wang, X.-D. The recent advances in C_{60} micro/nanostructures and their optoelectronic applications. *Org. Electron.* **2021**, *93*, 106142. [CrossRef]
11. Kianezhad, M.; Youzi, M.; Vaezi, M.; Nejat Pishkenari, H. Unidirectional motion of C_{60}-based nanovehicles using hybrid substrates with temperature gradient. *Sci. Rep.* **2023**, *13*, 1100. [CrossRef]
12. Vinnikov, N.; Dolbin, A.; Basnukaeva, R.; Gavrilko, V.; Eselson, V.; Buravtseva, L. Quantum effects in the low-temperature thermal expansion of fullerite C_{60} doped with a 4He impurity. *Low Temp. Phys.* **2022**, *48*, 791–797. [CrossRef]
13. Elistratova, M.; Zakharova, I. Temperature-dependent photoluminescence of thin tetraphenylporphyrin-based thin films and their composites with C_{60} fullerene. *J. Mater. Sci. Mater. Electron.* **2022**, *33*, 15554–15562. [CrossRef]
14. Sivkov, D.V.; Petrova, O.V.; Nekipelov, S.V.; Vinogradov, A.S.; Skandakov, R.N.; Bakina, K.A.; Isaenko, S.I.; Ob'edkov, A.M.; Kaverin, B.S.; Vilkov, I.V.; et al. Quantitative Characterization of Oxygen-Containing Groups on the Surface of Carbon Materials: XPS and NEXAFS Study. *Appl. Sci.* **2022**, *12*, 7744. [CrossRef]
15. Lin, I.H.; Lu, Y.H.; Chen, H.T. Nitrogen-doped C_{60} as a robust catalyst for CO oxidation. *J. Comput. Chem.* **2017**, *38*, 2041–2046. [CrossRef]
16. Tang, K.; Chen, X.; Ding, X.; Yu, X.; Yu, X. MoS2/graphene oxide/C_{60}-OH nanostructures deposited on a quartz crystal microbalance transducer for humidity sensing. *ACS Appl. Nano Mater.* **2021**, *4*, 10810–10818. [CrossRef]
17. Esrafili, M.D.; Janebi, H. B-, N-doped and BN codoped C_{60} heterofullerenes for environmental monitoring of NO and NO_2: A DFT study. *Mol. Phys.* **2020**, *118*, e1631495. [CrossRef]
18. Tabari, L.; Farmanzadeh, D. Interesting adsorption behavior of $C_{60}O$ fullerene oxide isomers toward O_3 and CO molecules: A DFT study. *Appl. Surf. Sci.* **2019**, *479*, 569–575. [CrossRef]
19. Dattani, R.; Gibson, K.F.; Few, S.; Borg, A.J.; DiMaggio, P.A.; Nelson, J.; Kazarian, S.G.; Cabral, J.T. Fullerene oxidation and clustering in solution induced by light. *J. Colloid Interface Sci.* **2015**, *446*, 24–30. [CrossRef] [PubMed]
20. Sanchís, J.; Freixa, A.; López-Doval, J.C.; Santos, L.H.; Sabater, S.; Barceló, D.; Abad, E.; Farré, M. Bioconcentration and bioaccumulation of C_{60} fullerene and C_{60} epoxide in biofilms and freshwater snails (*Radix* sp.). *Environ. Res.* **2020**, *180*, 108715. [CrossRef] [PubMed]
21. Jones, M.A.; Britz, D.A.; Morton, J.J.; Khlobystov, A.N.; Porfyrakis, K.; Ardavan, A.; Briggs, G.A.D. Synthesis and reactivity of N@$C_{60}O$. *Phys. Chem. Chem. Phys.* **2006**, *8*, 2083–2088. [CrossRef]
22. Mondal, T.; Tripathi, A.; Tiwari, A.; Zhang, J.; Shripathi, T.; Shinohara, H.J.J.o.A.P. Temperature and pressure induced Raman studies of C_{60} oxide. *J. Appl. Phys.* **2018**, *124*, 195105. [CrossRef]
23. Gao, B.; Chen, G. CO oxidization catalyzed by B, N, and their co-doped fullerenes: A first-principles investigation. *RSC. Adv.* **2019**, *9*, 21626–21636. [CrossRef]
24. Palotás, J.; Martens, J.; Berden, G.; Oomens, J. Laboratory IR Spectra of the Ionic Oxidized Fullerenes $C_{60}O^+$ and $C_{60}OH^+$. *J. Phys. Chem. A* **2022**, *126*, 2928–2935. [CrossRef] [PubMed]
25. Schulz, P.; Kelly, L.L.; Winget, P.; Li, H.; Kim, H.; Ndione, P.F.; Sigdel, A.K.; Berry, J.J.; Graham, S.; Brédas, J.L.; et al. Tailoring Electron-Transfer Barriers for Zinc Oxide/C_{60} Fullerene Interfaces. *Adv. Funct. Mater.* **2014**, *24*, 7381–7389. [CrossRef]
26. Tanaka, Y.; Kanai, K.; Ouchi, Y.; Seki, K. Oxygen effect on the interfacial electronic structure of C_{60} film studied by ultraviolet photoelectron spectroscopy. *Chem. Phys. Lett.* **2007**, *441*, 63–67. [CrossRef]

27. Wang, X.-B.; Woo, H.-K.; Kiran, B.; Wang, L.-S. Photoelectron Spectroscopy and Electronic Structures of Fullerene Oxides: $C_{60}O_x^-$ (x = 1–3). *J. Phys. Chem. A* **2005**, *109*, 11089–11092. [CrossRef] [PubMed]
28. Lim, J.; Park, J.; Yeom, G.Y. Interfacial electronic structure of molybdenum oxide on the fullerene layer, a potential hole-injecting layer in inverted top-emitting organic light-emitting diodes. *Curr. Appl. Phys.* **2013**, *13*, 1037–1041. [CrossRef]
29. Pan, F.; Ni, K.; Xu, T.; Chen, H.; Wang, Y.; Gong, K.; Liu, C.; Li, X.; Lin, M.-L.; Li, S.; et al. Long-range ordered porous carbons produced from C_{60}. *Nature* **2023**, *614*, 95–101. [CrossRef]
30. Tropin, T.V.; Karpets, M.L.; Kosiachkin, Y.; Gapon, I.V.; Gorshkova, Y.E.; Aksenov, V.L. Evaluation of fullerenes C_{60}/C_{70} layers in polystyrene thin films by neutron and X-ray reflectometry. *Fuller. Nanotub. Carbon Nanostructures* **2021**, *29*, 819–824. [CrossRef]
31. Asad, K.; Stergiou, A.; Kourtellaris, A.; Tagmatarchis, N.; Chronakis, N. First Synthesis of the Inherently Chiral Trans-4′ Bisadduct of $C_{59}N$ Azafullerene by Using Cyclo-[2]-dodecylmalonate as a Tether. *Chem. Eur. J.* **2021**, *27*, 13879–13886. [CrossRef]
32. Couto, R.C.; Kjellsson, L.; Ågren, H.; Carravetta, V.; Sorensen, S.L.; Kubin, M.; Bülow, C.; Timm, M.; Zamudio-Bayer, V.; von Issendorff, B.; et al. The carbon and oxygen K-edge NEXAFS spectra of CO^+. *Phys. Chem. Chem. Phys.* **2020**, *22*, 16215–16223. [CrossRef]
33. de Kock, S.; Skudler, K.; Matsidik, R.; Sommer, M.; Müller, M.; Walter, M. NEXAFS spectra of model sulfide chains: Implications for sulfur networks obtained from inverse vulcanization. *Phys. Chem. Chem. Phys.* **2023**, *25*, 20395–20404. [CrossRef] [PubMed]
34. Wasowicz, T.; Ljubić, I.; Kivimäki, A.; Richter, R. Core–shell excitation of isoxazole at the C, N, and O K-edges–an experimental NEXAFS and theoretical TD-DFT study. *Phys. Chem. Chem. Phys.* **2022**, *24*, 19302–19313. [CrossRef] [PubMed]
35. Cole, J.; Henderson, Z.; Thomas, A.G.; Compeán-González, C.L.; Greer, A.J.; Hardacre, C.; Venturini, F.; Garzon, W.Q.; Ferrer, P.; Grinter, D.C.; et al. Near-ambient pressure XPS and NEXAFS study of a superbasic ionic liquid with CO_2. *J. Phys. Chem. C* **2021**, *125*, 22778–22785. [CrossRef]
36. Braglia, L.; Tavani, F.; Mauri, S.; Edla, R.; Krizmancic, D.; Tofoni, A.; Colombo, V.; D'angelo, P.; Torelli, P. Catching the Reversible Formation and Reactivity of Surface Defective Sites in Metal–Organic Frameworks: An Operando Ambient Pressure-NEXAFS Investigation. *J. Phys. Chem. Lett.* **2021**, *12*, 9182–9187. [CrossRef] [PubMed]
37. Johnson, M.; Hawly, T.; Zhao, B.; Halik, M.; Nefedov, A.; Fink, R. Field-induced modification of the electronic structure in BTBT-based organic thin films observed by NEXAFS spectroscopy. *Appl. Phys. Lett.* **2022**, *121*, 183503. [CrossRef]
38. Couto, R.C.; Hua, W.; Lindblad, R.; Kjellsson, L.; Sorensen, S.L.; Kubin, M.; Bülow, C.; Timm, M.; Zamudio-Bayer, V.; von Issendorff, B.; et al. Breaking inversion symmetry by protonation: Experimental and theoretical NEXAFS study of the diazynium ion, N_2H^+. *Phys. Chem. Chem. Phys.* **2021**, *23*, 17166–17176. [CrossRef] [PubMed]
39. Lever, F.; Mayer, D.; Metje, J.; Alisauskas, S.; Calegari, F.; Düsterer, S.; Feifel, R.; Niebuhr, M.; Manschwetus, B.; Kuhlmann, M.; et al. Core-level spectroscopy of 2-thiouracil at the sulfur L_1-and $L_{2,3}$-edges utilizing a SASE free-electron laser. *Molecules* **2021**, *26*, 6469. [CrossRef] [PubMed]
40. Klein, B.P.; Hall, S.J.; Maurer, R.J. The nuts and bolts of core-hole constrained ab initio simulation for K-shell x-ray photoemission and absorption spectra. *J. Phys. Condens. Matter* **2021**, *33*, 154005. [CrossRef]
41. Armillotta, F.; Bidoggia, D.; Biasin, P.; Annese, A.; Cossaro, A.; Verdini, A.; Floreano, L.; Peressi, M.; Vesselli, E. Spectroscopic fingerprints of iron-coordinated cobalt and iron porphyrin layers on graphene. *Cell Rep. Phys. Sci.* **2023**, *4*, 101378. [CrossRef]
42. Creegan, K.M.; Robbins, J.L.; Robbins, W.K.; Millar, J.M.; Sherwood, R.D.; Tindall, P.J.; Cox, D.M.; McCauley, J.P.; Jones, D.R. Synthesis and characterization of $C_{60}O$, the first fullerene epoxide. *J. Am. Chem. Soc.* **1992**, *114*, 1103–1105. [CrossRef]
43. Lebedkin, S.; Ballenweg, S.; Gross, J.; Taylor, R.; Krätschmer, W. Synthesis of $C_{120}O$: A new dimeric [60] fullerene derivative. *Tetrahedron Lett.* **1995**, *36*, 4971–4974. [CrossRef]
44. Weisman, R.B.; Heymann, D.; Bachilo, S.M. Synthesis and characterization of the "missing" oxide of C_{60}:[5,6]-open $C_{60}O$. *J. Am. Chem. Soc.* **2001**, *123*, 9720–9721. [CrossRef] [PubMed]
45. Erbahar, D.; Susi, T.; Rocquefelte, X.; Bittencourt, C.; Scardamaglia, M.; Blaha, P.; Guttmann, P.; Rotas, G.; Tagmatarchis, N.; Zhu, X.; et al. Spectromicroscopy of C_{60} and azafullerene $C_{59}N$: Identifying surface adsorbed water. *Sci. Rep.* **2016**, *6*, 35605. [CrossRef] [PubMed]
46. Deng, Y.; Gao, B.; Deng, M.; Luo, Y. A comparative theoretical study on core-hole excitation spectra of azafullerene and its derivatives. *J. Chem. Phys.* **2014**, *140*, 124304. [CrossRef] [PubMed]
47. Momma, K.; Izumi, F. VESTA: A three-dimensional visualization system for electronic and structural analysis. *J. Appl. Crystallogr.* **2008**, *41*, 653–658. [CrossRef]
48. Mohammadi, M.D.; Abdullah, H.Y.; Louis, H.; Etim, E.E.; Edet, H.O. Evaluating the detection potential of $C_{59}X$ fullerenes (X= C, Si, Ge, B, Al, Ga, N, P, and As) for H_2SiCl_2 molecule. *J. Mol. Liq.* **2023**, *387*, 122621. [CrossRef]
49. Sood, P.; Kim, K.C.; Jang, S.S. Electrochemical and electronic properties of nitrogen doped fullerene and its derivatives for lithium-ion battery applications. *J. Energy Chem.* **2018**, *27*, 528–534. [CrossRef]
50. Nattagh, F.; Hosseini, S.; Esrafili, M.D. Effects of B and N doping/codoping on the adsorption behavior of C_{60} fullerene towards aspirin: A DFT investigation. *J. Mol. Liq.* **2021**, *342*, 117459. [CrossRef]
51. Käämbre, T.; Qian, L.; Rubensson, J.-E.; Guo, J.-H.; Såthe, C.; Nordgren, J.; Palmqvist, J.-P.; Jansson, U. Study of oxygen-C_{60} compound formation by NEXAFS and RIXS. *Eur. Phys. J. D* **2001**, *16*, 357–360. [CrossRef]
52. Schulte, K.; Wang, L.; Moriarty, P.; Prassides, K.; Tagmatarchis, N. Resonant processes and Coulomb interactions in $(C_{59}N)_2$. *J. Chem. Phys.* **2007**, *126*, 184707. [CrossRef] [PubMed]

53. Tan, Z.; Zhang, D.; Tian, H.-R.; Wu, Q.; Hou, S.; Pi, J.; Sadeghi, H.; Tang, Z.; Yang, Y.; Liu, J.; et al. Atomically defined angstrom-scale all-carbon junctions. *Nat. Commun.* **2019**, *10*, 1748. [CrossRef] [PubMed]
54. Wang, S.; Zhao, R.; Zheng, T.; Lu, Z.; Fang, Y.; Xie, H.; Wang, W.; Xue, W. Rational Design of a Low-Dimensional and Metal-free Heterostructure for Efficient Water Oxidation: DFT and Experimental Studies. *Langmuir* **2022**, *38*, 12562–12569. [CrossRef]
55. Wohlers, M.; Werner, H.; Herein, D.; Schedel-Niedrig, T.; Bauer, A.; Schlögl, R. Reaction of C_{60} and C_{70} with molecular oxygen. *Synth. Met.* **1996**, *77*, 299–302. [CrossRef]
56. Frisch, M.J.; Schlegel, H.B.; Scuseria, G.E.; Robb, M.A.; Scalmani, G.; Barone, V.; Petersson, G.A.; Li, X.; Caricato, M.; Marenich, A.V.; et al. *Gaussian 16, Revision A.03*; Gaussian Inc.: Wallingford, CT, USA, 2016.
57. Triguero, L.; Pettersson, L.; Ågren, H. Calculations of near-edge X-ray-absorption spectra of gas-phase and chemisorbed molecules by means of density-functional and transition-potential theory. *Phys. Rev. B* **1998**, *58*, 8097. [CrossRef]
58. Delesma, F.A.; Van den Bossche, M.; Grönbeck, H.; Calaminici, P.; Köster, A.M.; Pettersson, L.G. A chemical view on X-ray photoelectron spectroscopy: The ESCA molecule and surface-to-bulk XPS shifts. *ChemPhysChem* **2018**, *19*, 169–174. [CrossRef]
59. Hua, W.; Gao, B.; Luo, Y. First-principle simulation of soft X-ray spectroscopy. *Prog. Chem.* **2012**, *24*, 964.
60. Nyberg, M.; Luo, Y.; Triguero, L.; Pettersson, L.G.; Ågren, H. Core-hole effects in x-ray-absorption spectra of fullerenes. *Phys. Rev. B* **1999**, *60*, 7956. [CrossRef]
61. Bassan, A.; Nyberg, M.; Luo, Y. Identifying isomers of C_{78} by means of x-ray spectroscopy. *Phys. Rev. B* **2002**, *65*, 165402. [CrossRef]
62. Gao, B.; Liu, L.; Wang, C.; Wu, Z.; Luo, Y. Spectral identification of fullerene C_{82} isomers. *J. Chem. Phys.* **2007**, *127*, 164314. [CrossRef] [PubMed]
63. Brena, B.; Luo, Y. Electronic structures of azafullerene $C_{48}N_{12}$. *J. Chem. Phys.* **2003**, *119*, 7139–7144. [CrossRef]
64. Qi, J.; Hua, W.; Gao, B. Theoretical study of two I_h-symmetry-breaking C_{60} isomers and their chlorinated species in core-excited and ground states. *Chem. Phys. Lett.* **2012**, *539*, 222–228. [CrossRef]
65. Pettersson, L.G.; Wahlgren, U.; Gropen, O. Effective core potential parameters for first-and second-row atoms. *J. Chem. Phys.* **1987**, *86*, 2176–2184. [CrossRef]
66. Becke, A.D. Density-functional exchange-energy approximation with correct asymptotic behavior. *Phys. Rev. A* **1988**, *38*, 3098. [CrossRef] [PubMed]
67. Perdew, J.P. Density-functional approximation for the correlation energy of the inhomogeneous electron gas. *Phys. Rev. B* **1986**, *33*, 8822. [CrossRef] [PubMed]
68. Hermann, K.; Pettersson, L.; Casida, M.; Daul, C.; Goursot, A.; Koester, A.; Proynov, E.; St-Amant, A.; Salahub, D.R.; Carravetta, V.; et al. *StoBe-deMon v.3.3*; Royal Institute of Technology: Stockholm, Sweden; Berlin, Germany, 2014.
69. Vall-Llosera, G.; Gao, B.; Kivimäki, A.; Coreno, M.; Ruiz, J.; de Simone, M.; Ågren, H.; Rachlew, E. The C 1s and N 1s near edge x-ray absorption fine structure spectra of five azabenzenes in the gas phase. *J. Chem. Phys.* **2008**, *128*, 044316. [CrossRef]
70. Gao, B.; Jiang, J.; Liu, K.; Luo, Y. *Bionano-Lego, Version 2.0*; Royal Institute of Technology: Stockholm, Sweden, 2008.
71. Brena, B.; Carniato, S.; Luo, Y. Functional and basis set dependence of K-edge shake-up spectra of molecules. *J. Chem. Phys.* **2005**, *122*, 184316. [CrossRef]
72. Brena, B.; Luo, Y.; Nyberg, M.; Carniato, S.; Nilson, K.; Alfredsson, Y.; Åhlund, J.; Mårtensson, N.; Siegbahn, H.; Puglia, C. Equivalent core-hole time-dependent density functional theory calculations of carbon 1s shake-up states of phthalocyanine. *Phys. Rev. B* **2004**, *70*, 195214. [CrossRef]
73. Brena, B.; Luo, Y. Time-dependent DFT calculations of core electron shake-up states of metal-(free)-phthalocyanines. *Radiat. Phys. Chem.* **2006**, *75*, 1578–1581. [CrossRef]
74. Gao, B.; Wu, Z.; Luo, Y. A density functional theory study of shake-up satellites in photoemission of carbon fullerenes and nanotubes. *J. Chem. Phys.* **2008**, *128*, 234704. [CrossRef]
75. Kutzelnigg, W.; Fleischer, U.; Schindler, M. The IGLO-method: Ab initio calculation and interpretation of NMR chemical shifts and magnetic susceptibilities. *Deuterium Shift Calc.* **1990**, *23*, 165–262.
76. Zhang, B.-B.; Lin, J.; Song, X.-N.; Wang, C.-K.; Hua, W.; Ma, Y. Identification of oxidation states in γ-graphyne by computational XPS and NEXAFS spectra. *Appl. Surf. Sci.* **2023**, *609*, 155134. [CrossRef]
77. Ming, J.; Zhang, J.-R.; Song, X.-N.; Li, X.; Hua, W.; Ma, Y. First-principles simulation of X-ray spectra of graphdiyne and graphdiyne oxides at the carbon K-edge. *Phys. Chem. Chem. Phys.* **2023**, *25*, 32421–32429. [CrossRef] [PubMed]

molecules

MDPI

Article

Quantum Chemical Investigation into the Structural Analysis and Calculated Raman Spectra of Amylose Modeled with Linked Glucose Molecules

Dapeng Zhang and Naoki Kishimoto *

Department of Chemistry, Graduate School of Science, Tohoku University, 6-3, Aoba, Aramaki, Aoba-ku, Sendai 980-8578, Japan; zhang.dapeng.c5@tohoku.ac.jp
* Correspondence: kishimoto@tohoku.ac.jp; Tel.: +81-22-795-6576

Abstract: This study presents a quantum chemical investigation into the structural analysis and calculated Raman spectra of modeled amylose with varying units of linked glucose molecules. We systematically examined the rotation of hydroxymethyl groups and intramolecular hydrogen bonds within these amylose models. Our study found that as the number of linked glucose units increases, the linear structure becomes more complex, resulting in curled, cyclic, or helical structures facilitated by establishing various intramolecular interactions. The hydroxymethyl groups were confirmed to form interactions with oxygen atoms and with hydroxymethyl and hydroxyl groups from adjacent rings in the molecular structures. We identified distinct peaks and selected specific bands applicable in various analytical contexts by comparing their calculated Raman spectra. Representative vibrational modes within selected regions were identified across the different lengths of amylose models, serving as characteristic signatures for linear and more coiled structural conformations. Our findings contribute to a deeper understanding of amylose structures and spectroscopic signatures, with implications for theoretical studies and potential applications. This work provides valuable reference points for the detailed assignment of Raman peaks of amylose structure, facilitating their application in broader research on carbohydrate structures and their associated spectroscopic properties.

Keywords: amylose; calculated Raman spectra; glucose linkage; quantum chemical calculation; structural analysis

1. Introduction

Starch, a ubiquitous component of carbohydrate-rich foods, serves as a primary energy source for human metabolism [1–5] and holds considerable promise for the development of biopolymer composite materials [6–8], such as bioplastics [9,10]. Offering a sustainable alternative to conventional petroleum-based plastics, these bioplastics thereby contribute to the mitigation of environmental concerns associated with non-biodegradable synthetic polymers [11–13]. As an abundant and biodegradable polysaccharide endowed with a chemically modifiable structure, starch has attracted substantial attention as a promising raw resource for advanced functional materials [14]. However, further exploration and exploitation of starch necessitates a comprehensive elucidation of its intricate structural features across multiple length scales, particularly those of its amylose and amylopectin components. Amylose is a linear polysaccharide composed of α-1,4 glycosidic linkages of glucose units, forming a helical structure stabilized by intramolecular hydrogen bonds.

In contrast, amylopectin features a branched structure, where the linear glucose backbone is interconnected by α-1,4 glycosidic linkages interconnected with additional α-1,6 glycosidic points [15]. The relative proportions of amylose and amylopectin in starch vary, and their distinct assembled features contribute to the potential for forming diverse structures, thereby rendering the structural composition of starch replete with various conformations arising from the interplay between these components, enabling the formation of intricate polymerized architectures [16–18]. Given the pivotal role of α-1,4 glycosidic linkages in amylose and amylopectin, a comprehensive structural analysis of the linear glucose linkages becomes imperative. Elucidating the intricate details of these structural features across multiple length scales will facilitate the understanding and development of tailor-made starch-based materials with customized functionalities, unlocking their potential for various applications across diverse domains.

Raman spectroscopy is a versatile and powerful analytical technique for investigating structural information and intramolecular interactions of carbohydrate polymers [19]. This technique can be complemented by quantum chemical calculations, which enable the examination and detailed assignment of the observed Raman spectra of compounds, including amylose and amylopectin [20,21]. In this study, we perform quantum chemical calculations to investigate the structural features of linear glucose models and their corresponding calculated Raman spectra. By combining conformational analysis and assignment of the calculated Raman spectra, we aim to explain the structural details of amylose and amylopectin, thereby facilitating the understanding of their polymerization properties and potential bio-inspired applications.

2. Calculation Details

The investigation commenced with the initial optimization of amylose models, consisting of four glucose units (4Glc), utilizing the B3LYP/3-21G level of theory. Uniform orientation of two out of four hydroxymethyl groups was presumed throughout the calculations, performed using the GRRM (Global Reaction Route Mapping) program [22,23] interfaced with Gaussian 16 [24]. Subsequent refinement encompassed further optimization and Raman spectra calculations at the CAM-B3LYP/6-311+G(d,p) level of theory [25–29] using Gaussian 16. To address potential conformational influences on the Raman spectra, a conformational exploration was conducted employing the PM6 semi-empirical method [30] implemented through the ADDF algorithm [31–34] within the GRRM software (GRRM 17). This algorithm was applied exclusively to the ten lowest local minima (NLowest = 10). A constrained exploration of local minima was performed to minimize computational costs by bypassing transition state optimizations. Tracking the seven largest anharmonic downward distortions (ADDs) along the reaction coordinates of the computed equilibrium (EQ) structures (l-ADDF7) was conducted. The NoBondRearrange option was employed to expedite the search for structures without bond rearrangements. Subsequently, the ten most stable structures identified in the conformational search were optimized along with their corresponding Raman spectra at the CAM-B3LYP/6-311+G(d,p) level of theory using Gaussian 16. Additionally, the amylose structures comprising two, six, and eight glucose units (2Glc, 6Glc, and 8Glc) were investigated for comparison, beginning with initial optimization via the B3LYP/3-21G method utilizing the GRRM program, followed by successive optimization and Raman spectra calculation employing the CAM-B3LYP/6-311+G(d,p) level of theory for 2Glc and 6Glc, while for 8Glc the CAM-B3LYP/6-31G(d) approach [35–44] within Gaussian 16. The dihedral angles related to the hydroxymethyl groups of each glucose ring were manually adjusted among the 2Glc, 4Glc, 6Glc, and 8Glc models to analyze the formation of intramolecular hydrogen bonds.

Given the potential for intramolecular interactions between hydroxymethyl and hydroxy groups to induce coiling of the linear structure, we additionally examined the conformational landscape without imposing manual selection of molecular dihedral angles. Instead, the ADDF algorithm generated a diverse ensemble of potential structural configurations. This systematic computational algorithm explored various molecular structures, enabling the automatic generation of plausible models while minimizing potential biases that could arise from manual adjustments. The ADDF calculation was initiated from an optimized 4Glc model, wherein the ten lowest local minima (NLowest = 10) and only the equilibrium (EQ) structures were traced. In order to achieve computational efficiency, the seven largest ADDs (l-ADD7) surrounding each stable minima point were followed. The conformational exploration was conducted using the semi-empirical PM6 method, yielding 169 stable structures. The ten most stable structures found in the exploration were re-optimized at the CAM-B3LYP/6-311+G(d,p) level of theory, and their Raman spectra were calculated at the same level.

3. Results and Discussion

3.1. Structural Analysis and Calculated Raman Spectra of 4Glc

The initial models were constructed by manipulating the dihedral angles between the O5 and C5 atoms within the ring, as well as those between the C6 and O6 atoms of hydroxymethyl groups linked to the ring (O5-C5-C6-O6), as shown in Figure 1. Detailed parameters governing these adjustments are outlined in Supporting Information, Tables S1 and S2. Subsequent structural optimization resulted in the convergence of certain structures toward similar conformations, which can be attributed to the interplay of hydroxymethyl groups forming intramolecular hydrogen bonds, as the structures depicted in Figure 2 (with calculated Raman spectra in Figure 3) and Figure 4 (with calculated Raman spectra in Figure 5), and Tables 1 and 2. Two distinct types of intramolecular hydrogen bonds are confirmed: O5· · · H-O6′, formed between a ring and the hydroxymethyl groups at the succeeding ring, and O6· · · H-O6′, formed between the hydroxymethyl groups of two adjacent rings. Notably, *Om*· · · H-O*n*′ indicates that the hydrogen bonds were formed between the *Om* atom of one ring and the H-O*n*′ group from another nearby ring, where the apostrophe (′) denotes a different glucose ring. While these interactions are comparatively weaker than those established between amylose during polymerization, they engender diverse conformational arrangements within the molecular framework.

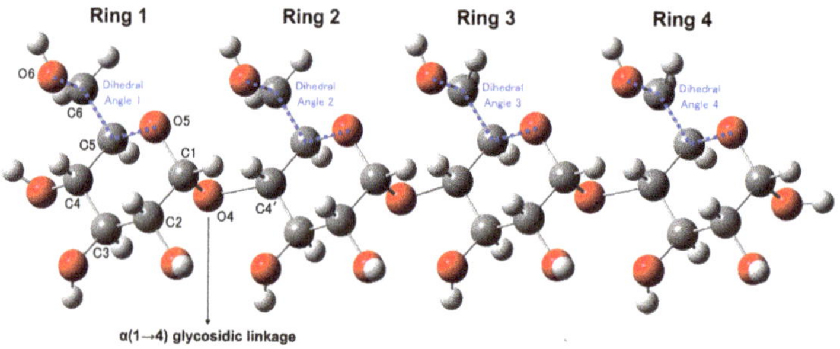

Figure 1. Atom labeling and the dihedral angles examined within four rings in a 4Glc structure. Red balls represent O atoms, white balls represent H atoms, and gray balls represent C atoms. The remaining figures are consistent with the representation.

Table 1. Intramolecular hydrogen bonds formed by the rotation of hydroxymethyl groups within four rings (4Glc1-4Glc10 structures), showing dihedral angles (in degrees) and hydrogen bond lengths (in Å).

Structure	O5 (Ring 1)· · ·H-O6′ (Ring 2)	O5 (Ring 2)· · ·H-O6′ (Ring 3)	O5 (Ring 3)· · ·H-O6′ (Ring 4)	O6 (Ring 3)· · ·H-O6′ (Ring 4)
4Glc1	158.39, 2.06	159.73, 2.10	161.46, 2.09	-
4Glc2	158.21, 2.05	159.93, 2.10	161.29, 2.10	-
4Glc3	158.95, 2.06	160.10, 2.10	-	-
4Glc4	163.07, 2.08	-	-	164.08, 1.89
4Glc5	163.60, 2.11	-	-	163.98, 1.89
4Glc6	-	162.69, 2.09	161.40, 2.09	-
4Glc7	159.34, 1.94	160.02, 2.12	161.23, 2.09	-
4Glc8	-	-	-	158.83, 1.98
4Glc9	160.63, 1.95	-	-	165.88, 1.89
4Glc10	157.96, 1.95	-	154.90, 1.84	158.17, 1.95

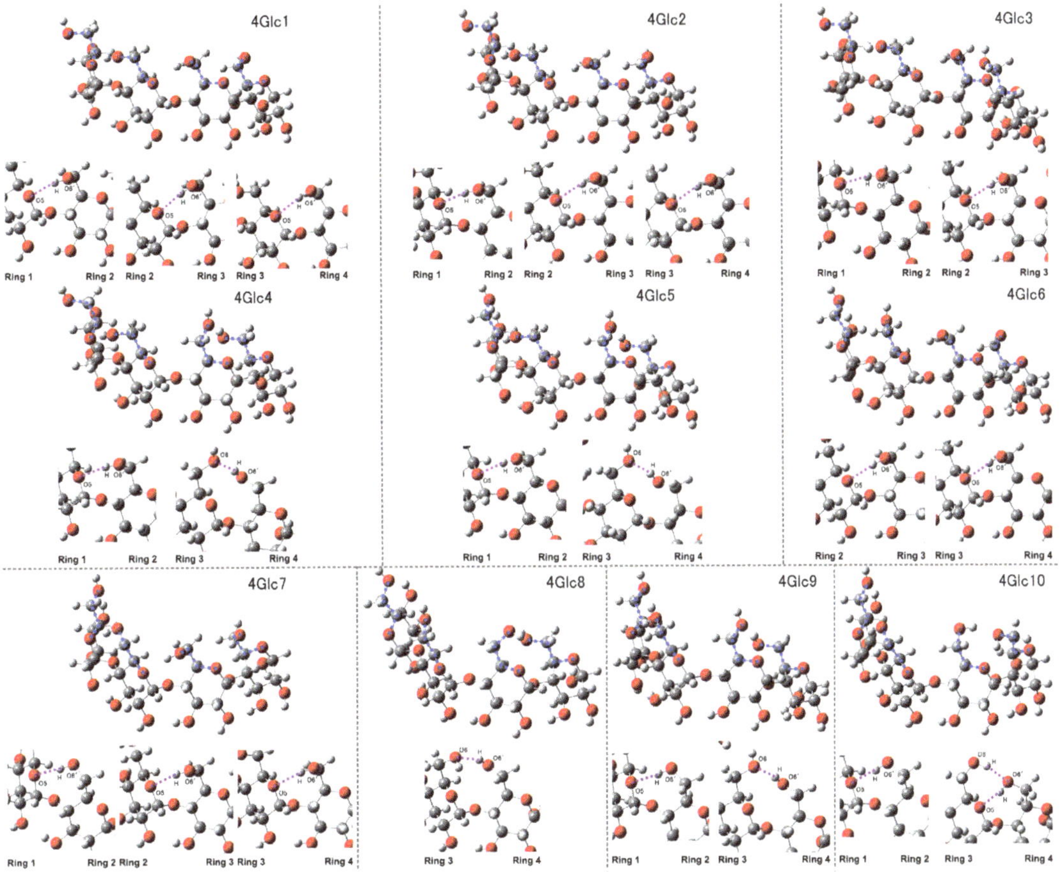

Figure 2. Structural changes observed in 4Glc1-10 models, with dihedral angles highlighted by blue dashed lines and weak intramolecular hydrogen bonds denoted by purple dashed lines.

Table 2. Intramolecular hydrogen bonds formed by the rotation of hydroxymethyl groups within four rings (4Glc11-4Glc18 structures), showing dihedral angles (in degrees) and hydrogen bond lengths (in Å).

Structure	O6 (Ring 1)···HOCH2 (Ring 2)	O6 (Ring 2)···H-O6′ (Ring 3)	O5 (Ring 3)···HOCH2 (Ring 4)	O6 (Ring 3)···HOCH2 (Ring 4)
4Glc11	156.42, 2.00	168.22, 1.99	159.78, 2.08	-
4Glc12	156.48, 2.00	168.32, 1.99	159.69, 2.09	
4Glc13	161.70, 1.95	167.75, 1.96	-	-
4Glc14	167.11, 1.87	173.30, 1.85	-	162.54, 1.84
4Glc15	-	158.29, 1.97	162.81, 1.96	-
4Glc16	162.60, 1.92	158.74, 1.89	164.22, 1.95	-
4Glc17	168.32, 1.86	-	-	167.68, 1.88
4Glc18	164.86, 1.86	-	-	161.44, 1.94

4Glc1 and 4Glc2 show analogous conformations, characterized by the presence of three intramolecular hydrogen bonds (O5· · · H-O6′ type) linking ring 1 to ring 2, ring 2 to ring 3, and ring 3 to ring 4. These weak interactions exhibit bond angles of approximately 158.21–161.46 degrees and lengths spanning 2.05–2.10 angstroms. The hydroxymethyl group of ring 4 in 4Glc3 exhibited a different rotational conformation compared to that in 4Glc1 and 4Glc2, with a dihedral angle of 78.30 degrees instead of around −81 to −82 degrees. This variance resulted in the disruption of the hydrogen bond between rings 3 and 4. Additionally, intramolecular hydrogen bonds (O5· · · H-O6′ type) were observed between rings 1 and 2, as well as between rings 2 and 3. 4Glc4 and 4Glc 5 exhibit similar intramolecular interactions, with hydrogen bonds of the O5· · · H-O6′ type observed between ring 1 and ring 2, while another type (O6· · · H-O6′) is confirmed between ring 3 and ring 4. This result is accompanied by the disruption of hydrogen bonds between ring 2 and ring 3. Interestingly, the dihedral angle 1 for 4Glc4 is −82.18 degrees, while that of 4Glc5 is −177.38 degrees, indicating significant differences in the conformational rotation of hydroxymethyl groups at ring 1.

4Glc6 exhibits unique weak interactions, characterized by two hydrogen bonds of the O5· · · H-O6′ type, between ring 2 and ring 3, as well as between ring 3 and ring 4. The presence of three hydrogen bonds (O5· · · H-O6′ type) across rings 1 to 2, 2 to 3, and 3 to 4 in 4Glc1-2 was also observed in 4Glc7. Nonetheless, notable conformational disparities were identified in dihedral angle 1 (−66.70 and −66.67 degrees for 4Glc1-2, −174.49 degrees for 4Glc7) and dihedral angle 2 (−81.12 and −81.09 degrees for 4Glc1-2, 168.55 degrees for 4Glc7). In 4Glc8, we observed a single hydrogen bond of the O6· · · H-O6′ type, formed between ring 3 and ring 4, with bond angles of 158.83 degrees and a bond length of 1.98 angstroms. The dihedral angles 1 and 2 exhibited similar values in 4Glc9 and 4Glc10, measuring −174.92 and −173.37 degrees and 168.24 and 170.75 degrees, respectively. In both 4Glc9 and 4Glc10, an O5· · · H-O6′ type hydrogen bond was observed between ring 1 and ring 2, while no hydrogen bond interaction was observed between ring 2 and ring 3. Although both structures exhibited an O6· · · H-O6′ type hydrogen bond between ring 3 and ring 4, the difference in dihedral angles 3 (72.53 and 94.97 degrees for 4Glc9 and 4Glc10, respectively) and dihedral angles 4 (148.49 and 171.18 degrees for 4Glc9 and 4Glc10, respectively) resulted in the simultaneous presence of an O5· · · H-O6′ type hydrogen bond between ring 3 and ring 4 in 4Glc10.

The Raman spectra calculated for 4Glc1-10 exhibited noticeable disparities in relative intensity and subtle shifts in band positions across the spectrum. Notably, relative intensity comparisons were conducted within each designated molecule region rather than between different structures. These findings are shown in Figure 3. It is important to note that all the calculated results presented in this work are from the Raman activity spectrum. Our analysis specifically targeted four distinct regions to elucidate the conformational characteristics in depth. The calculated Raman spectra of 4Glc1 and 4Glc2 exhibit similarities attributable to their shared conformations. However, discernible disparities were observed, particularly in 4Glc2, wherein a distinct band emerged at approximately 1103 cm^{-1} within region C, alongside an intensified band at 1304 cm^{-1} within region D. By contrast, 4Glc3 displayed significant alternations, particularly characterized by a diminished relative intensity at 852 cm^{-1} within region 2, a reduction in relative intensity of the 1121 cm^{-1} band within region 3, as well as attenuated signals at 1374 and 1469 cm^{-1} within region 4, in addition to the attenuation observed at 1152 cm^{-1} within region 4. 4Glc4 and 4Glc5 exhibited a novel peak around 506 cm^{-1} within region 1, while 4Glc4 displayed notably increased intensities at the 870 cm^{-1} band within region 2, as well as at the 1309 and 1515 cm^{-1} bands within region 4. Moreover, an increased intensity was observed around the 1446 or 1448 cm^{-1} bands in region 4, for both 4Glc4 and 4Glc5.

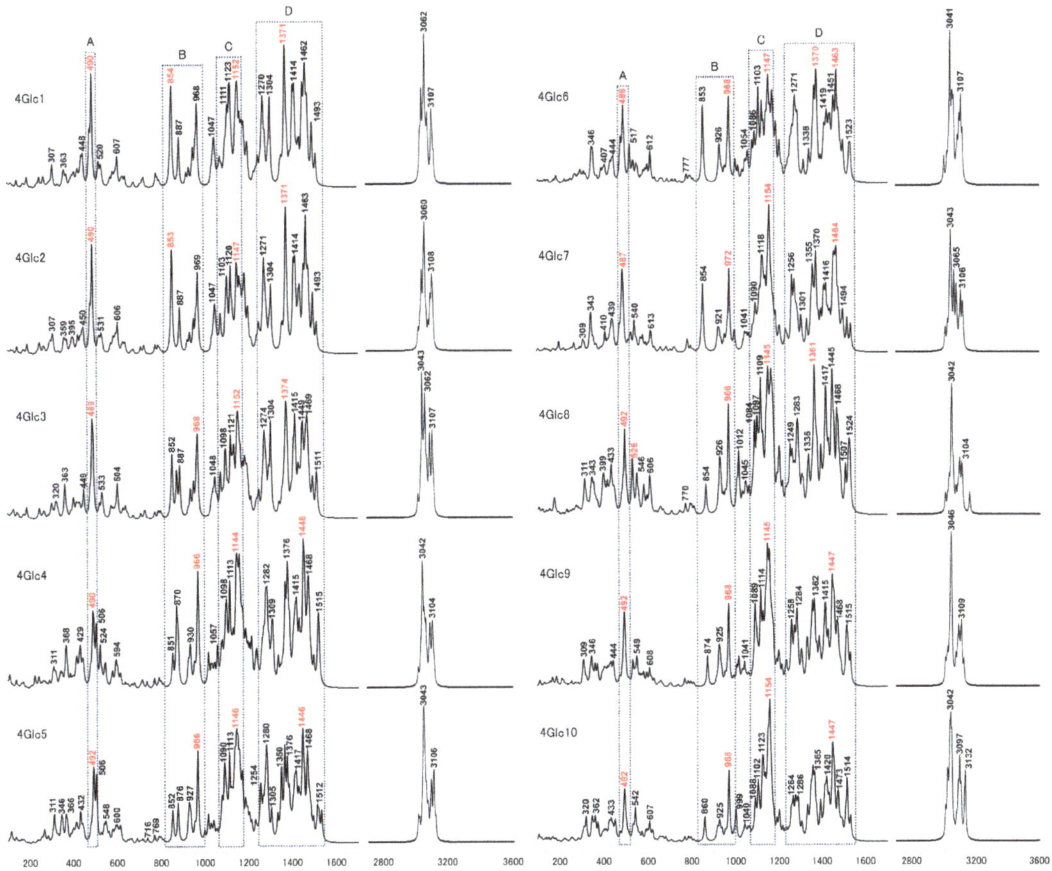

Figure 3. Calculated Raman spectra for 4Glc1-10 models, highlighting bands with the highest intensity in four delineated regions, A, B, C, and D, denoted in red.

Although 4Glc6 and 4Glc7 share similar parameters of dihedral angles 1, 3, and 4, the latter exhibited a hydrogen bond (O5· · · H-O6′ type) between ring 1 and ring 2, as depicted in Figure 2. The calculated Raman spectra of both models are nearly identical in regions 1 and 2. However, in region C, the intensity of the 1154 cm^{-1} peak in 4Glc7 significantly dominated. In contrast, in 4Glc6, the intensity differences between the most intensified band of 1147 cm^{-1} within region C and the two strong peaks (1370 and 1463 cm^{-1}) in region D are less pronounced. In the B region of 4Glc8-10, the bands exhibited decreased intensities at 854, 874, and 860 cm^{-1}, respectively. In the C region of 4Glc8, the peak intensities at 1109 and 1145 cm^{-1} are similar, whereas, in 4Glc9 and 4Glc10, only the band at 1145 or 1154 cm^{-1} showed a prominent peak. Furthermore, the strongest peak in the C region of 4Glc8 (1145 cm^{-1}) is comparable in intensity to the strongest peak in the D region (1361 cm^{-1}), whereas, in 4Glc9 and 4Glc10, the intensity of this peak notably decreased. Additionally, the intensity of the 1286 cm^{-1} bands in 4Glc10 decreased compared to 4Glc9, and the difference in intensity between the 1286 cm^{-1} and 1264 cm^{-1} bands is less pronounced. These three structures exhibited prominent peaks at 1524, 1515, and 1514 cm^{-1}, similar to the independent sharp peak observed at 1515 cm^{-1} in 4Glc4.

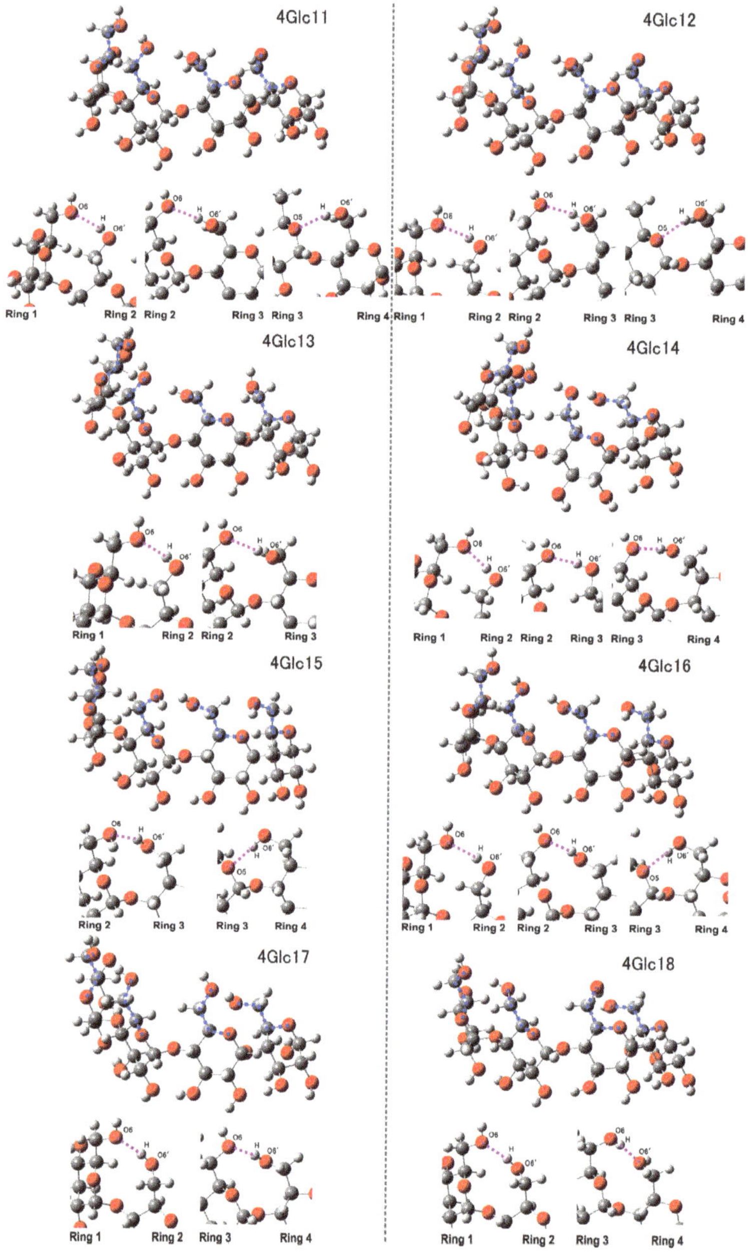

Figure 4. Structural changes observed in 4Glc11-18 models, with dihedral angles highlighted by blue dashed lines and weak intramolecular hydrogen bonds denoted by purple dashed lines.

The conversion of dihedral angles 1 and 2 to positive values resulted in distinct conformations within the interconnected four-glucose molecules, as illustrated in Figure 4. Both 4Glc11 and 4Glc12 exhibited akin structures characterized by three intramolecular hydrogen bonds. Specifically, hydrogen bonds of the O6$\cdots$H-O6′ type were established between the hydroxymethyl groups of ring 1 and ring 2 and between ring 2 and ring 3.

Concurrently, the O5· · · H-O6′ type was observed to form between ring 3 and ring 4. In contrast, a significant deviation in the dihedral angle 4 of 4Glc13 was observed, measuring 77.92 degrees, in contrast to those of 4Glc11 and 4Glc12 (−81.33 and −82.14 degrees, respectively). This alteration resulted in the hydroxymethyl group in ring 4 relocating farther from ring 3, disrupting the intermolecular interactions with ring 3. Notably, only two hydrogen bonds of the O6· · · H-O6′ type were evident, occurring between ring 1 and ring 2, as well as between ring 2 and ring 3, respectively. In 4Glc14, three intramolecular hydrogen bonds of the O6· · · H-O6′ type were identified, attributed to rotations of hydroxymethyl groups between ring 1 and ring 2, between ring 2 and ring 3, and between ring 3 and ring 4, respectively. 4Glc15 and 4Glc16 displayed similarity in dihedral angles 1, 3, and 4, while a notable disparity arose in dihedral angle 2, exhibiting values of 57.27 and 96.93 degrees, respectively. This discrepancy implied that conformational variations primarily stem from the rotational dynamics of the hydroxymethyl group within ring 2. Notably, 4Glc15 manifested intramolecular hydrogen bonds formed between ring 2 and ring 3 (O6· · · H-O6′ type) and between ring 3 and ring 4 (O5· · · H-O6′ type). Additionally, in 4Glc16, alongside these aforementioned interactions, the hydroxymethyl groups within ring 1 and ring 2 facilitated the formation of weak bonds of the O6· · · H-O6′ type. In both 4Glc 17 and 4Glc 18, analogous interactions between the hydroxymethyl groups within rings 1 and 2, as well as between rings 3 and 4, led to the formation of O6· · · H-O6′ type hydrogen bonds. However, due to differences in dihedral angles 2 and 3, with values of 139.65 and 73.00 degrees, respectively, for the former structure and 151.75 and 64.18 degrees, respectively, for the latter, these molecules failed to establish interactions between rings 2 and 3.

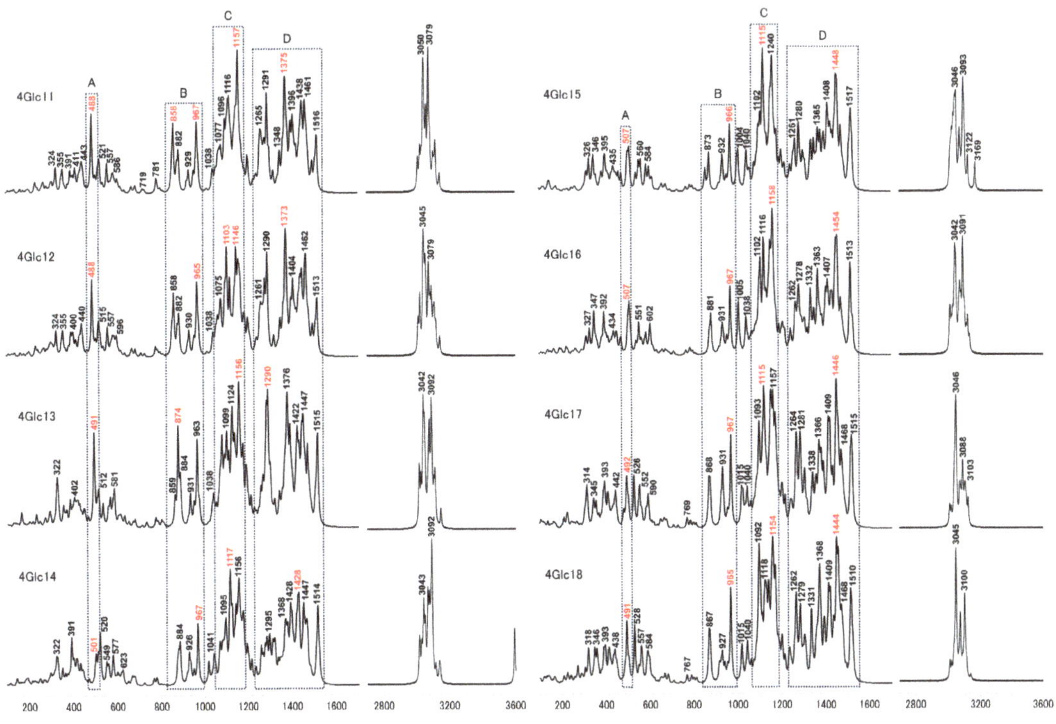

Figure 5. Calculated Raman spectra for 4Glc11-18 models, highlighting bands with the highest intensity in four delineated regions, A, B, C, and D, denoted in red.

Conformational similarities between 4Glc11 and 4Glc12 were confirmed, complemented by comparable calculated Raman spectra in regions A and B, as shown in Figure 5. However, in the case of 4Glc11, the dominance of the 1157 cm^{-1} peaks in region C exceeded even the most prominent band in region D (1375 cm^{-1}). In addition, region C of 4Glc11 showed two notable bands at 1103 and 1146 cm^{-1}, albeit with relative intensities lower than the predominant band within region D (1373 cm^{-1}). Conversely, 4Glc12 experienced considerable alternations in dihedral angle 4, as indicated by a notable amplification of the 874 cm^{-1} band in its region B, surpassing the intensity of the 963 cm^{-1} bands. Notably, the 1156 cm^{-1} band in region C of 4Glc13 mirrored the behavior in 4Glc11. Additionally, region D of 4Glc12 featured two strong peaks at 1290 and 1376 cm^{-1}, accompanied by a significant increase in the intensity of the 1515 cm^{-1} band. In region A of 4Glc14, the peak at 501 cm^{-1} exhibited relatively low intensity, while region B shared similarity with the corresponding regions of 4Glc15 and 4Glc16. Within region C, two peaks were discerned at 1117 and 1156 cm^{-1}, mirroring the spectral characteristics observed in 4Glc15 (which manifested peaks at 1115 and 1240 cm^{-1}). In contrast, in 4Glc16, a solitary dominant peak appeared at 1158 cm^{-1} within the same spectral region. Contrasted with 4Glc14, regions D of 4Glc15 and 4Glc16 showcased robust peaks aligning with the 1448 or 1454 cm^{-1} bands. Region C of 4Glc17 revealed the presence of three discernible peaks at 868, 931, and 967 cm^{-1}, with escalating relative intensities. By contrast, the intensity of the 927 cm^{-1} band in 4Glc18 is noticeably diminished compared to those at 867 and 965 cm^{-1}. Additionally, distinctions between the two compounds emerged in region C, wherein the 1093 cm^{-1} peak exhibited reduced intensity relative to the 1115 cm^{-1} peak in 4Glc17, while conversely observed in 4Glc18 (1092 to 1118 cm^{-1} bands). Additionally, an augmentation in the intensity of the 1368 cm^{-1} peak in region D became evident in 4Glc18.

The ten most stable structures derived from the conformational exploration and calculated Raman spectra are shown in Figures 6 and 7. In contrast to their 4Glc1-18 counterparts, these structures exhibited entirely distinct sets of four dihedral angles, as detailed in Table S3. The emergence of coiled conformations prompted notable shifts in intramolecular hydrogen network, involving eight distinct types: O3 (ring 1)···H-O6′ (ring 1), O4 (ring 1)···H-O6′ (ring 2), O4 (ring 1)···H-O6′ (ring 4), O6 (ring 1)···H-O6′ (ring 2), O6 (ring 1)···H-O6′ (ring 4), O6 (ring 2)···H-O6′ (ring 3), O6 (ring 2)···H-O6′ (ring 4), and O6 (ring 3)···H-O6′ (ring 4). The intramolecular hydrogen bonding patterns observed for each conformation are delineated in Table 3, with the parameters of the dihedral angles provided in Supporting Information Table S3.

Table 3. Intramolecular hydrogen bonds formed by the rotation of hydroxymethyl groups within four rings (the explored structures), showing dihedral angles (in degrees) and hydrogen bond lengths (in Å).

Structure	O3 (Ring 1)···H-O6′ (Ring 1)	O4 (Ring 1)···H-O6′ (Ring 2)	O4 (Ring 1)···H-O6′ (Ring 4)	O6 (Ring 1)···H-O6′ (Ring 2)	O6 (Ring 1)···H-O6′ (Ring 4)	O6 (Ring 2)···H-O6′ (Ring 3)	O6 (Ring 2)···H-O6′ (Ring 4)	O6 (Ring 3)···H-O6′ (Ring 4)
EQ21		-	167.38, 1.82	-	-	-	-	164.81, 1.84
EQ145	153.45, 1.91	-	-	-	-	150.32, 1.93	150.43, 1.90	-
EQ52		-	-	161.65, 1.92	154.06, 1.78	-	-	162.14, 1.86
EQ54		-	-	161.64, 1.92	154.06, 1.78	-	-	162.13, 1.86
EQ141	165.17, 1.84	175.28, 1.88	-	-	-	155.49, 1.93	-	164.93, 1.78
EQ67		-	-	156.21, 1.80	170.57, 1.77	166.54, 1.80	-	154.00, 1.82
EQ9		-	166.67, 1.84	-	-	-	-	163.91, 1.87
EQ112		-	-	-	-	150.48, 1.93	152.49, 1.88	-
EQ32		-	-	-	-	151.75, 1.92	155.48, 1.86	-
EQ134		-	168.25, 1.82	-	-	164.33, 2.00	-	156.73, 1.82

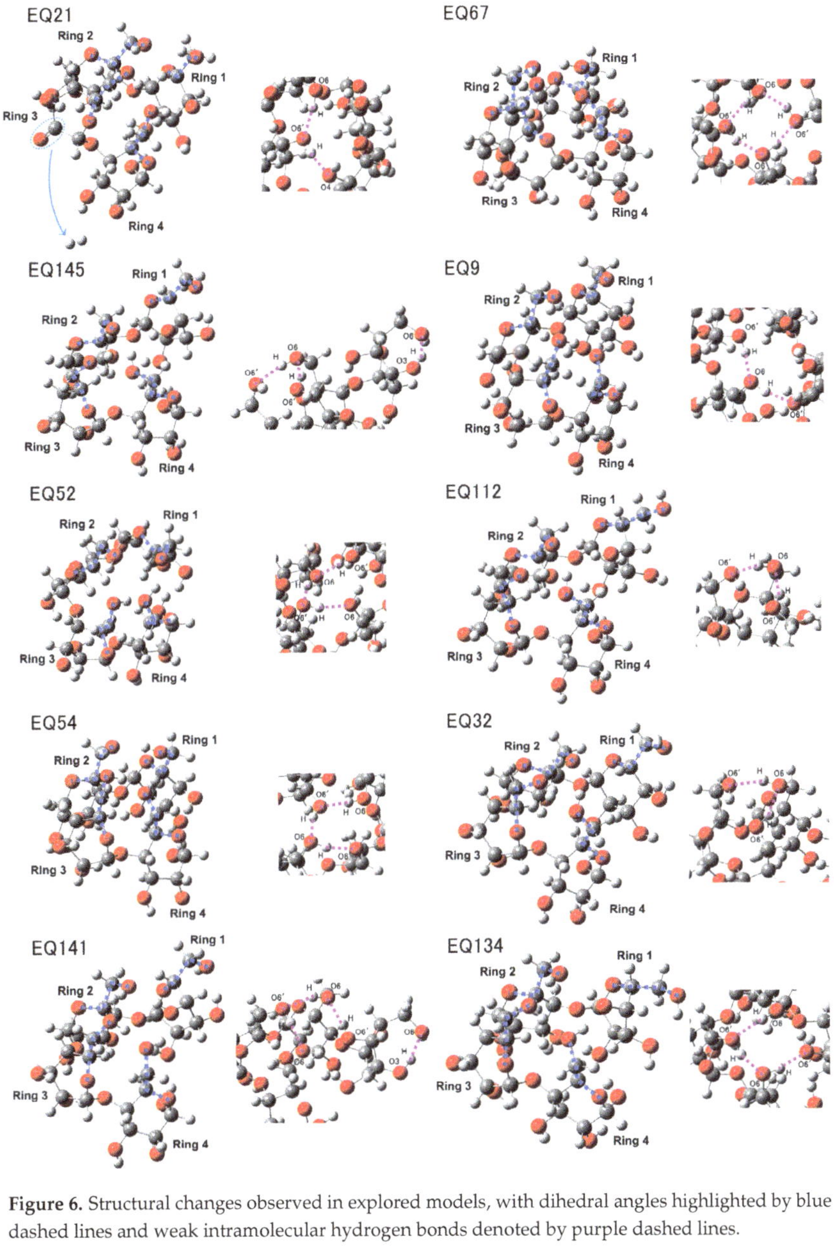

Figure 6. Structural changes observed in explored models, with dihedral angles highlighted by blue dashed lines and weak intramolecular hydrogen bonds denoted by purple dashed lines.

For EQ21, ring 3 underwent dehydrogenation, there is an O4· · · H-O6′ hydrogen bond between ring 1 and ring 4, and an O6· · · H-O6′ hydrogen bond between ring 3 and ring 4, although it maintained the α-1,4 linked structure, its stability decreased compared to others; In EQ145, there is a special O3-H· · · O6′ hydrogen bond (153.45°, 1.91 Å) within ring 1, and the O6 of ring 2 forms hydrogen bonds with the O6 of ring 3 and 4, respectively; In EQ52 and EQ54, the O6 of ring 1 forms hydrogen bonds with the O6 of ring 2 and 4, respectively, and there is an O6· · · H-O6′ hydrogen bond between ring 3 and ring 4; In EQ141, there is an O3-H· · · O6′ (165.17°, 1.84 Å) within ring 1 and a relatively strong O4· · · H-O6′ (175.28°,

1.88 Å) hydrogen bond between ring 1 and ring 2, and there are also O6· · · H-O6′ hydrogen bonds between rings 2 and 3, and between rings 3 and 4; In EQ67, there are O6· · · H-O6′ hydrogen bonds between ring 1 and rings 2 and 4, ring 2 and rings 3 and 4, and ring 3 and ring 4; In EQ9, there is a hydrogen bond (166.67°, 1.84 Å) between the O4 of ring 1 and the O6 of ring 4, and a hydrogen bond (163.91°, 1.87 Å) between the O6 of ring 3 and the O6 of ring 4; In EQ112 and EQ32, the O6 of ring 2 forms similar hydrogen bonds with the O6 of ring 3 and 4 (for EQ112: 150.48°, 1.93 Å and 152.49°, 1.88 Å, for EQ32: 151.75°, 1.92 Å and 155.48°, 1.86 Å); In EQ134, there are hydrogen bonds between the O4 of ring 1 and the O6 of ring 4 (168.25°, 1.82 Å), the O6 of ring 2 and the O6 of ring 3 (164.33°, 2.00 Å), and the O6 of ring 3 and the O6 of ring 4 (156.73°, 1.82 Å).

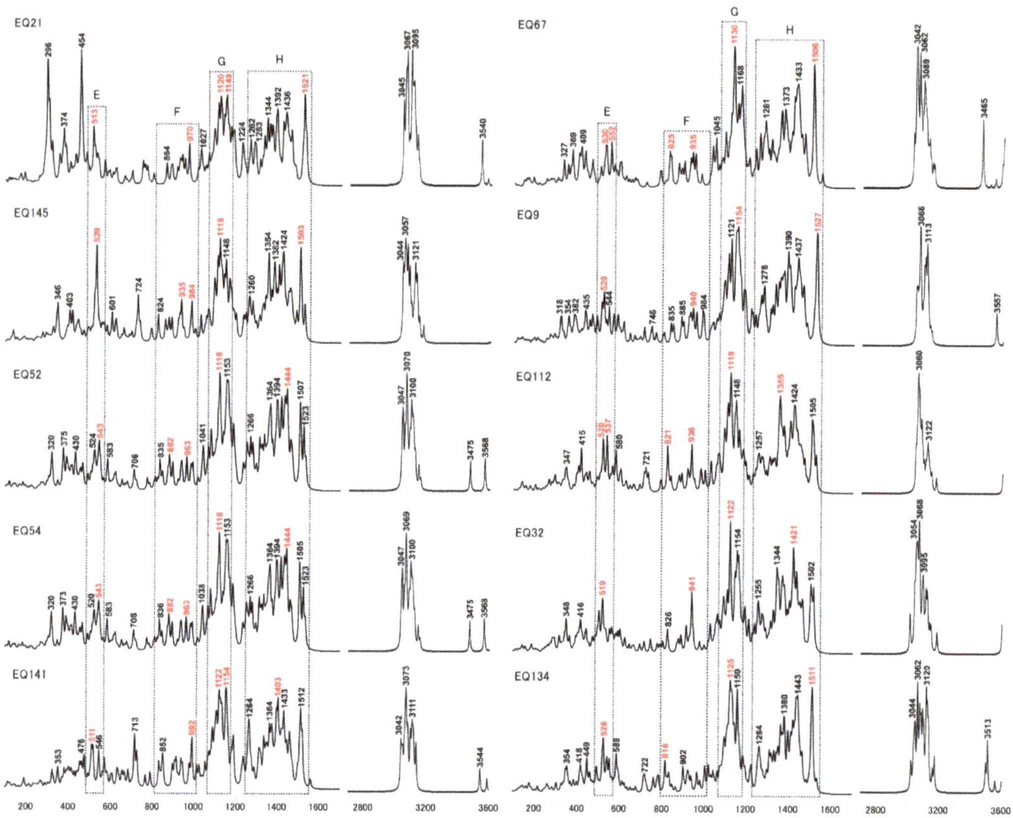

Figure 7. Calculated Raman spectra for explored models, highlighting bands with the highest intensity in four delineated regions, E, F, G, and H, denoted in red.

We selected and focused our analysis on four distinct peak regions in the calculated Raman spectra of the coiled glucose models, denoted as E, F, G, and H. Compared to the uncoiled conformation exhibiting the A, B, C, and D regions, the spectral features within each of the newly observed E, F, G, and H regions exhibited discernible alterations. These changes likely arise from forming more intramolecular interactions induced by the conformational changes. Due to dehydrogenation, the EQ21 displayed two prominent bands at 296 and 454 cm^{-1}, albeit the four selected regions of interest remained broadly analogous to other structures. In region E, a distinct singular feature emerged at 513 cm^{-1}. The spectral region F was characterized by a peak at 970 cm^{-1} and a relatively feeble signal at 864 cm^{-1}. Region G showed two peaks of comparable intensities at 1120 and

1149 cm^{-1}, while region H featured an intense band at 1521 cm^{-1}. Region E of the EQ145 was likewise characterized by a prominent singular band at 529 cm^{-1}, exhibiting a higher relative intensity than the features observed in region F.

In contrast to EQ21, the spectral signature of EQ145 in the region was dominated by an intense peak in region G at 1118 cm^{-1}, representing the most pronounced signal across the four regions under investigation. Conversely, in region H, the dominant band for EQ145 was analogous to that of EQ21, located at 1503 cm^{-1}. The EQ52 and EQ54 exhibited a comparable distribution of spectral features. In region E, two distinct bands were discernible at 524 and 543 cm^{-1} for EQ52, while EQ54 bands were located at slightly different wavenumbers of 520 and 543 cm^{-1}. Region F was characterized by multiple weak peaks. The spectral region G manifested two prominent bands of analogous intensities situated at 1118 and 1153 cm^{-1}. The most intense peak across region H occurred at 1444 cm^{-1}. The calculated Raman spectra of EQ141 revealed two distinct bands in region E at 511 and 546 cm^{-1}, respectively. The region F exhibited the anticipated two prominent peaks at 852 and 992 cm^{-1}. In region G, the two bands' relative intensities at 1122 and 1154 cm^{-1} were comparable. The spectral region H was characterized by the most intense peak across region H, located at 1403 cm^{-1}, accompanied by two additional well-defined sharp peaks at 1264 and 1512 cm^{-1}.

The calculated spectrum of EQ67 revealed the presence of two peaks in region E, exhibiting comparable intensities at 530 and 552 cm^{-1}. Region F was characterized by two peaks of analogous intensities at 825 and 935 cm^{-1}. In region G, a distinct, intense band was observed at 1130 cm^{-1}. The spectral region H exhibited a series of four well-defined peaks with gradually increasing intensities, located at 1281, 1373, 1433, and 1506 cm^{-1}. The spectral region E of the EQ9 was characterized by the presence of two bands of relatively low intensity at 520 and 544 cm^{-1}. The band at 544 cm^{-1} exhibited a comparatively lower intensity and was in close proximity to the prominent peak at 940 cm^{-1} in region F. Region G manifested two distinct bands of substantial intensity at 1121 and 1154 cm^{-1}. The spectral region H displayed the most intense peak around 1527 cm^{-1}, notably sharp and pronounced compared to the other peaks within the same region. For the case of EQ112, region E was characterized by two well-defined bands at 520 and 537 cm^{-1}. Region F exhibited two distinct, prominent peaks at 821 and 936 cm^{-1}. In region G, a pronounced intense band was observed at 1119 cm^{-1}, showing a significant enhancement in intensity relative to the other regions. Region H revealed three peaks with gradually diminishing intensities at 1355, 1424, and 1505 cm^{-1}. In the case of EQ32, region E displayed a peak at 519 cm^{-1}, which is close to a peak at 941 cm^{-1} in region F. However, the peak at 826 cm^{-1} in region F exhibited reduced intensity. Region G presented the most intense peak at 1122 cm^{-1}, comparable to EQ67 and EQ112, while region H showed the most prominent peak at 1421 cm^{-1}.

In comparison to EQ32, the spectral region E of EQ134 was characterized by a relatively intense band at 526 cm^{-1}, unlike region F, which did not exhibit the anticipated two distinct prominent peaks. The spectral region G revealed the presence of the most intense band across the four regions at 1125 cm^{-1}, accompanied by an adjacent band of comparable intensity situated at 1159 cm^{-1}. Furthermore, region H displayed a pronounced peak at 1511 cm^{-1}, exhibiting an intensity analogous to the two bands observed in region G.

3.2. Band Assignments for 4Glc Structures

The structural investigation into the structural characteristics of 4Glc1-18 and the coiled structures after conformational changes revealed that the rotational behavior of hydroxymethyl groups across the four rings can induce the formation of diverse intramolecular hydrogen bonds, resulting in distinct features within the corresponding vibrational properties. Despite these structural variants exhibiting fluctuations in relative intensity across the two sets of four regions (A, B, C, and D, as well as E, F, G, and H), the positional attributes within similar conformational structures remained notably stable. Specifically, spectral signatures within regions A and B, and corresponding E and F, exhibited relatively simple patterns, facilitating straightforward determination. In contrast, those within regions C and D, and corresponding G and H presented heightened intricacy.

The comprehensive assignment of calculated Raman spectra presents challenges attributed to vibrational modes' structural complexity and overlapping nature. In response, our investigation prioritized the examination of vibrational signatures associated with hydroxymethyl groups and α-1,4 glycosidic linkages. Notably, the vibrational modes were classified into stretching (υ), bending (β), and ring stretching and breathing (ρ), with subscripts indicating the vibration of specific groups occurring at ring n. Particular attention was given to discernible bands corresponding to simpler vibrational modes, thereby facilitating the identification of more distinguishable bands applicable to a broader range of cases. The band assignments for the primary vibrations of two spectral parts (100–1700 cm^{-1} and 2700–3600 cm^{-1}) are detailed in Table 4, while the corresponding vibrational animations can be found in the Supporting Information.

Table 4. Raman band assignments for selected 4Glc models with corresponding structure designation following the band wavenumber (cm^{-1}).

Part	Wavenumber (Structure Name)	Band Assignment (Primary Vibration) [1]
II	3092 (4Glc14)	υ (4, -CH)
	3046 (4Glc17)	υ (2, -CH)
I	1527 (EQ9)	β (3, 4, -OH), β (1, 2, 3, 4, -CH_2-OH)
	1462 (4Glc1)	β (2, -OH), β (2, 3, -CH), ρ (2, 3, 4)
	1417 (4Glc8)	υ (2, -C-O-C-), β (1, 2, 3, 4, -CH), β (4, -OH), ρ (3, 4)
	1371 (4Glc1)	β (2, 3, 4, -CH), β (4, -OH), ρ (2, 3, 4)
	1304 (4Glc1)	β (1, -CH), β (1, -OH), β (1, -CH_2-OH), ρ (1)
	1290 (4Glc13)	β (3, 4, -CH_2-OH), β (3, 4, -CH), β (4, -OH)
	1280 (4Glc5)	β (1, -CH_2-OH), β (1, 2, 3, 4, -CH), β (1, 2, 3, 4 -OH)
	1154 (4Glc10)	υ (1, -CH-OH), υ (1, -C-O-C-), β (2, -CH_2-OH), β (1, 2, 4, -CH), β (1, 2, -OH)
	1123 (4Glc1)	υ (1, 2, 3, 4, -CH-OH), β (1, 2, 3, 4, -CH), β (1, 2, 3, 4 -OH), ρ (1, 2, 3, 4)
	968 (4Glc1)	β (1, 2, 3, 4, -CH_2-OH), β (1, 2, 3, 4, -CH), β (1, 2, 3, 4, -OH), ρ (1, 2, 3, 4)
	926 (4Glc8)	β (1, 2, 3, 4, -CH_2-OH), β (1, 2, 3, 4, -CH), β (1, 2, 3, 4, -OH), ρ (1, 2, 3, 4)
	887 (4Glc1)	β (1, -CH_2-OH), β (1, -CH), ρ (1)
	854 (4Glc1)	β (3, 4, -CH_2-OH), β (3, 4, -CH), ρ (3, 4)
	529 (EQ145)	β (2, 3, 4, -CH_2-OH), β (1, 3, -OH), ρ (1, 2, 3)
	490 (4Glc1)	β (1, 2, 3, -CH_2-OH), β (2, 3, -CH), ρ (1, 2, 3, 4)

[1] υ—stretching, β—bending, ρ—ring stretching and breathing.

3.3. *Structural Conformations and Calculated Raman Spectra of 2Glc, 6Glc, and 8Glc*

The 2Glc structures exhibit a relatively simple conformation, with rotations of the hydroxymethyl group observed in one ring (ring 1). In contrast, the other ring (ring 2) maintains its orientation, as illustrated by the 2Glc1 and 2Glc2 structures in Figure 8. The dihedral angles are provided in Supporting Information Table S4. As the number of amylose units increases to form the 6Glc structures, distortions and curling emerge, and the structures gradually adopt a coiled conformation. Notably, although six amylose units per turn are typically sufficient in starch to form a helical structure, the formation of helical structures is not necessarily observed in our calculated models. This discrepancy may arise from the lack of intramolecular interactions that could restrain the conformational arrangements in these model systems. Further research broadening the perspective to incorporate surrounding molecular movements and environments may be necessary for determining the underlying factors driving the formation of helical structures.

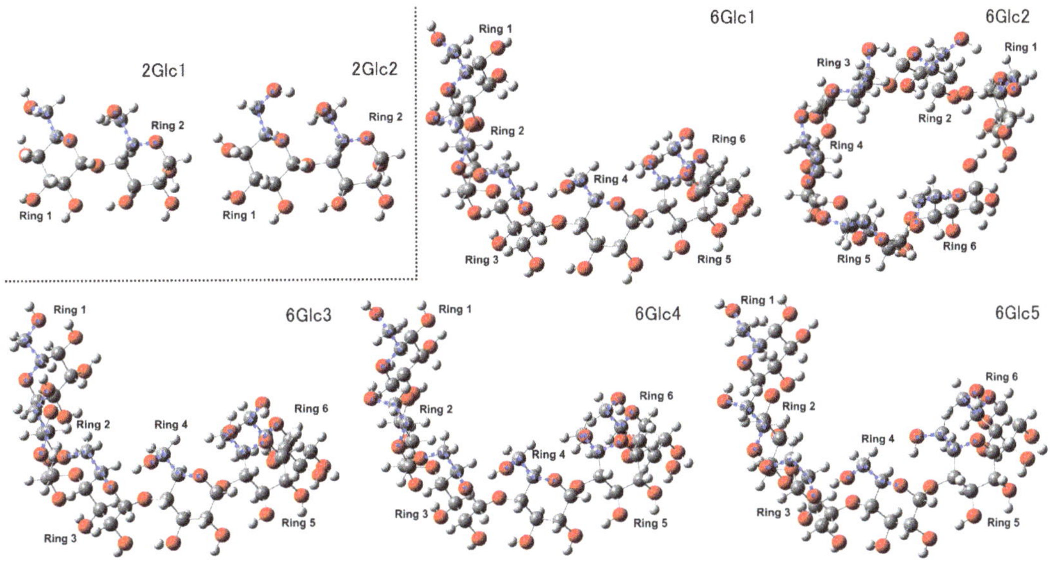

Figure 8. Structural changes observed in 2Glc1-2 and 6Glc1-5 models, highlighting the dihedral angles between specific atoms with blue dashed lines.

As the number of amylose units increases to eight (8Glc), the structures exhibit more intricate intramolecular hydrogen bonding networks, leading to observations of twisting, curling, or the formation of coiled conformations. The rotations of the hydroxymethyl, represented by the dihedral angles provided in Supporting Information Table S4, facilitate the formation of these weak interactions while simultaneously inducing varying degrees of coiling or partial inversion. To this end, these rotations give rise to diverse coiled or quasi-coiled curved structures, as shown in Figure S1. Notably, the 4Glc structures are intermediate between fully linear and helical conformations. Therefore, investigating the structural properties of 4Glc oligomers offers acceptable computational costs and provides valuable insights into the amylose linkage formation process toward a more complicated system.

Representative calculated Raman spectra for the 2Glc and 6Glc models were presented in Figure 9, with their band assignments of the primary vibrational modes in Table 5, enabling a comparative analysis of their primary vibrational modes. The Raman spectra for selected 8Glc structures, calculated using a different basis set, were provided in Supporting Information Figures S1 and S2. In analyzing the calculated Raman spectra for 2Glc and 6Glc structures, we identified several key differences in the major peaks. For 2Glc1 and 2Glc2, the primary distinctions are the relative intensity of the peak at 492 and 499 cm^{-1} and the relative changes in the bands at 854 and 952 cm^{-1} and 855 and 950 cm^{-1}. In both 6Glc1 and 6Glc2, there is a noticeable enhancement of the band at 489 cm^{-1}. Specifically, for 6Glc1, the band at 851 cm^{-1} is significantly stronger than at 972 cm^{-1}, whereas in 6Glc2, the bands at 851 and 977 cm^{-1} are comparable.

A strong peak at 1154 cm^{-1} is observed in 6Glc2 but not in 6Glc1. However, 6Glc1 and 6Glc2 exhibit their strongest bands near 1371 and 1353 cm^{-1}, respectively. This selection aims to provide further insights for various applications of band assignments.

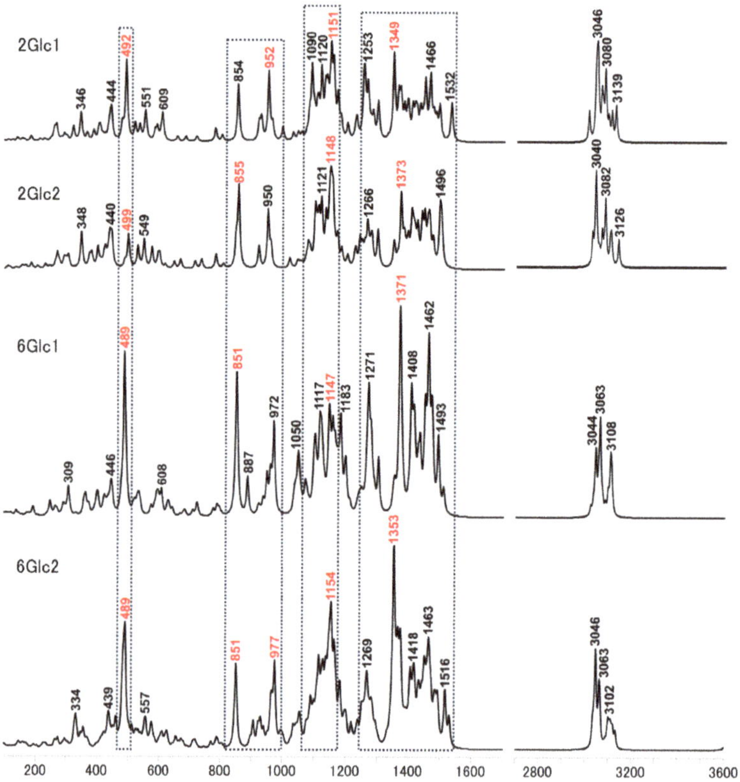

Figure 9. Calculated Raman spectra for representative models of 2Glc and 6Glc, highlighting bands with the highest intensity in comparable regions denoted in red.

Table 5. Raman band assignments for selected 2Glc and 6Glc models, with corresponding structure designation following the band wavenumber (cm^{-1}).

Part	Wavenumber (Structure Name)	Band Assignment (Primary Vibration) [1]
II	3046 (6Glc2)	υ (1, 3, 4, -CH), υ (1, 2, 3, $-CH_2$-OH)
I	1371 (6Glc1)	υ (2, $-CH_2$-OH), β (2, 3, 4, -CH), β (2, 3, 4 -OH), ρ (2, 3, 4)
	1353 (6Glc2)	υ (1, 4, 5, $-CH_2$-OH), β (1, 2, 4, 5, 6, -CH), β (1, 4, 5, 6, -OH), ρ (1, 2, 3, 4)
	1154 (6Glc2)	υ (1, 2, 3, 4, 5, 6, $-CH_2$-OH), β (1, 2, 3, 4, 5, 6, $-CH_2$-OH), β (1, 2, 3, 4, 5, 6, -CH), β (1, 2, 3, 4, 5, 6, -OH), ρ (1, 2, 3, 4, 5, 6)
	977 (6Glc2)	υ (2, 3, 4, 5, 6, $-CH_2$-OH), β (1, 2, 3, 4, 5, 6, -CH), β (1, 2, 3, 4, 5, 6, -OH), ρ (1, 2, 3, 4, 5, 6)
	851 (6Glc1)	β (2, 3, 4, 5, 6, $-CH_2$-OH), β (2, 3, 4, 6, -CH), ρ (2, 3, 4, 6)
	489 (6Glc1)	β (2, 3, 4, 5, 6, $-CH_2$-OH), ρ (1, 2, 3, 4, 5, 6)

[1] υ—stretching, β—bending, ρ—ring stretching and breathing.

4. Conclusions

In this comprehensive quantum chemical exploration, we investigated the structural features and calculated Raman spectra of modeled structures of amylose, encompassing varying units of linked glucose molecules (4Glc, 2Glc, 6Glc, and 8Glc). Our study focused on systematically analyzing 4Glc structures and applied the band feature to other models. The hydroxymethyl group rotation and intramolecular hydrogen bonds were analyzed, revealing that as the number of linked glucose units increases, the linear configuration of linked amylose becomes progressively more intricate, giving rise to curled, cyclic, or helical conformations. Various intramolecular interactions drive these structural transformations, resulting in different vibrational modes of constituent rings in linked glucose structures. The intricate coiling process of amylose is further complicated by the formation of intramolecular hydrogen bonding interactions between various functional groups. These interactions can occur between two hydroxymethyl groups in neighboring rings, an oxygen atom in one ring and a hydroxymethyl group from another ring, or hydroxyl groups from adjacent rings.

The detailed comparison of calculated Raman spectra, emphasizing distinctive vibrational bands, provides critical reference points for reliably assigning vibrational modes across various analytical applications. Despite the intricate structures of amylose molecules, the analyzed vibrational mode can be used as a reference to assign the observed Raman peaks. This enhances our theoretical grasp of carbohydrate structures and broadens research into their spectroscopic properties. By unraveling the spectroscopic fingerprints of linked glucose molecules, this study paves the way for more advanced analysis of complex linked glucose architectures governed by intramolecular hydrogen bond interactions. Importantly, the spectroscopic insights gained herein provide a molecular-level understanding of the formation of biopolymeric structures derived from glucose monomers. The conformational analysis and band assignments lay a foundation for further theoretical and experimental investigations, enabling more accurate interpretation of spectroscopic data for structural analysis of biomolecular systems involving carbohydrate moieties.

Supplementary Materials: The following supporting information can be downloaded at: https://www.mdpi.com/article/10.3390/molecules29122842/s1, Table S1: Dihedral angles (in degrees) describing the orientations of hydroxymethyl groups in each glucose ring for the 4Glc1-4Glc10 structural models; Table S2: Dihedral angles (in degrees) describing the orientations of hydroxymethyl groups in each glucose ring for the 4Glc11-4Glc18 structural models; Table S3: Dihedral angles (in degrees) describing the orientations of hydroxymethyl groups in each glucose ring for the explored structures; Table S4: Dihedral angles (in degrees) describing the orientations of hydroxymethyl groups in each glucose ring for the 2Glc1-2 and 6Glc1-5 structural models; Table S5: Dihedral angles (in degrees) describing the orientations of hydroxymethyl groups in each glucose ring for the 8Glc1-10 structural models; Figure S1: Structural changes observed in 8Glc1-10 models, highlighting the dihedral angles between specific atoms with blue dashed lines; Figure S2. Calculated Raman spectra for 8Glc1-10 models, with the most intense bands in four selected regions highlighted in red (calculated at the CAM-B3LYP/6-31G(d) level of theory); Appendix SA: Detailed diagrams of hydrogen bonding formations in the linear models; Vibrational animations can be found in Band_Assignment.rar.

Author Contributions: Conceptualization, D.Z. and N.K.; investigation, formal analysis, visualization, and writing—original draft preparation, D.Z.; resources, project administration, and writing—review and editing, N.K. All authors have read and agreed to the published version of the manuscript.

Funding: N.K. received research funds from IQCE research funding and KAKENHI (21K04806, 24K08182).

Institutional Review Board Statement: Not applicable.

Informed Consent Statement: Not applicable.

Data Availability Statement: Data are contained within the article and Supplementary Materials.

Acknowledgments: We would like to thank Wulan Intan Sari for her basic research relative to this study. We express our gratitude to Shin-ichi Morita from Tohoku University regarding the Raman spectroscopy analysis.

Conflicts of Interest: The authors declare no conflicts of interest.

References

1. Pérez, S.; Bertoft, E. The molecular structures of starch components and their contribution to the architecture of starch granules: A comprehensive review. *Starch/Stärke* **2010**, *62*, 389–420. [CrossRef]
2. Zhu, J.; Bai, Y.; Gilbert, R.G. Effects of the Molecular Structure of Starch in Foods on Human Health. *Foods* **2023**, *12*, 2263. [CrossRef] [PubMed]
3. Svihus, B.; Hervik, A.K. Digestion and metabolic fates of starch, and its relation to major nutrition-related health problems: A review. *Starch/Stärke* **2016**, *68*, 302–313. [CrossRef]
4. Bertoft, E. Understanding Starch Structure: Recent Progress. *Agronomy* **2017**, *7*, 56. [CrossRef]
5. Salimi, M.; Channab, B.-E.; Idrissi, A.E.; Zahouily, M.; Motamedi, E. A comprehensive review on starch: Structure, modification, and applications in slow/controlled-release fertilizers in agriculture. *Carbohydr. Polym.* **2023**, *322*, 121326. [CrossRef] [PubMed]
6. Ogunsona, E.; Ojogbo, E.; Mekonnen, T. Advanced material applications of starch and its derivatives. *Eur. Polym. J.* **2018**, *108*, 570–581. [CrossRef]
7. Zarski, A.; Bajer, K.; Kapuśniak, J. Review of the Most Important Methods of Improving the Processing Properties of Starch toward Non-Food Applications. *Polymers* **2021**, *13*, 832. [CrossRef] [PubMed]
8. Compart, J.; Singh, A.; Fettke, J.; Apriyanto, A. Customizing Starch Properties: A Review of Starch Modifications and Their Applications. *Polymers* **2023**, *15*, 3491. [CrossRef] [PubMed]
9. Jayarathna, S.; Andersson, M.; Andersson, R. Recent Advances in Starch-Based Blends and Composites for Bioplastics Applications. *Polymers* **2022**, *14*, 4557. [CrossRef]
10. Marichelvam, M.K.; Jawaid, M.; Asim, M. Corn and Rice Starch-Based Bio-Plastics as Alternative Packaging Materials. *Fibers* **2019**, *7*, 32. [CrossRef]
11. Zhang, H.; Su, Z.; Wang, X. Starch-Based Rehealable and Degradable Bioplastic Enabled by Dynamic Imine Chemistry. *ACS Sustain. Chem. Eng.* **2022**, *10*, 8650–8657. [CrossRef]
12. Xie, D.; Zhang, R.; Zhang, C.; Yang, S.; Xu, Z.; Song, Y. A novel, robust mechanical strength, and naturally degradable double crosslinking starch-based bioplastics for practical applications. *Int. J. Biol. Macromol.* **2023**, *253*, 126959. [CrossRef] [PubMed]
13. Tan, S.X.; Andriyana, A.; Ong, H.C.; Lim, S.; Pang, Y.L.; Ngoh, G.C. A Comprehensive Review on the Emerging Roles of Nanofillers and Plasticizers towards Sustainable Starch-Based Bioplastic Fabrication. *Polymers* **2022**, *14*, 664. [CrossRef] [PubMed]
14. Xie, F. Natural polymer starch-based materials for flexible electronic sensor development: A review of recent progress. *Carbohydr. Polym.* **2024**, *337*, 122116. [CrossRef] [PubMed]
15. Seung, D. Amylose in starch: Towards an understanding of biosynthesis, structure and function. *New Phytol.* **2020**, *228*, 1490–1504. [CrossRef] [PubMed]
16. Cummings, S.; Zhang, Y.; Smeets, N.; Cunningham, M.; Dubé, M.A. On the Use of Starch in Emulsion Polymerizations. *Processes* **2019**, *7*, 140. [CrossRef]
17. Sarder, R.; Piner, E.; Rios, D.C.; Chacon, L.; Artner, M.A.; Barrios, N.; Argyropoulos, D. Copolymers of starch, a sustainable template for biomedical applications: A review. *Carbohydr. Polym.* **2022**, *278*, 118973. [CrossRef] [PubMed]
18. Pesek, S.; Silaghi-Dumitrescu, R. The Iodine/Iodide/Starch Supramolecular Complex. *Molecules* **2024**, *29*, 641. [CrossRef]
19. Wang, Z.; He, M.; Sari, W.I.; Kishimoto, N.; Morita, S.-I. Bio-Raman research using principal component analysis and non-negative matrix factorization on rice grains: Detections of ordered and disordered states of starch in the cooking process. *Jpn. J. Appl. Phys.* **2021**, *60*, 060903. [CrossRef]
20. Wiercigroch, E.; Szafraniec, E.; Czamara, K.; Pacia, M.Z.; Majzner, K.; Kochan, K.; Kaczor, A.; Baranska, M.; Malek, K. Raman and infrared spectroscopy of carbohydrates: A review. *Spectrochim. Acta A* **2017**, *185*, 317–335. [CrossRef]
21. Bağcıoğlu, M.; Zimmermann, B.; Kohler, A. A Multiscale Vibrational Spectroscopic Approach for Identification and Biochemical Characterization of Pollen. *PLoS ONE* **2015**, *10*, e0137899. [CrossRef]
22. Maeda, S.; Harabuchi, Y.; Sumiya, Y.; Takagi, M.; Suzuki, K.; Hatanaka, M.; Osada, Y.; Taketsugu, T.; Morokuma, K.; Ohno, K. GRRM17. Available online: http://iqce.jp/GRRM/index_e.shtml (accessed on 20 August 2021).
23. Maeda, S.; Ohno, K.; Morokuma, K. Systematic Exploration of the Mechanism of Chemical Reactions: The Global Reaction Route Mapping (GRRM) Strategy by the ADDF and AFIR Methods. *Phys. Chem. Chem. Phys.* **2013**, *15*, 3683–3701. [CrossRef]
24. Frisch, M.J.; Trucks, G.W.; Schlegel, H.B.; Scuseria, G.E.; Robb, M.A.; Cheeseman, J.R.; Scalmani, G.; Barone, V.; Petersson, G.A.; Nakatsuji, H.; et al. *Gaussian 16, Revision B.01*; Gaussian, Inc.: Wallingford, CT, USA, 2016.
25. Yanai, T.; Tew, D.; Handy, N. A new hybrid exchange–correlation functional using the Coulomb-attenuating method (CAM-B3LYP). *Chem. Phys. Lett.* **2004**, *393*, 51–57. [CrossRef]
26. McLean, A.D.; Chandler, G.S. Contracted Gaussian basis sets for molecular calculations. I. Second row atoms, Z=11–18. *J. Chem. Phys.* **1980**, *72*, 5639–5648. [CrossRef]
27. Raghavachari, K.; Binkley, J.S.; Seeger, R.; Pople, J.A. Self-consistent molecular orbital methods. XX. A basis set for correlated wave functions. *J. Chem. Phys.* **1980**, *72*, 650–654. [CrossRef]
28. Clark, T.; Chandrasekhar, J.; Spitznagel, G.W.; Schleyer, P.V.R. Efficient diffuse function-augmented basis sets for anion calculations. III.† The 3-21+G basis set for first-row elements. Li–F. *J. Comput. Chem.*. [CrossRef]
29. Frisch, M.J.; Pople, J.A.; Binkley, J.S. Self-consistent molecular orbital methods 25. Supplementary functions for Gaussian basis sets. *J. Chem. Phys.* **1984**, *80*, 3265–3269. [CrossRef]

30. Stewart, J.J.P. Optimization of parameters for semiempirical methods. V. Modification of NDDO approximations and application to 70 elements. *J. Mol. Model.* **2007**, *13*, 1173–1213. [CrossRef] [PubMed]
31. Maeda, S.; Taketsugu, T.; Morokuma, K.; Ohno, K. Anharmonic Downward Distortion Following for Automated Exploration of Quantum Chemical Potential Energy Surfaces. *Bull. Chem. Soc. Jpn.* **2014**, *87*, 1315–1334. [CrossRef]
32. Ohno, K.; Maeda, S. A Scaled Hypersphere Search Method for the Topography of Reaction Pathways on the Potential Energy Surface. *Chem. Phys. Lett.* **2004**, *384*, 277–282. [CrossRef]
33. Maeda, S.; Ohno, K. Global Mapping of Equilibrium and Transition Structures on Potential Energy Surfaces by the Scaled Hypersphere Search Method: Applications to Ab Initio Surfaces of Formaldehyde and Propyne Molecules. *J. Phys. Chem. A* **2005**, *109*, 5742–5753. [CrossRef]
34. Ohno, K.; Maeda, S. Global Reaction Route Mapping on Potential Energy Surfaces of Formaldehyde, Formic Acid, and their Metal Substituted Analogues. *J. Phys. Chem. A* **2006**, *110*, 8933–8941. [CrossRef] [PubMed]
35. Ditchfield, R.; Hehre, W.J.; Pople, J.A. Self-Consistent Molecular Orbital Methods. 9. Extended Gaussian-type basis for molecular-orbital studies of organic molecules. *J. Chem. Phys.* **1971**, *54*, 724–728. [CrossRef]
36. Hehre, W.J.; Ditchfield, R.; Pople, J.A. Self-Consistent Molecular Orbital Methods. 12. Further extensions of Gaussian-type basis sets for use in molecular-orbital studies of organic-molecules. *J. Chem. Phys.* **1972**, *56*, 2257–2261. [CrossRef]
37. Hariharan, P.C.; Pople, J.A. Influence of polarization functions on molecular-orbital hydrogenation energies. *Theor. Chem. Acc.* **1973**, *28*, 213–222. [CrossRef]
38. Hariharan, P.C.; Pople, J.A. Accuracy of AH equilibrium geometries by single determinant molecular-orbital theory. *Mol. Phys.* **1974**, *27*, 209–214. [CrossRef]
39. Gordon, M.S. The isomers of silacyclopropane. *Chem. Phys. Lett.* **1980**, *76*, 163–168. [CrossRef]
40. Francl, M.M.; Pietro, W.J.; Hehre, W.J.; Binkley, J.S.; DeFrees, D.J.; Pople, J.A.; Gordon, M.S. Self-Consistent Molecular Orbital Methods. 23. A polarization-type basis set for 2nd-row elements. *J. Chem. Phys.* **1982**, *77*, 3654–3665. [CrossRef]
41. Binning, R.C., Jr.; Curtiss, L.A. Compact contracted basis-sets for 3rd-row atoms—GA-KR. *J. Comp. Chem.* **1990**, *11*, 1206–1216. [CrossRef]
42. Blaudeau, J.-P.; McGrath, M.P.; Curtiss, L.A.; Radom, L. Extension of Gaussian-2 (G2) theory to molecules containing third-row atoms K and Ca. *J. Chem. Phys.* **1997**, *107*, 5016–5021. [CrossRef]
43. Rassolov, V.A.; Pople, J.A.; Ratner, M.A.; Windus, T.L. 6-31G* basis set for atoms K through Zn. *J. Chem. Phys.* **1998**, *109*, 1223–1229. [CrossRef]
44. Rassolov, V.A.; Ratner, M.A.; Pople, J.A.; Redfern, P.C.; Curtiss, L.A. 6-31G* Basis Set for Third-Row Atoms. *J. Comp. Chem.* **2001**, *22*, 976–984. [CrossRef]

molecules

MDPI

Article

Synergistic Effects of *Pyrrosia lingua* Caffeoylquinic Acid Compounds with Levofloxacin Against Uropathogenic *Escherichia coli*: Insights from Molecular Dynamics Simulations, Antibiofilm, and Antimicrobial Assessments

Yan Zhang [1,†], Fangfang Jiao [2,†], Derong Zeng [1], Xiang Yu [1], Yongqiang Zhou [1], Juan Xue [1], Wude Yang [1,*] and Jingjing Guo [2,*]

[1] College of Pharmacy, Guizhou University of Traditional Chinese Medicine, Guiyang 550025, China; zhangyan0003@gzy.edu.cn (Y.Z.); zengderogs@foxmail.com (D.Z.); yuxiang@gzy.edu.cn (X.Y.); zhouxiaoqiang1988@126.com (Y.Z.); xuejuan062@gzy.edu.cn (J.X.)

[2] Centre in Artificial Intelligence Driven Drug Discovery, Faculty of Applied Sciences, Macao Polytechnic University, Macao, China; p2212238@mpu.edu.mo

* Correspondence: yangwude476@gzy.edu.cn (W.Y.); jguo@mpu.edu.mo (J.G.)

† These authors contributed equally to this work.

Abstract: Urinary tract infections (UTIs), primarily caused by uropathogenic *Escherichia coli* (UPEC), have high morbidity and recurrence rates. Resistance to levofloxacin hydrochloride (LEV), a commonly used treatment for UTIs, is increasingly problematic, exacerbated by biofilm formation mediated by interactions between cyclic di-GMP (c-di-GMP or CDG) and YcgR. In this study, we identified three caffeoylquinic acid compounds from *Pyrrosia lingua*—chlorogenic acid (CGA), sibiricose A5 (SiA5), and 3-*O*-caffeoylquinic acid methyl ester (CAM)—that target YcgR through molecular docking. Biological assays revealed that combining these compounds with levofloxacin hydrochloride significantly enhanced antibacterial activity against standard UPEC strains in a concentration-dependent manner and clinically isolated UPEC strains. Notably, chlorogenic acid and sibiricose A5, when used with levofloxacin hydrochloride, enhanced intracellular c-di-GMP levels and swimming motility, significantly reduced YcgR gene expression, and effectively inhibited biofilm formation of UPEC at multiple time points. Additionally, molecular dynamics simulations elucidated the strong binding of these compounds to YcgR, underscoring the critical roles of residues, such as Arg118 and Asp145. This research serves as a foundation for tackling antibiotic resistance and developing innovative therapeutics for UTIs.

Keywords: UTIs; UPEC; *Pyrrosia lingua*; levofloxacin; molecular simulation

1. Introduction

Urinary tract infections (UTIs) are the second most common infectious disease, primarily caused by uropathogenic *Escherichia coli* (UPEC). UPEC accounts for 70–95% of UTIs [1,2]. Globally, approximately 150 million people suffer from UTIs each year, and at least one-third of them will develop recurrent infections [3]. The high morbidity and recurrence rates lead to great burdens on society and public health [4]. Currently, antibiotics are mainly used to treat UTIs; however, antibiotic resistance has been a global problem. Biofilms serve as reservoirs of infection, making it difficult to kill bacteria even at concentrations far beyond the minimum inhibitory concentration (MIC) [5]. When UPEC forms biofilms, it becomes much more difficult to eradicate with antibiotics, leading to recurring UTIs that are challenging to treat effectively [6].

Cyclic diguanosine monophosphate (c-di-GMP or CDG), serving as a ubiquitous second messenger across bacterial species, plays a key role in regulating biofilm dispersal and formation. An elevated cellular concentration of c-di-GMP binds the C terminal conserved PilZ domain of the only flagellar brake protein YcgR of UPEC. This interaction allows the N-terminal domain of YcgR (YcgR-N) to stably interact with motor proteins MotA, thereby inhibiting the MotA-FliG association and reducing bacterial motility. As a result, it increases matrix component production and promotes biofilm formation. However, a low cellular concentration of c-di-GMP promotes biofilm dispersal through increased bacterial motility. The YcgR mutant with an Arg118Asp substitution cannot associate with c-di-GMP and, therefore, cannot interact with the FliG protein. The mechanism by which an elevated cellular concentration of c-di-GMP induces a structural rearrangement of the PilZ and YcgR-N domains of YcgR is not clear, yet it is vital for biofilm formation and dispersal [7–10]. This understanding is crucial, because the use of antibiotics alone can easily lead to the development of antimicrobial drug resistance.

Natural products and synthetic compounds alone and combined with antibiotics including ciprofloxacin, such as citrus peel extract, cinnamaldehyde, resveratrol, disperazol, and zingerone, inhibit bacterial biofilm formation by regulating c-di-GMP concentrations and enhancing antibiotic efficacy [11–15]. *Pyrrosia lingua* has diuretic and strangury-relieving properties and has been widely used to treat UTIs in combination with antibiotics. The herbal remedy has advantages in terms of low side effects, low risk of drug resistance, and high efficacy [16–18]. According to previous studies, *Pyrrosia lingua* contains high levels of caffeoylquinic acid compounds, such as chlorogenic acid (CGA), sibiricose A5 (Si-A5), and 3-*O*-caffeoylquinic acid methyl ester (CAM) [19,20]. These compounds have anti-inflammatory, antimicrobial, antioxidant, antitumor, hypolipidemic, and hypoglycemic activities [21–23]. Despite the potential benefits, there are limited reports on the use of chlorogenic acid analogs in combination with antibiotics to enhance their antibacterial and anti-biofilm activities regulating c-di-GMP [24,25].

Our molecular docking studies indicate that these three compounds bind well with YcgR [20,26,27]. Given the promising results from our docking studies and the known properties of chlorogenic acid, sibiricose A5, and 3-*O*-caffeoylquinic acid methyl ester, we aimed to further investigate their combined effects with antibiotics against UPEC. We employed methods such as disk diffusion, broth microdilution, microdilution checkerboard, crystal violet assays, swimming motility assay, enzyme-linked immunosorbent assay (ELISA) and quantitative real-time PCR (qRT-PCR) to evaluate their synergistic antibacterial and anti-biofilm activities. Additionally, we explored the regulatory mechanism of these compounds on the interaction between YcgR and c-di-GMP through molecular dynamics simulations.

Experimental results suggest that they exhibit additive or synergistic effects when used in combination with antibiotics. This combination approach can help mitigate the issues associated with the use of antibiotics alone, such as the development of antimicrobial drug resistance. Our findings based on simulations suggest that these compounds may regulate the interaction between YcgR and c-di-GMP, potentially reducing antibiotic resistance and preventing recurrent UTIs. This underscores the potential of combining herbal remedies with conventional antibiotics to enhance treatment efficacy and combat drug-resistant infections. Additionally, understanding the synergistic mechanisms of caffeoylquinic acid compounds with antibiotics is crucial for reducing UTI recurrence and antibiotic resistance, providing significant insights and guiding value for future treatments.

2. Results and Discussion

2.1. Enhanced Antibacterial Activity of Caffeoylquinic Acid Compounds Combined with Levofloxacin Hydrochloride at Various Concentrations

To identify potential antibacterial compounds, molecular docking studies were performed on the compounds previously isolated from *Pyrrosia lingua* by our research group. As a result, three caffeoylquinic acid compounds exhibited strong binding with the active site of YcgR, such as chlorogenic acid, sibiricose A5, and 3-*O*-caffeoylquinic acid methyl ester, the docking scores were −10.344, −10.156, and −9.935 kcal/mol, and the corresponding ligand efficiency (LE) were −0.282, −0.382, and −0.414 kcal/(mol·heavy atom), respectively (Figure 1). These three compounds were selected for further evaluation of their antibacterial activity independently and in combination with levofloxacin hydrochloride (LEV).

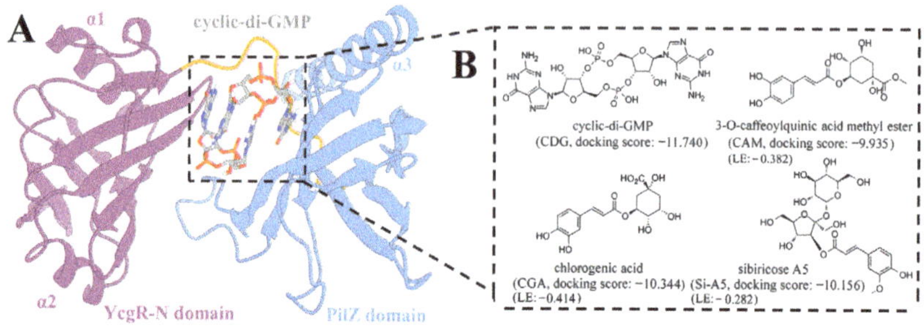

Figure 1. System information. (**A**) A cartoon representation of the YcgR-CDG complex structure. The YcgR-N domain is colored purple, and the PilZ domain is colored blue. The two c-di-GMP molecules are shown as sticks, with carbon atoms in gray, oxygen atoms in red, nitrogen atoms in blue, and phosphorus atoms in orange. The binding site of c-di-GMP is highlighted with a dashed box. (**B**) Chemical structures of c-di-GMP and three caffeoylquinic acid compounds studied in this work with their docking scores (kcal/mol) and ligand efficiency (LE, kcal/(mol·heavy atom)).

As detailed in Table 1, the antibacterial efficacy of chlorogenic acid, sibiricose A5, and 3-*O*-caffeoylquinic acid methyl ester, both as standalone agents and in combination with LEV, was assessed against UPEC using the disk diffusion method at concentrations of 25, 50, 100, 200, and 400 μg/mL, respectively. Levofloxacin hydrochloride alone inhibited *E. coli* ATCC 10389 with zone diameters ranging from 15.5 to 27.7 mm. When combined with the caffeoylquinic acid compounds, the inhibition zones increased, measuring between 16.4 and 30.6 mm. Additionally, the antibacterial activity of levofloxacin hydrochloride and its combinations improved as the concentration increased.

Table 1. Inhibition zone diameters for caffeoylquinic acid compounds and levofloxacin hydrochloride alone and in combination against *E. coli* ATCC 10389.

Concentration (μg/mL)	Inhibition Zone Diameters (mm) [a]						
	CGA	Si-A5	CAM	LEV	LEV + CGA	LEV + Si-A5	LEV + CAM
25 [b]	-	-	-	15.5 ± 0.1	17.1 ± 0.5 **	16.6 ± 0.2 **	16.4 ± 0.2 **
50	-	-	-	16.5 ± 0.4	19.3 ± 0.8 **	19.7 ± 0.4 **	17.0 ± 0.2
100	-	-	-	20.4 ± 0.5	23.6 ± 0.2 **	23.4 ± 0.2 **	24.7 ± 0.7 **
200	-	-	-	26.3 ± 0.3	27.7 ± 0.3 *	27.0 ± 0.3	27.8 ± 1.1 *
400	-	-	-	27.7 ± 0.2	29.3 ± 0.1	30.6 ± 1.8 **	29.6 ± 0.5 *
MIC [c]	-	-	-	16.3 ± 0.4	16.4 ± 0.4	17.9 ± 0.7 *	19.2 ± 0.7 **

[a] Values were the mean ± SD of three replicates. [b] Under the concentrations of LEV (25 μg/mL) and caffeoylquinic acid compounds (25 μg/mL). [c] Under the MIC concentrations of LEV (6.25 μg/mL) and caffeoylquinic acid compounds (250 μg/mL). */** The single (*) and double (**) asterisks represented $p < 0.05$ and $p < 0.01$, compared with LEV, respectively. - No activity.

As shown in Table 1, when administered alongside levofloxacin hydrochloride, chlorogenic acid, sibiricose A5, and 3-*O*-caffeoylquinic acid methyl ester achieved inhibition zone diameters (IZDs) of 16.4 mm, 17.9 mm, and 19.2 mm, respectively. For comparison, levofloxacin hydrochloride alone produced an IZD of 16.3 mm. Notably, all combinations exhibited superior antibacterial activity compared to levofloxacin hydrochloride alone, with 3-*O*-caffeoylquinic acid methyl ester showing the most significant enhancement of the synergistic antibacterial effects.

2.2. Synergetic Antibacterial and Antibiofilm Activities of Caffeoylquinic Acid Compounds and Levofloxacin Hydrochloride

The results of the synergistic antibacterial activities of caffeoylquinic acid compounds with levofloxacin hydrochloride, assessed using the broth microdilution method, are presented in Table 2. These compounds exhibited synergistic effects with levofloxacin hydrochloride against the standard strain ATCC 10389 and clinically isolated UPEC 64222 and 55758 strains, with the fractional inhibitory concentration index (FIC) ranging from 0.25 to 0.75. Specifically, the minimum inhibitory concentrations (MICs) of the antibiotic had a 2–8-fold reduction when used in combination. Chlorogenic acid achieved a 4–16-fold reduction, while both sibiricose A5 and 3-*O*-caffeoylquinic acid methyl ester demonstrated an 8-fold and 4–16-fold reduction in MIC, respectively. Moreover, compared with clinically isolated UPEC strains, caffeoylquinic acid compounds and levofloxacin hydrochloride in single and combined use showed better antimicrobial efficacy against the standard strain. This synergy not only enhanced the antimicrobial efficacy but also reduced the likelihood of resistance development, addressing a critical clinical healthcare challenge.

Table 2. Fractional inhibitory concentration (FIC) index of antimicrobial combinations of caffeoylquinic acid compounds and levofloxacin hydrochloride against UPEC.

Strain	MIC (μg/mL)			FIC	MIC (μg/mL)		FIC	MIC (μg/mL)		FIC
	LEV	CGA	LEV + CGA		Si-A5	LEV + Si-A5		CAM	LEV + CAM	
E. coli ATCC 10389	6.25	250	0.78 + 62.50	0.38	250	0.78 + 31.25	0.25	250	0.78 + 31.25	0.25
E. coli 64222	31.25	≥1000	15.63 + 125	0.63	≥1000	3.91 + 125	0.38	1000	7.81 + 62.50	0.31
E. coli 55758	31.25	≥1000	15.63 + 62.50	0.56	≥1000	15.63 + 125	0.63	≥1000	15.63 + 250	0.75

Based on the better antimicrobial efficacy of caffeoylquinic acid compounds and levofloxacin hydrochloride in single and combined use, the effects of the antibiotic and these compounds on biofilm formation across the standard strain were analyzed. Levofloxacin hydrochloride significantly inhibited biofilm formation at multiple time points, including 24 h, 36 h, 3 days, 5 days, and 7 days (Figure 2). Additionally, the combinations of chlorogenic acid or sibiricose A5 with levofloxacin hydrochloride markedly reduced biofilm formation across the standard strain at different time points. In contrast, 3-*O*-caffeoylquinic acid methyl ester did not inhibit biofilm formation. While increasing the concentration of levofloxacin hydrochloride slightly enhanced its ability to inhibit biofilm formation, decreasing the concentration of 3-*O*-caffeoylquinic acid methyl ester was beneficial for biofilm inhibition. Furthermore, higher concentrations of chlorogenic acid were particularly effective in reducing biofilm formation.

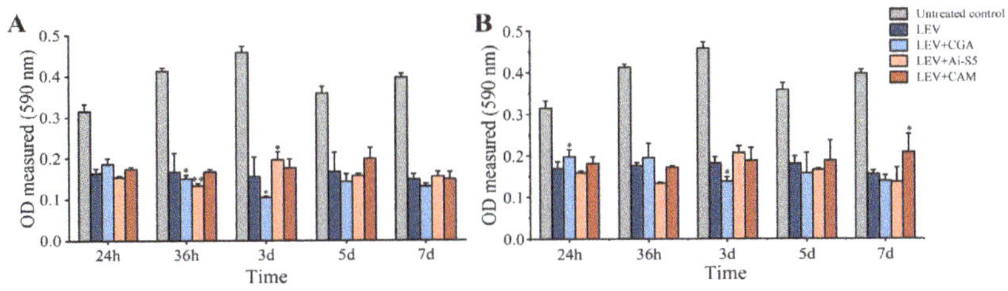

Figure 2. OD values of levofloxacin hydrochloride in combination with caffeoylquinic acid compounds against *E. coli* ATCC 10389 at multiple time points and concentrations. (**A**) OD values of levofloxacin hydrochloride and caffeoylquinic acid compounds at MIC concentrations; (**B**) OD values of levofloxacin hydrochloride and caffeoylquinic acid compounds at 1/2 MIC concentrations. The single (*) and double (**) asterisks represent $p < 0.05$ and $p < 0.01$, compared with levofloxacin hydrochloride, respectively.

2.3. Swimming Motility Inhibition Activities of Caffeoylquinic Acid Compounds and Levofloxacin Hydrochloride

As illustrated in Figure 3A, the swimming motility of *E. coli* ATCC 10389, when treated with levofloxacin hydrochloride at 1/2 MIC concentration, did not significantly increase compared with the untreated control group. However, the combinations of levofloxacin hydrochloride with chlorogenic acid or sibiricose A5 resulted in notable enhancements in swimming ability, especially sibiricose A5 ($p < 0.05$). In contrast, levofloxacin hydrochloride combined with 3-*O*-caffeoylquinic acid methyl ester led to a reduction in swimming ability. Consistently, the combinations of chlorogenic acid or sibiricose A5 with levofloxacin hydrochloride markedly reduced biofilm formation, while its combination with 3-*O*-caffeoylquinic acid methyl ester did not inhibit biofilm formation. These findings suggested that levofloxacin hydrochloride combined with chlorogenic acid or sibiricose A5 enhanced the swimming motility of *E. coli* ATCC 10389 and reduced biofilm formation, whereas its combination with 3-*O*-caffeoylquinic acid methyl ester reduced the swimming motility of *E. coli* ATCC 10389 and did not inhibit biofilm formation.

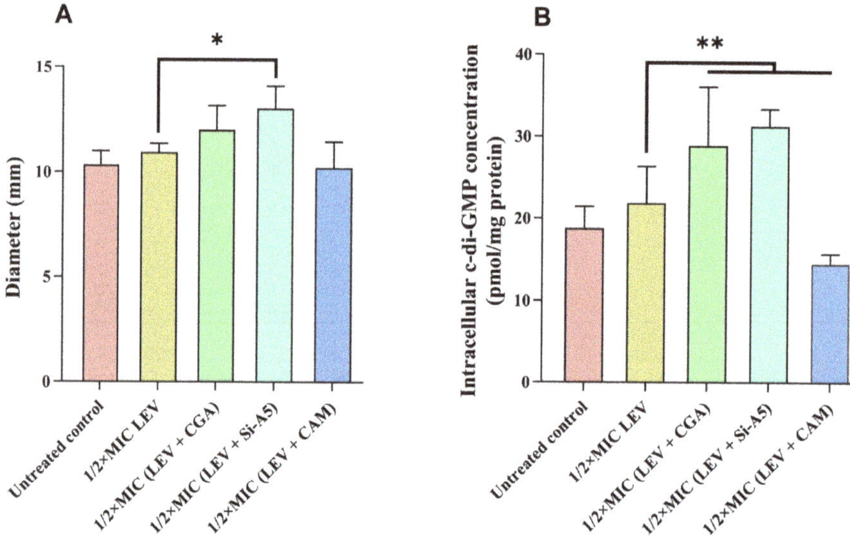

Figure 3. Effects of levofloxacin hydrochloride and caffeoylquinic acid compounds at 1/2 MIC concentrations on swimming motility and intracellular c-di-GMP concentrations in *E. coli* ATCC 10389. (**A**) Swimming motility inhibition activities of *E. coli* ATCC 10389 treated with levofloxacin hydrochloride and caffeoylquinic acid compounds at 1/2 MIC concentrations; (**B**) Intracellular c-di-GMP concentrations in *E. coli* ATCC 10389 treated with levofloxacin hydrochloride and caffeoylquinic acid compounds at 1/2 MIC concentrations. The data are expressed as the mean ± SD of at least three independent experiments. The single (*) and double (**) asterisks represent $p < 0.05$ and $p < 0.01$, compared with 1/2 MIC concentration of levofloxacin hydrochloride, respectively.

2.4. Intracellular c-di-GMP Concentrations Regulated by Caffeoylquinic Acid Compounds and Levofloxacin Hydrochloride

To study the correlation between biofilm formation, swimming motility, and intracellular c-di-GMP levels of *E. coli* ATCC 10389, intracellular c-di-GMP concentrations in *E. coli* ATCC 10389 were measured when treated with levofloxacin hydrochloride and caffeoylquinic acid compounds at 1/2 MIC concentrations. As shown in Figure 3B, levofloxacin hydrochloride alone did not significantly affect the intracellular c-di-GMP content of *E. coli* ATCC 10389. However, the c-di-GMP content of *E. coli* ATCC 10389 was significantly increased after levofloxacin hydrochloride in combination with chlorogenic acid or sibiricose A5 ($p < 0.01$). In contrast, c-di-GMP content was decreased after levofloxacin hydrochloride in combination with 3-*O*-caffeoylquinic acid methyl ester ($p < 0.01$). These results suggested that levofloxacin hydrochloride in combination with chlorogenic acid or sibiricose A5 increased intracellular c-di-GMP content of *E. coli* ATCC 10389. However, the combination of levofloxacin hydrochloride and 3-*O*-caffeoylquinic acid methyl ester decreased c-di-GMP content. Thus, levofloxacin hydrochloride, in combination with chlorogenic acid or sibiricose A5, enhanced intracellular c-di-GMP levels, swimming motility, and reduced biofilm formation. In contrast, the combination of CAM and LEV resulted in similar swimming motility patterns as the control and the LEV-only treatments (Figure 3A), and the intracellular c-di-GMP concentration in the CAM + LEV system remains even lower than the levofloxacin-only and control groups.

2.5. *Significant Reduction in YcgR Flexibility Due to Ligand Binding*

Next, molecular dynamics (MD) simulations were performed to elucidate the binding interactions between the YcgR protein and CDG (PDB ID: 5Y6F [10]), as well as the three caffeoylquinic acid compounds. Here, we established seven systems: YcgR-apo, representing the unbound state of the YcgR protein; YcgR-CDG-dual, featuring YcgR complexed with two c-di-GMP molecules; YcgR-CDG-PilZ, showing YcgR with a c-di-GMP molecule close to the PilZ domain; and YcgR-CDG-N domain, with a c-di-GMP molecule close to the YcgR-N domain. In addition, YcgR-CDG-dual-PilZ refers to the complex formed by a c-di-GMP molecule located near the PilZ domain within the YcgR-CDG-dual complex. Conversely, the YcgR-CDG-dual-N domain denotes the complex formed by a c-di-GMP molecule situated near the N-domain within the YcgR-CDG-dual complex. Moreover, YcgR-CAM refers to the complex of YcgR bound with 3-*O*-caffeoylquinic acid methyl ester, YcgR-CGA represents the complex of YcgR with chlorogenic acid, and YcgR-Si-A5 stands for YcgR bound to sibiricose A5.

The root-mean-square deviations (RMSDs) of the Cα atoms of YcgR were calculated by aligning to the initial conformation to evaluate the structural stability of each simulated system. As illustrated in Figure 4, the RMSD values of these seven systems were stable after the 500 ns MD simulations, but the RMSD value (<12 Å) for the YcgR-apo system was larger than the other six YcgR-ligand complexes (<6 Å). In addition, YcgR-CDG-dual, YcgR-CDG-PilZ, and YcgR-Si-A5 exhibited the lowest RMSD values (~4 Å) among all simulated systems, which were attributed to their larger molecular skeletons acting as a bridge and stabilizing the conformation between the PilZ and YcgR-N domains (Figure 1A). Moreover, caffeoylquinic acid compounds significantly reduced the flexibility of PilZ and YcgR-N domains, especially chlorogenic acid and sibiricose A5. Hence, the presence of ligands between the PilZ domain and the YcgR-N domain should be beneficial to the stability of YcgR.

2.6. *The Binding Mechanism of YcgR and Caffeoylquinic Acid Compounds*

To further evaluate the binding mechanism between YcgR and ligands, the MM-GBSA method was employed to calculate the binding free energy for the YcgR-ligand complexes over the last 50 ns MD trajectories. Entropic contribution ($T\Delta S$) was not considered here due to the high resource consumption. As listed in Tables 3 and 4, the average binding free energies ($\Delta G_{MM/GBSA}$) were −29.59, −31.87, −42.48, −44.12, −33.30, −55.89, and −40.52 kcal/mol for complexes YcgR-CAM, YcgR-CGA, YcgR-Si-A5, YcgR-CDG-PilZ, YcgR-CDG-N domain, YcgR-CDG-dual-PilZ (a c-di-GMP molecule near the PilZ domain in YcgR-CDG-dual), and YcgR-CDG-dual-N domain (a c-di-GMP molecule near the N-domain in YcgR-CDG-dual), respectively. Two c-di-GMP were more favorable for the association between c-di-GMP and YcgR than one c-di-GMP. Moreover, the c-di-GMP adjacent to the PilZ domain was more favorable. Caffeoylquinic acid compounds exhibited strong binding with YcgR, especially chlorogenic acid and sibiricose A5. In detail, the van der Waals interaction (ΔE_{vdw}) and the direct electrostatic interaction (ΔE_{ele}) were the crucial driving forces for the binding between ligands and YcgR. In contrast, the polar component

($\Delta G_{pol,sol}$) was not conducive to the ligand binding. It is noteworthy that chlorogenic acid and sibiricose A5, which bound well with YcgR, were unfavorable for the association between c-di-GMP and YcgR. This led to enhanced intracellular c-di-GMP levels, swimming motility, and reduced biofilm formation. However, 3-*O*-caffeoylquinic acid methyl ester bound weakly with YcgR had weak effects on the association between YcgR and c-di-GMP, which led to reduced c-di-GMP levels and swimming motility. It is well known that levofloxacin hydrochloride targets DNA gyrase and thus has no obvious effects on c-di-GMP levels and swimming motility. CAM may exert its synergistic antimicrobial effects through alternative mechanisms, rather than the c-di-GMP-mediated motility regulation observed with CGA and Si-A5.

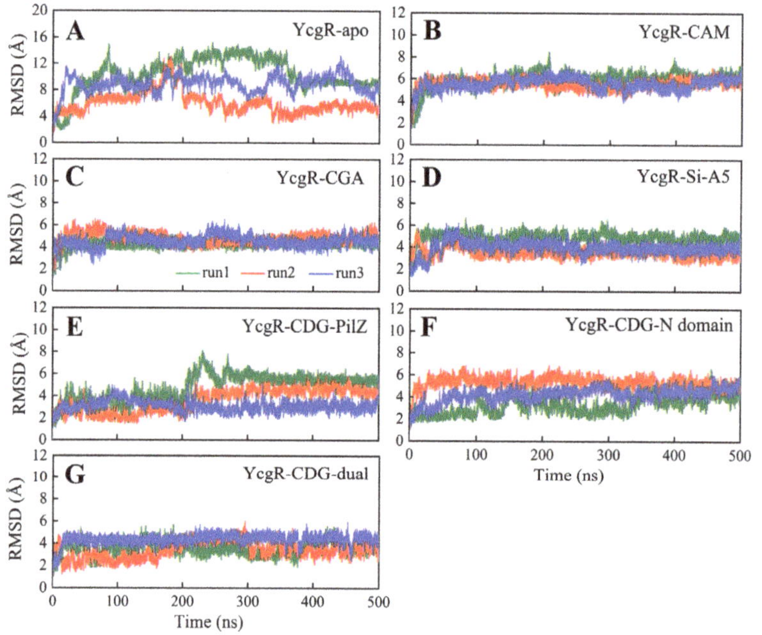

Figure 4. The RMSDs of all simulated systems during the MD simulations: (**A**) YcgR-apo system; (**B**) YcgR-CAM system; (**C**) YcgR-CGA system; (**D**) YcgR-Si-A5 system; (**E**) YcgR-CDG-PilZ system; (**F**) YcgR-CDG-N domain system; (**G**) YcgR-CDG-dual system. Each system is carried out in triplicate.

Table 3. Components of the binding free energy between YcgR and three ligands averaged in the last 50 ns of the three replicates (kcal/mol).

Contribution	YcgR-CAM	YcgR-CGA	YcgR-Si-A5
ΔE_{vdw}	−30.29 ± 2.69	−26.57 ± 0.63	−40.46 ± 2.57
ΔE_{ele}	−67.49 ± 4.92	−217.26 ± 5.85	−110.06 ± 31.49
$\Delta G_{pol,sol}$	72.99 ± 3.73	216.30 ± 5.39	115.10 ± 27.97
$\Delta G_{npol,sol}$	−4.80 ± 0.34	−4.35 ± 0.08	−7.06 ± 0.36
ΔE_{MM}	−97.77 ± 7.11	−243.82 ± 5.23	−150.52 ± 29.20
ΔG_{sol}	68.19 ± 3.41	211.95 ± 5.44	108.04 ± 27.76
$\Delta G_{MM/GBSA}$	−29.59± 3.70	−31.87 ± 0.49	−42.48 ± 1.45

Table 4. Components of the binding free energy between YcgR and CDG averaged in the last 50 ns of the three replicates (kcal/mol).

Contribution	YcgR-CDG-PilZ	YcgR-CDG-N Domain	YcgR-CDG-dual-PilZ	YcgR-CDG-dual-N Domain
ΔE_{vdw}	−55.25 ± 3.57	−43.13 ± 5.82	−58.16 ± 0.85	−44.58 ± 0.47
ΔE_{ele}	−103.98 ± 36.09	−140.00 ± 28.44	−134.06 ± 1.04	−166.02 ± 7.96
$\Delta G_{pol,sol}$	121.10 ± 33.93	155.29 ± 29.87	143.18 ± 1.97	176.22 ± 5.26
$\Delta G_{npol,sol}$	−5.99 ± 0.41	−5.45 ± 0.58	−6.39 ± 0.03	−6.14 ± 0.03
ΔE_{MM}	−159.23 ± 39.27	−183.13 ± 28.71	−192.67 ± 1.11	−210.60 ± 8.38
ΔG_{sol}	115.11 ± 33.62	149.83 ± 29.48	136.79 ± 1.98	170.08 ± 5.24
$\Delta G_{MM/GBSA}$	−44.12 ± 5.97	−33.30 ± 1.50	−55.89 ± 1.04	−40.52 ± 3.20

Owing to the effects of the association of caffeoylquinic acid compounds and YcgR on c-di-GMP levels, swimming motility, and biofilm formation, the relative expression of YcgR in *E. coli* ATCC 10389 was examined with either levofloxacin hydrochloride alone or levofloxacin hydrochloride combined with caffeoylquinic acid compounds at a concentration of 1/4 MIC for 4 h. As shown in Figure 5, levofloxacin hydrochloride combined with caffeoylquinic acid compounds significantly reduced YcgR gene expression in UPEC. Specifically, the LEV + CGA and LEV + Si-A5 groups resulted in 1.14-fold and 1.08-fold reductions in YcgR expression, respectively. In contrast, the LEV + CAM group exhibited a 1.36-fold reduction in YcgR expression. Consistently, CGA and Si-A5 with YcgR had more stable binding while CAM with YcgR had weak association. Therefore, the combination of levofloxacin hydrochloride and caffeoylquinic acid compounds exerted an inhibitory effect on the expression of the YcgR-1 gene through the binding of caffeoylquinic acid compounds and YcgR.

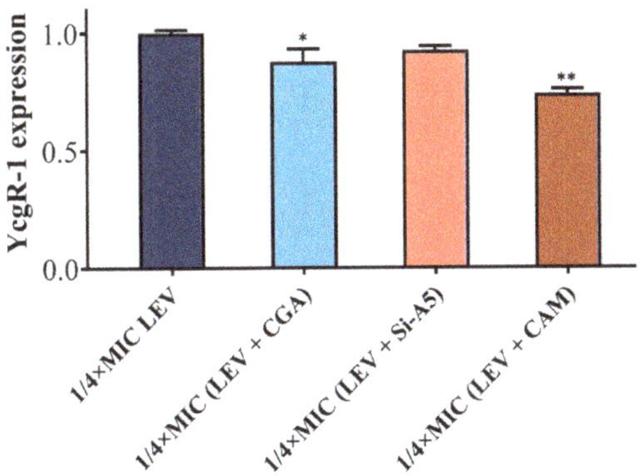

Figure 5. Effects of levofloxacin hydrochloride and caffeoylquinic acid compounds on YcgR gene expression in *E. coli* ATCC 10389 at 1/4 MIC concentration. The single (*) and double (**) asterisks represented $p < 0.05$ and $p < 0.01$, compared with 1/4 MIC concentration of levofloxacin hydrochloride, respectively.

The identified caffeoylquinic acid compounds significantly inhibited the interaction between c-di-GMP and YcgR, disrupting biofilm formation pathways that are key in antibiotic resistance and bacterial persistence. These findings offer a deeper understanding of how these compounds could be applied to resolve biofilm-associated infections. In addition, as natural products obtained from *Pyrrosia lingua*, these compounds have advantages such as low toxicity and compatibility with existing antibiotics. They strengthened the contributions of traditional herbal medicine to modern therapeutic solutions.

2.7. The Hotspot Residues in YcgR-Ligand Binding

The binding free energies were decomposed at the per-residue level to identify the critical residues involved in YcgR-ligand interactions. Residues with a per-residue contribution lower than −1 kcal/mol were considered hotspot residues. As shown in Figure 6, Figure S1 and S2, two different interaction patterns were identified in all simulated systems. In detail, YcgR-CAM, YcgR-CGA, YcgR-CDG-PilZ, and YcgR-CDG-dual-PilZ mainly bound to the upper region of the PilZ domain (Figure S3), and residue Y116, D145, S147, G150, and R212 were identified as the common hotspot residues in these four complexes, especially residue D145 exhibited high interaction energy among all the key residues. In addition, YcgR-Si-A5, YcgR-CDG-N domain, and YcgR-CDG-dual-N domain bound to the lower region of the PilZ domain and YcgR-N domain (Figure S4), and residue R208 or R113 was identified as one of the common hotspot residues in these three systems. Additionally, hotspot residues in these three systems changed significantly. For example, R118 was favorable for the binding of caffeoylquinic acid compounds. However, it was unfavorable for the binding of the two c-di-GMP molecules and showed no significant contribution to the one c-di-GMP system.

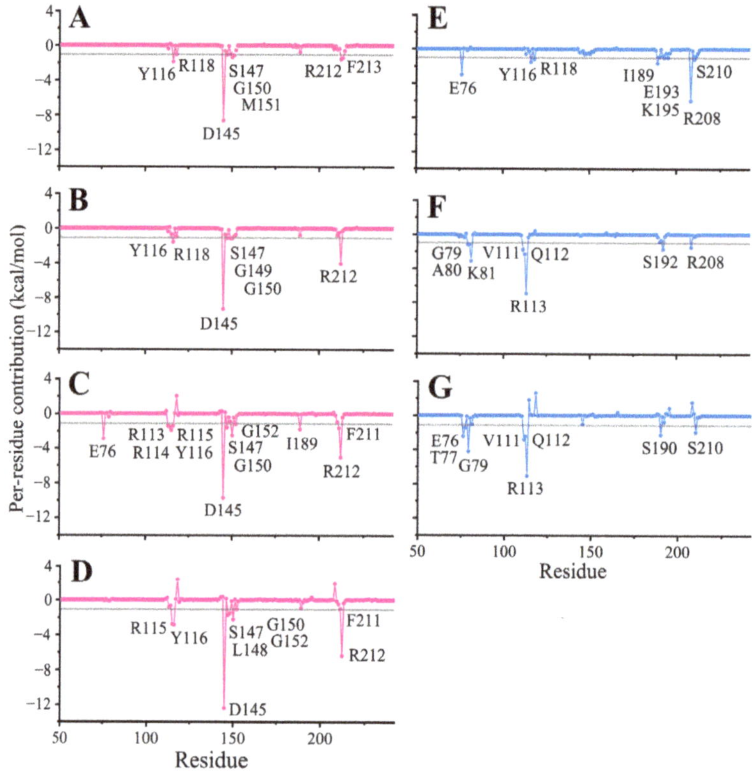

Figure 6. Residue contribution to the binding of ligands: (**A**) YcgR-CAM system; (**B**) YcgR-CGA system; (**C**) YcgR-CDG-PilZ system; (**D**) YcgR-CDG-dual-PilZ system; (**E**) YcgR-Si-A5 system; (**F**) YcgR-CDG-N domain system; (**G**) YcgR-CDG-dual-N domain system.

2.8. The Binding Mechanisms Between YcgR and Ligands

Based on the further structural analysis of the YcgR-CDG-dual and apo YcgR, the V-shape conformation of apo YcgR derived from the experimental structure of CDG-YcgR complex became linear during MD simulation, with the two YcgR subdomains moving to their farthest separation (Figure 7). However, c-di-GMP binding led to the PilZ domain being closer to the YcgR-N domain. As a result, YcgR exhibited a compact V-shape conformation, and the structure of YcgR-N domain was almost invariant. Similarly, YcgR-CDG-dual and YcgR-CAM had compact V-shape structures, and the PilZ and YcgR-N domains of YcgR-CDG-dual were nearer. Consistent with previous studies [7–10], an elevated cellular concentration of c-di-GMP binds the PilZ domain and reduces bacterial motility through its interaction with the YcgR-N domain, thereby facilitating biofilm formation. The ligand primarily binds to the PilZ domain, bringing the PilZ and YcgR-N domains into close proximity. Thus, the flexibility of YcgR and bacterial motility will be reduced.

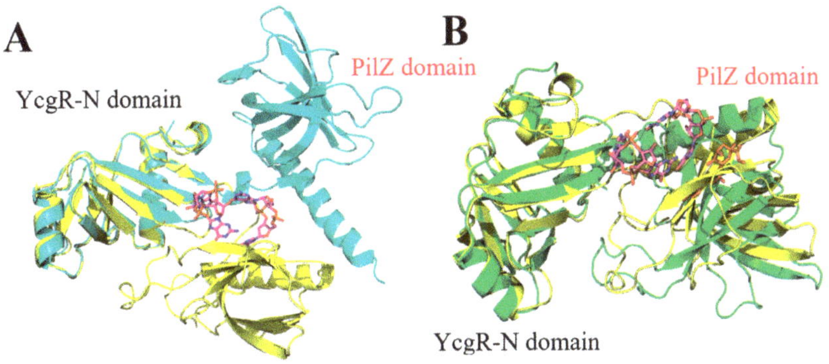

Figure 7. A cartoon representation of YcgR-CDG-dual (yellow) and apo YcgR (cyan). C-di-GMP was shown in pink (**A**). A cartoon representation of YcgR-CDG-dual (yellow) and YcgR-CAM (green) (**B**). C-di-GMP and 3-*O*-caffeoylquinic acid methyl ester were shown in stick representation, with carbon atoms colored pink (C-di-GMP) or orange (3-caffeoylquinic acid methyl ester), oxygen atoms in red, nitrogen atoms in blue, and phosphorus atoms in orange.

Next, the 2D representations of the binding mode of ligands with YcgR are depicted in Figure 8. As can be seen, YcgR-CAM, YcgR-CGA, YcgR-CDG-PilZ, and YcgR-CDG-dual-PilZ displayed a similar binding mode, forming hydrogen bonds with residue Asp145. Interestingly, Arg212 was identified as a common key residue from the residue energy decomposition analysis, but it played various contributions for these four complexes. In detail, the hydrogen bond was observed between ARG212 and YcgR-CAM, YcgR-CGA, and YcgR-CDG-PilZ and attractive charges were found in ARG212 with YcgR-CDG-dual-PilZ. In addition, hydrogen bonds with residue ARG208 were observed in YcgR-Si-A5, YcgR-CDG-N domain, and YcgR-CDG-dual-N domain systems. Besides, van der Waals, carbon-hydrogen bond, Pi-cation, and Alkyl were also observed in the simulated systems. Additionally, the hydrogen bonds observed in CDG-bound systems (Arg113, Arg114, Arg115, Asp145, Ser147, and Arg208) were consistent with the previous study [10]. It is worth noting that Arg118 had favorable interactions with the benzene ring or hydroxyl group of caffeoylquinic acid compounds, while it had unfavorable interactions with the nitrogen atom of c-di-GMP by electrostatic repulsion. An in-depth analysis of Arg118 and the residues around it was performed. Conformation changes of Arg118 were induced by electrostatic interactions of Asp145, which is one of the key residues driving ligand binding with YcgR. Consistent with previous studies [9,10], the YcgR^{Arg118Asp} mutant is incapable of associating with c-di-GMP and interacting with the FliG protein.

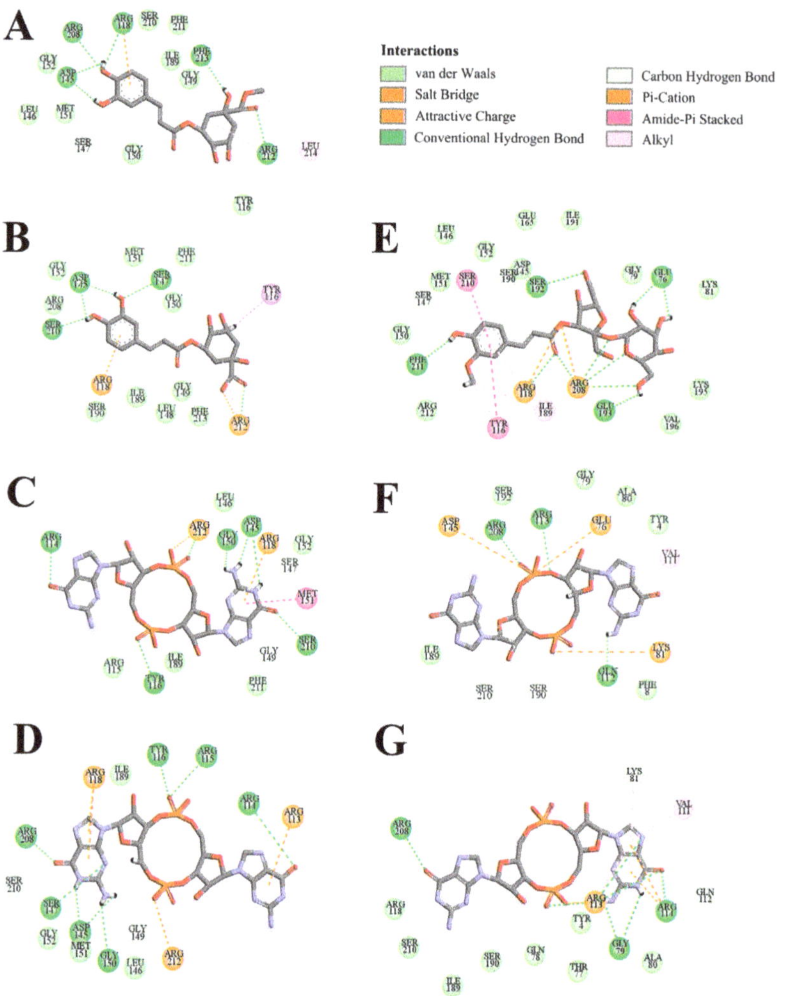

Figure 8. 2D representations of the binding mode of YcgR with the docking ligands: (**A**) CAM; (**B**) CGA; (**C**) CDG-PilZ; (**D**) CDG-dual-PilZ; (**E**) Si-A5; (**F**) CDG-N domain; (**G**) CDG-dual-N domain. The ligands are shown in stick representation, with carbon atoms colored gray, oxygen atoms in red, nitrogen atoms in blue, and phosphorus atoms in orange.

3. Materials and Methods

3.1. Molecular Docking Study

The X-ray crystallographic structure of the YcgR protein in complex with c-di-GMP from *Escherichia coli* (PDB ID: 5Y6F [10]) was retrieved from the Protein Data Bank. Then, the structure was prepared using the Protein Preparation Wizard [28] in the Maestro interface of the Schrödinger 2023 suite, which involved adding hydrogen atoms, filling in the missing structures (residues 123–124, 156–160, and 195–203), and assigning charges and protonation states at pH 7.0. Subsequently, the structure was minimized using the OPLS-2005 force field [29]. The binding site for each compound was defined within a 10 Å × 10 Å × 10 Å cube centered on the mass center of the co-crystallized ligand, c-di-GMP, using the receptor grid generation tool in Glide, Schrödinger 2018. Sixty compounds previously isolated from *Pyrrosia lingua* were used for molecular docking. The 3D structures of these compounds were generated using the LigPrep module of Schrödinger 2018 based on their initial chirality. Protonation states and tautomers were assigned using Epik at a pH range of 7.0 ± 2.0, and up to 32 stereoisomers were allowed for each compound, with all other parameters set to default. In addition, the ligand efficiency (LE) for each compound based on the docking scores and the number of non-hydrogen atoms were also calculated based on the LE = (docking score)/(number of heavy atoms). According to good docking scores and key binding interactions, three compounds, including chlorogenic acid, sibiricose A5, and 3-*O*-caffeoylquinic acid methyl ester, were selected and purchased for further bioactivity validation.

3.2. *In Vitro Antibacterial Assay*

3.2.1. Instruments and Chemicals

The following instruments were used: Haloes Caliper Circle of Inhibition Meter (Haloes Caliper, Shanghai, China), Hve-50 autoclave, and HCB-1300V medical ultra clean table (Haier, Qingdao, China). The following chemicals and materials were applied. Levofloxacin hydrochloride injection was purchased from CSPC OUYI Pharmaceutical Co., Ltd. (Taizhou, China), and chlorogenic acid, sibiricose A5, and 3-*O*-caffeoylquinic acid methyl ester were all purchased from Vicci Biotech Co., Ltd. (Chengdu, China). LB agar, powder, LB Broth, powder, 1% crystal violet ammonium oxalate solution, and phosphate buffer saline (PBS) were all purchased from Beijing Solepol Science and Technology Co., Ltd. (Beijing, China). Blank drug-sensitive paper tablets were purchased from Shanghai Qiyin Biotechnology Co., Ltd. (Shanghai, China). Ninety-six-well cell culture plates were purchased from Wuhan Seville Technology Co., Ltd. (Wuhan, China). UPEC (ATCC 10389) was provided by the laboratory of the College of Life Sciences, Guizhou University (Guiyang, China). Clinical isolates (*E. coli* 64222 and *E. coli* 55758) were provided by the Second Clinical Affiliated Hospital of Guizhou University of Traditional Chinese Medicine, and both were resistant to levofloxacin hydrochloride.

3.2.2. Disk Diffusion Method

The concentrations of caffeoylquinic acid compounds, levofloxacin hydrochloride, and their combinations were set as 400 μg/mL, 200 μg/mL, 100 μg/mL, 50 μg/mL, and 25 μg/mL, respectively. The blank drug-sensitive paper (6 mm × 1 mm) was soaked in the above-mentioned compound solutions. Bacterial suspensions with a concentration of 1.5×10^8 CFU/mL were evenly inoculated onto LB agar plates using sterile cotton swabs. The sterile water was used as the blank control. UPEC was placed in an incubator at 37 °C for 12 h. The experiments were repeated three times. Inhibition zone diameters (mm) were read by the Circle of Inhibition Measuring Instrument [30–32]. In addition, the MIC of levofloxacin hydrochloride in combination with caffeoylquinic acid compounds was determined using the same disk diffusion method mentioned above, which was used to evaluate the antibacterial activity of the combinations.

3.2.3. Minimum Inhibitory Concentration

MIC is the lowest concentration of a drug or chemical that inhibits bacterial growth under in vitro conditions. The antimicrobial activities of each caffeoylquinic acid compound and antibiotic were determined by broth microdilution method with the 96-well plate [33,34]. UPEC was selected as the test organism, and incubated in a shaking incubator at 37 °C, 150 rpm for 12 h to revive the strain. LB Broth powder was used to culture bacteria. Serial two-fold dilutions for each caffeoylquinic acid compound and antibiotic were carried out ranging from 2000 μg/mL to 15.6 μg/mL and from 800 μg/mL to 6.25 μg/mL in a 96-well plate, respectively. A total of 100 μL bacterial suspension (1.5×10^8 CFU/mL) and 2 μL TTC (0.5%) were added to each well. LB Broth, a powder containing UPEC, was set as a negative control. UPEC was cultured at 37 °C for 24 h. All the experiments were repeated three times.

3.2.4. Microdilution Checkerboard Method

The combined effects of the caffeoylquinic acid compounds and the antibiotic were evaluated using the microdilution checkerboard method. Specifically, the activity of each combination was measured in a two-dimensional checkerboard fashion (8 × 8). The concentration changes were monitored by the MICs measured above. Each drug was used at seven increasing (two-fold) concentrations (1/32 × to 2 × MIC). A total of 100 μL bacterial suspension (1.5 × 10^8 CFU/mL) was added. The microtiter plates were cultured at 37 °C for 24 h.

The fractional inhibitory concentration (FIC) index was calculated according to the formula as follows:

$$FIC = \frac{MIC_{(\text{A in the presence of B})}}{MIC_{(\text{A alone})}} + \frac{MIC_{(\text{B in the presence of A})}}{MIC_{(\text{B alone})}}$$

The effects of the combinations of caffeoylquinic acid compounds and antibiotics were evaluated based on the FIC index: the FIC index ≤ 0.5 (Synergistic), $0.5 <$ FIC index ≤ 1 (additive), $1 <$ FIC index ≤ 2 (indifferent), FIC index > 2.0 (antagonistic), respectively [35,36].

3.2.5. Crystal Violet Assay

The antibiofilm activity of each sample was determined using crystal violet assay in 96-well plates according to the method of Bai J et al. [37–39]. A total of 100 μL of the drug solutions at the concentrations of the MIC and 1/2 MIC were added to the 96-well plate, and then 100 μL of the bacterial suspension at a concentration of 1.5 × 10^8 CFU/mL was added to each well. UPEC was incubated for 24 h, 36 h, 3 d, 5 d, and 7 d at 37 °C. All the experiments were performed three times. A total of 100 μL of LB Broth, powder and 100 μL of bacterial suspension were set as the negative control group. A total of 200 μL of LB Broth powder without the drugs was set as the blank control group. Each well was washed three times with PBS, and then fixed with methanol for 15 min. After drying, 1% crystal violet was added to each well and incubated for 15 min. Each well was washed with PBS. The plates were dried at 55 °C and 200 μL of 33% acetic acid was added. After 15 min, the absorbance at 590 nm was measured using an enzyme counter.

3.2.6. Swimming Motility Assay

The swimming motility assay was performed according to Gao et al. with slight modifications [40]. In brief, a volume of 2.5 μL of the bacterial suspension (OD_{600} = 0.1) was inoculated into a swim plate containing (0.5% tryptone, 0.5% sodium chloride, and 0.3% agar), whose final drug concentration was set as 1/2 MIC. After 16 h of incubation at 37 °C, the diameter of the colonies occupied by each strain was measured. All experiments were conducted at least three times.

3.2.7. Extraction and Quantification of c-di-GMP

C-di-GMP was extracted from UPEC (ATCC 10389) using the method previously described [41]. Following the cultivation of the bacteria during the late exponential growth phase, the cell suspension was diluted 50-fold in sterile LB broth containing the drugs at a concentration of half MIC. This mixture was then incubated at 37 °C with shaking at 200 rpm until the optical density at 600 nm (OD_{600}) reached approximately 0.6 to 0.8, indicating mid-log phase growth. Two milliliters of planktonic culture were centrifuged at 10,000× *g* for 5 min at 4 °C. Bacterial cells were harvested and resuspended in 2 mL ice-cold PBS, incubated at 100 °C for 5 min, and sonicated for 15 min (power, 100%; frequency, 37 kHz) in an ice-water bath. After centrifugation, the supernatant containing c-di-GMP was collected, and the pellet was resuspended in 2 mL of ice-cold PBS and reextracted another two times. Intracellular c-di-GMP levels were quantified using a c-di-GMP ELISA kit (Mlbio, Shanghai, China). The total protein concentration in the supernatant was measured using a Pierce bicinchoninic acid (BCA) protein assay kit (Solarbio, Beijing, China). Intracellular c-di-GMP levels were reported as pmol/mg protein.

3.2.8. qRT-PCR Analysis

UPEC was cultured with shaking at 37 °C until reaching the logarithmic phase (OD_{600} = 0.4) and subsequently incubated with drugs at a concentration of 1/4 MIC for 4 h. Total RNA was extracted using the M5 EASY spin plus kit (Mei5bio, Beijing, China). Subsequently, the RNA was reverse transcribed to cDNA using the M5 Sprint qPCR RT kit with gDNA remover (Mei5bio, Beijing, China). Real-time PCR was conducted using the Q2000C (LongGene instruments, Hangzhou, China), employing a mixture of primers and 2 × M5 HiPer SYBR Premix EsTaq (Mei5bio, Beijing, China). The 16S rRNA gene was utilized as the reference gene for the strain (Table 5). The Ct value for each sample was measured in triplicate, and the $2^{-\Delta\Delta Ct}$ method was employed to calculate the mRNA expression levels of YcgR.

Table 5. Sequences of primers used in this study.

Primer Name	Sequence (5′-3′)	Reference
16SrRNA	F:GGCTGAAAAGCTGCATTACC R:CATCAGGCCGATGTTACCTT	[42]
YcgR-1	F:GCGCATTACTGGAAACAGC R:TTCATTCTTGCCATCAATCACT	This study

3.3. *Molecular Dynamics Simulations*

Table 2 shows that three compounds, including chlorogenic acid, sibiricose A5, and 3-O-caffeoylquinic acid methyl ester, achieve significant synergistic effects with levofloxacin hydrochloride, as evidenced by their FIC values under 0.5. To further explore antibiofilm mechanisms, three docked structures (YcgR_CAM, YcgR_CGA, and YcgR_Si-A5) were derived from the molecular docking process detailed in the Section 3. Additionally, various configurations of the YcgR protein, including the unbound YcgR-apo, YcgR-CDG-dual with two c-di-GMP molecules (PDB ID: 5Y6F [10]), YcgR-CDG-PilZ with one c-di-GMP molecule near the PilZ domain, and YcgR-CDG-N-domain with one c-di-GMP molecule near the YcgR-N domain, were constructed. These configurations result in seven systems: YcgR-apo, YcgR-CDG-dual, YcgR-CDG-PilZ, YcgR-CDG-N domain, YcgR-CAM, YcgR-CGA, and YcgR-Si-A5.

Three independent 500-ns MD simulations for the seven systems were conducted using the AMBER 20 package [43]. The resp charges for ligands were generated utilizing Gaussian 16 combined with the Antechamber module in the AMBER 20 package. The protein YcgR and ligands were described by the ff14SB [44] and GAFF [45] force fields, respectively. Then, each system was neutralized with an appropriate amount of counterion (Cl^- or Na^+), followed by solvation with TIP3P [46] water molecules in an orthogonal box, leaving at least 10 Å between the solute atoms and the box boundary. The cutoff distance for nonbonded interactions was set to 10 Å [47] In addition, the particle mesh Ewald (PME) [47] summation method was implemented to deal with Coulombic interactions, and the Shake [48] algorithm was applied to restrain bond vibrations involving hydrogen atoms. Four steps, including energy minimization, heating, equilibration, and production runs, were performed for the MD simulations. At first, four minimizations were carried out by employing the steepest descent and the conjugate gradient algorithm through 27,000 steps with the force constants of 5.0, 2.0, 0.1, and 0 kcal/mol/$Å^2$ to all heavy atoms, protein backbone, and C_α atoms, respectively. Following the minimization steps, each system was heated from 0 K to 310 K with a force constant of 5.0 kcal/mol/$Å^2$ for all heavy atoms. Then, all systems were equilibrated through 3000 steps with harmonic restraints of 1.0, 0.5, and 0.1 kcal/mol/$Å^2$ on all heavy atoms, protein backbone, and C_α atoms, respectively. Next, a long 10-ns equilibration was conducted with a harmonic restraint of

0.1 kcal/mol/Å^2. Finally, three independent 500-ns unrestrained MD simulations were simulated with randomized initial atomic velocities for each system.

3.4. Molecular Mechanics Generalized Born Surface Area Calculation

To evaluate the binding affinity of ligands on YcgR, the binding free energy and per-residue free energy decomposition were calculated using the MM/GBSA [49] method for each system. At first, water molecules and counter ions were extracted from the MD trajectory, and then 500 snapshots at time intervals of 25 ps were taken from the last 50-ns MD simulation. For each snapshot, the binding free energy (ΔG_{bind}) was calculated using the following formula:

$$\Delta G_{\text{bind}} = \Delta G_{\text{complex}} - \Delta G_{\text{receptor}} - \Delta G_{\text{ligand}}$$

where

$$\Delta G_{\text{bind}} = \Delta E_{\text{MM}} + \Delta G_{\text{sol}} - T\,\Delta S$$

ΔE_{MM} is the change in the molecular mechanics energy upon complexation in the gas phase, ΔG_{sol} is the change in solvation free energy, and $T\,\Delta S$ is the change of conformational entropy associated with inhibitor binding. However, the entropy contribution is not considered here because our goal is not to obtain the absolute Gibbs energy. $\Delta G_{\text{MM/GBSA}}$ was averaged over the three parallel MD simulations, and per-residue free energy decomposition was determined by summing the energy components of each residue involved in the stabilization of the complex.

3.5. Statistical Analysis

In this study, all data were recorded as means ± standard deviations (SD), and a one-way analysis of variance (ANOVA) was conducted using SPSS Statistics version 26 software. $p < 0.05$ was considered significant. All experiments were performed at least in triplicate.

4. Conclusions

In summary, we illustrated the synergistic antibacterial and anti-biofilm mechanism of chlorogenic acid, sibiricose A5, and 3-*O*-caffeoylquinic acid methyl ester of *Pyrrosia lingua* combined with levofloxacin hydrochloride against UPEC. Our results showed that the combinations of the three compounds with levofloxacin hydrochloride enhanced the antibacterial activity against standard UPEC strain as the concentration increased. Moreover, the combined use led to MIC of levofloxacin hydrochloride and three caffeoylquinic acid compounds with a maximum reduction of 16-fold against the standard and clinically isolated UPEC strains. Chlorogenic acid and sibiricose A5 inhibited the biofilm formation of UPEC with levofloxacin hydrochloride at multiple time points. In addition, levofloxacin hydrochloride combined with chlorogenic acid or sibiricose A5 significantly increased the motility and c-di-GMP contents of UPEC, while the combined use of 3-*O*-caffeoylquinic acid methyl ester and levofloxacin hydrochloride had the opposite effects. The combination of levofloxacin hydrochloride and caffeoylquinic acid compounds exerted an inhibitory effect on the expression of the YcgR-1 gene. Two c-di-GMP within the binding site enhanced the stability of the complex, resulting in a more favorable and compact structure that promotes biofilm formation. Conversely, the presence of a single c-di-GMP molecule appeared to weaken this binding, thereby increasing protein's motility. Notably, the interaction involving the c-di-GMP adjacent to the PilZ domain was more favorable compared to that near the YcgR-N domain. Additionally, the introduction of chlorogenic acid and sibiricose A5 was observed to decrease the motility of the YcgR protein, which in turn facilitated a more stable association with YcgR and was unfavorable for the association between c-di-GMP and YcgR. The van der Waals and direct electrostatic interaction were crucial driving forces for the binding between ligands and YcgR. These caffeoylquinic acid analogs have the potential to be used as antibiotic adjuvants for the treatment of urinary tract infections in the future. The study provides the foundation for solving the problem of antibiotic resistance and developing novel drugs for treating urinary tract infections.

Supplementary Materials: The following supporting information can be downloaded at https://www.mdpi.com/article/10.3390/molecules29235679/s1, Figure S1: Residue contribution to the binding of ligands considering three replicates: (A) YcgR-CAM; (B) YcgR-CGA; (C) YcgR-CDG-PilZ; (D) YcgR-CDG-dual-PilZ; Figure S2: Residue contribution to the binding of ligands considering three replicates: (A) YcgR-Si-A5; (B) YcgR-CDG-N domain; (C) YcgR-CDG-dual-N domain; Figure S3: A cartoon representation of YcgR-CAM (green), YcgR-CGA (cyan), YcgR-CDG-PilZ (magenta), and YcgR-CDG-dual-PilZ (yellow). The ligands were highlighted by the black box; Figure S4: A cartoon representation of YcgR-Si-A5 (pink), YcgR-CDG-N domain (silver), and YcgR-CDG-dual-N domain (purple). The ligands were highlighted by the black box.

Author Contributions: Conceptualization, Y.Z. (Yan Zhang), J.G. and W.Y.; methodology, D.Z., F.J., Y.Z. (Yan Zhang) and J.G.; software, J.G.; validation, Y.Z. (Yan Zhang), F.J. and X.Y.; formal analysis, D.Z., F.J., J.X. and Y.Z. (Yan Zhang); investigation, W.Y. and X.Y.; resources, W.Y., Y.Z. (Yongqiang Zhou) and J.G.; data curation, D.Z. and F.J.; writing—original draft preparation, D.Z., Y.Z. (Yan Zhang) and F.J.; writing—review and editing, Y.Z. (Yan Zhang), J.G and W.Y.; supervision, Y.Z. (Yan Zhang) and J.G. All authors have read and agreed to the published version of the manuscript.

Funding: This study was supported by the Science and Technology Program of Guizhou Province [QKHF-ZK(2021)G539], the Science and Technology Program of Guizhou Province [QKHF-ZK(2023)G426], the Science and Technology Program of Guizhou Province [QKHF-ZK(2021)G518], the Science and Technology Program of Guizhou Province [QKHF-ZK(2023)G420], Guizhou University of Traditional Chinese Medicine [Gui Ke He Xue Shu Xin Miao [2023]-10], and Macao Polytechnic University (RP/CAI-01/2022).

Institutional Review Board Statement: Not applicable.

Informed Consent Statement: Not applicable.

Data Availability Statement: Data are available within the manuscript or the Supporting Materials.

Conflicts of Interest: The authors declare no conflicts of interest.

References

1. Naziri, Z.; Kilegolan, J.A.; Moezzi, M.S.; Derakhshandeh, A. Biofilm formation by uropathogenic *Escherichia coli*: A complicating factor for treatment and recurrence of urinary tract infections. *J. Hosp. Infect.* **2021**, *117*, 9–16. [CrossRef] [PubMed]
2. Boya, B.R.; Lee, J.H.; Lee, J. Antibiofilm and antimicrobial activities of chloroindoles against uropathogenic *Escherichia coli*. *Front. Microbiol.* **2022**, *13*, 872943. [CrossRef] [PubMed]
3. Murray, B.O.; Flores, C.; Williams, C.; Flusberg, D.A.; Marr, E.E.; Kwiatkowska, K.M.; Charest, J.L.; Isenberg, B.C.; Rohn, J.L. Recurrent Urinary Tract Infection: A Mystery in Search of Better Model Systems. *Front. Cell. Infect. Microbiol.* **2021**, *11*, 691210. [CrossRef]
4. Aziminia, N.; Hadjipavlou, M.; Philippou, Y.; Pandian, S.S.; Malde, S.; Hammadeh, M.Y. Vaccines for the prevention of recurrent urinary tract infections: A systematic review. *BJU Int.* **2019**, *123*, 753–768. [CrossRef] [PubMed]
5. Whelan, S.; O'Grady, M.C.; Corcoran, G.D.; Finn, K.; Lucey, B. Effect of sub-inhibitory concentrations of nitrofurantoin, ciprofloxacin, and trimethoprim on in vitro biofilm formation in uropathogenic *Escherichia coli* (UPEC). *Med. Sci.* **2022**, *11*, 1. [CrossRef] [PubMed]
6. Yang, Z.; Zuying, Z.; Lin, Z.; Zipeng, G.; Yueting, L.; Yang, J.; Yong, H.; Mingyan, C. Urinary tract infections caused by uropathogenic *Escherichia coli*: Mechanisms of infection and treatment options. *Int. J. Mol. Sci.* **2023**, *24*, 10537. [CrossRef]
7. Fang, X.; Gomelsky, M. A post-translational, c-di-gmp-dependent mechanism regulating flagellar motility. *Mol. Microbiol.* **2010**, *76*, 1295–1305. [CrossRef]
8. Han, Q.; Wang, S.F.; Qian, X.X.; Guo, L.; Shi, Y.F.; He, R.; Yuan, J.H.; Hou, Y.J.; Li, D.F. Flagellar brake protein ycgr interacts with motor proteins mota and flig to regulate the flagellar rotation speed and direction. *Front. Microbiol.* **2023**, *14*, 1159974. [CrossRef] [PubMed]
9. Wang, X.; Liu, M.; Yu, C.; Li, J.; Zhou, X. Biofilm formation: Mechanistic insights and therapeutic targets. *Mol. Biomed.* **2023**, *4*, 49. [CrossRef] [PubMed]
10. Hou, Y.J.; Yang, W.S.; Hong, Y.; Zhang, Y.; Wang, D.C.; Li, D.F. Structural insights into the mechanism of c-di-gmp-bound ycgr regulating flagellar motility in *Escherichia coli*. *J. Biol. Chem.* **2020**, *295*, 808–821. [CrossRef] [PubMed]
11. Topa, S.H.; Subramoni, S.; Palombo, E.A.; Kingshott, P.; Rice, S.A.; Blackall, L.L. Cinnamaldehyde disrupts biofilm formation and swarming motility of *Pseudomonas aeruginosa*. *Microbiology* **2018**, *164*, 1087–1097. [CrossRef]

12. Rasamiravaka, T.; Labtani, Q.; Duez, P.; El, J.M. The formation of biofilms by *Pseudomonas aeruginosa*: A review of the natural and synthetic compounds interfering with control mechanisms. *Biomed. Res. Int.* **2015**, *2015*, 759348. [CrossRef] [PubMed]
13. Hultqvist, L.D.; Andersen, J.B.; Nilsson, C.M.; Jansen, C.U.; Rybtke, M.; Jakobsen, T.H.; Nielsen, T.E.; Qvortrup, K.; Moser, C.; Graz, M.; et al. High efficacy treatment of murine *Pseudomonas aeruginosa* catheter-associated urinary tract infections using the c-di-gmp modulating anti-biofilm compound disperazol in combination with ciprofloxacin. *Antimicrob. Agents Chemother.* **2024**, *68*, e0148123. [CrossRef]
14. Qin, T.; Chen, K.; Xi, B.; Pan, L.; Xie, J.; Lu, L.; Liu, K. In vitro antibiofilm activity of resveratrol against *Aeromonas hydrophila*. *Antibiotics* **2023**, *12*, 4. [CrossRef] [PubMed]
15. Kim, H.S.; Ham, S.Y.; Ryoo, H.S.; Kim, D.H.; Yun, E.T.; Park, H.D.; Park, J.H. Inhibiting bacterial biofilm formation by stimulating c-di-gmp regulation using citrus peel extract from Jeju Island. *Sci. Total Environ.* **2023**, *872*, 162180. [CrossRef]
16. van Wietmarschen, H.; van Steenbergen, N.; van der Werf, E.; Baars, E. Effectiveness of herbal medicines to prevent and control symptoms of urinary tract infections and to reduce antibiotic use: A literature review. *Integr. Med. Res.* **2022**, *11*, 100892. [CrossRef] [PubMed]
17. Zuvairiya, A.; Lydia, M.A.; Gayathri, J.; Gayathri, N.; Kasipandi, M.; Saikumar, S.; Parimelazhagan, T. Evaluation of an edible polyherbal formulation against urinary tract infection pathogens, its antioxidant and anti-inflammatory potential. *Biocatal. Agric. Biotechnol.* **2021**, *35*, 102104.
18. Barrett, P.; Flower, A.; Lo, V. What's past is prologue: Chinese medicine and the treatment of recurrent urinary tract infections. *J. Ethnopharmacol.* **2015**, *167*, 86–96. [CrossRef] [PubMed]
19. Chang, J.; Sui, Y.; Sun, W.J.; He, H.J.; Hu, S.Y.; Tong, P.Z.; Yi, X.; Yang, W.D.; Long, Y. Effective Components and Mechanism of Miao Medicine Herb *Pyrrosia petiolosa* (Christ) Ching in Lowering Blood Glucose. *Pharmacol. Clin. Chin. Mater. Medica* **2023**, *39*, 75–81.
20. Hu, Y.; Chen, F.; Long, Y.; Yu, X.; Wu, J.; Luo, G.; Yang, W. Chemical Constituents from Promoting Diuresis and Relieving Stranguria Effective Part of *Pyrrosia petiolosa* in Guizhou (II). *J. Chin. Med. Mater.* **2021**, *44*, 1386–1391.
21. Xi, L.; Mu, T.; Sun, H. Progresses in the Research of Chlorogenic Acids. *J. Nucl. Agric. Sci.* **2014**, *28*, 292–301.
22. Gupta, A.; Atanasov, A.G.; Li, Y.; Kumar, N.; Bishayee, A. Chlorogenic acid for cancer prevention and therapy: Current status on efficacy and mechanisms of action. *Pharmacol. Res.* **2022**, *186*, 106505. [CrossRef] [PubMed]
23. Bai, P.; Zhang, C.; Xu, M.; Li, Y. Biological function of chlorogenic acid compounds and its application in animal husbandry production. *Feed Res.* **2022**, *45*, 127–130.
24. Chai, B.; Jiang, W.; Hu, M.; Wu, Y.; Si, H. In vitro synergistic interactions of protocatechuic acid and chlorogenic acid in combination with antibiotics against animal pathogens. *Synergy* **2019**, *9*, 100055. [CrossRef]
25. Chai, B.; Jiang, W.; Wang, L.; Hu, M.; Zhao, Y.; Wu, Y.; Si, H. Antibacterial effects of protocatechuic acid and chlorogenic acid combined with antibiotics against fish-source *Streptococcus*. *J. South. Agric.* **2018**, *49*, 580–585.
26. Chen, F.; Luo, G.; Yang, W. Chemical Constituents from Promoting Diuresis and Relieving Stranguria Effective Part of *Pyrrosia petiolosa* in Guizhou. *J. Chin. Med. Mater.* **2019**, *42*, 2822–2826.
27. Wu, X.; Hu, Y.; Zhang, Y.; Long, Y.; Wu, J.; Luo, G.; Yang, W. Chemical Constituents of the Effective Part of Promoting Diuresis and Relieving Stranguria of *Pyrrosia petiolosa* in Guizhou (III). *J. Chin. Med. Mater.* **2022**, *45*, 1861–1865.
28. Sastry, G.M.; Adzhigirey, M.; Day, T.; Annabhimoju, R.; Sherman, W. Protein and ligand preparation: Parameters, protocols, and influence on virtual screening enrichments. *J. Comput. Aided. Mol. Des.* **2013**, *27*, 221–234. [CrossRef] [PubMed]
29. Kaminski, G.A.; Friesner, R.A.; Tirado-Rives, J.; Jorgensen, W.L. Evaluation and reparametrization of the opls-aa force field for proteins via comparison with accurate quantum chemical calculations on peptides. *J. Phys. Chem. B* **2001**, *105*, 6474–6487. [CrossRef]
30. Ngamsurach, P.; Praipipat, P. Comparative antibacterial activities of *Garcinia cowa* and *Piper sarmentosum* extracts against *Staphylococcus aureus* and *Escherichia coli* with studying on disc diffusion assay, material characterizations, and batch experiments. *Heliyon* **2022**, *8*, e11704. [CrossRef]
31. Lazou, T.P.; Chaintoutis, S.C. Comparison of disk diffusion and broth microdilution methods for antimicrobial susceptibility testing of campylobacter isolates of meat origin. *J. Microbiol. Methods* **2023**, *204*, 106649. [CrossRef] [PubMed]
32. Carvalho, G.M.; Silva, B.A.; Xavier, R.; Zanon, I.P.; Vilela, E.G.; Nicolino, R.R.; Tavares, G.C.; Silva, R. Evaluation of disk diffusion method for testing the rifampicin, erythromycin, and tetracycline susceptibility of *Clostridioides* (prev. *Clostridium) difficile*. *Anaerobe* **2023**, *80*, 102720. [CrossRef] [PubMed]
33. Cordovana, M.; Ambretti, S. Antibiotic susceptibility testing of anaerobic bacteria by broth microdilution method using the micronaut-s anaerobes mic plates. *Anaerobe* **2020**, *63*, 102217. [CrossRef]
34. Lin, H.; Liang, Y.; Kaliaperumal, K.; Xiong, Q.; Duan, S.; Jiang, Y.; Zhang, J. Linoleic acid from the endophytic fungus *Diaporthe* sp. Ht-79 inhibits the growth of *Xanthomonas citri* subsp. citri by destructing the cell membrane and producing reactive oxygen species (ros). *Pest. Biochem. Physiol.* **2023**, *192*, 105423.
35. Yu, W.; Shen, P.; Bao, Z.; Zhou, K.; Zheng, B.; Ji, J.; Guo, L.; Huang, C.; Xiao, Y. In vitro antibacterial activity of fosfomycin combined with other antimicrobials against kpc-producing *Klebsiella pneumoniae*. *Int. J. Antimicrob. Agents* **2017**, *50*, 237–241. [CrossRef] [PubMed]

36. European Committee for Antimicrobial Susceptibility Testing (EUCAST) of the European Society of Clinical Microbiology and Infectious Diseases (ESCMID). Terminology Relating to Methods for the Determination of Susceptibility of Bacteria to Antimicrobial Agents. *Clin. Microbiol. Infect.* **2000**, *6*, 503–508. [CrossRef]
37. Bai, J.-R.; Zhong, K.; Wu, Y.-P.; Grosu, E.; Gao, H. Antibiofilm activity of shikimic acid against *Staphylococcus aureus*. *Food Control* **2019**, *95*, 327–333. [CrossRef]
38. Ashkezari, S.; Abtahi, M.S.; Sattari, Z.; Yaraki, M.T.; Hosseini, F.; Salehi, R.I.; Afzali, E.; Hajihosseini, S.; Mousavi-Niri, N. Antibiotic and inorganic nanoparticles co-loaded into carboxymethyl chitosan-functionalized niosome: Synergistic enhanced antibacterial and anti-biofilm activities. *J. Drug Deliv. Sci. Technol.* **2023**, *83*, 104386. [CrossRef]
39. Tan, Y.; Leonhard, M.; Moser, D.; Schneider-Stickler, B. Inhibition activity of lactobacilli supernatant against fungal-bacterial multispecies biofilms on silicone. *Microb. Pathog.* **2017**, *113*, 197–201. [CrossRef] [PubMed]
40. Gao, H.; Ma, L.; Qin, Q.; Qiu, Y.; Zhang, J.; Li, J.; Lou, J.; Diao, B.; Zhao, H.; Shi, Q.; et al. Fur represses vibrio cholerae biofilm formation via direct regulation of viesab, cdgd, vpsu, and vpsa-k transcription. *Front. Microbiol.* **2020**, *11*, 587159. [CrossRef] [PubMed]
41. Zhang, M.; Cai, L.; Luo, X.; Li, X.; Zhang, T.; Wu, F.; Zhang, Y.; Lu, R. Effect of sublethal dose of chloramphenicol on biofilm formation and virulence in *Vibrio parahaemolyticus*. *Front. Microbiol.* **2023**, *14*, 1275441. [CrossRef]
42. Yu, L.; Zhang, S.; Xu, Y.; Mi, X.; Xing, T.; Li, J.; Zhang, L.; Gao, F.; Jiang, Y. Acid resistance of *E. coli* o157:h7 and o26:h11 exposure to lactic acid revealed by transcriptomic analysis. *LWT Food Sci. Technol.* **2021**, *136 Pt 2*, 110352. [CrossRef]
43. Case, D.A.; Aktulga, H.M.; Belfon, K.; Ben-Shalom, I.; Brozell, S.R.; Cerutti, D.S.; Cheatham, T.E., III; Cruzeiro, V.W.D.; Darden, T.A.; Duke, R.E.; et al. *Amber 2021*; University of California: San Francisco, CA, USA, 2021.
44. Maier, J.A.; Martinez, C.; Kasavajhala, K.; Wickstrom, L.; Hauser, K.E.; Simmerling, C. Ff14sb: Improving the accuracy of protein side chain and backbone parameters from ff99sb. *J. Chem. Theory Comput.* **2015**, *11*, 3696–3713. [CrossRef] [PubMed]
45. Wang, J.; Wolf, R.M.; Caldwell, J.W.; Kollman, P.A.; Case, D.A. Development and testing of a general amber force field. *J. Comput. Chem.* **2004**, *25*, 1157–1174. [CrossRef] [PubMed]
46. Jorgensen, W.L.; Chandrasekhar, J.; Madura, J.D.; Impey, R.W.; Klein, M.L. Comparison of simple potential functions for simulating liquid water. *J. Chem. Phys.* **1983**, *79*, 926–935. [CrossRef]
47. Essmann, U.; Perera, L.; Berkowitz, M.L.; Darden, T.; Lee, H.; Pedersen, L.G. A smooth particle mesh ewald method. *J. Chem. Phys.* **1995**, *103*, 8577–8593. [CrossRef]
48. Ryckaert, J.; Ciccotti, G.; Berendsen, H.J. Numerical integration of the cartesian equations of motion of a system with constraints: Molecular dynamics of n-alkanes. *J. Comput. Phys.* **1977**, *23*, 327–341. [CrossRef]
49. Genheden, S.; Ryde, U. The mm/pbsa and mm/gbsa methods to estimate ligand-binding affinities. *Expert Opin. Drug Discov.* **2015**, *10*, 449–461. [CrossRef] [PubMed]

molecules

MDPI

Article

2,4-Dichlorophenoxyacetic Acid in the Gas and Crystal Phases and Its Intercalation in Montmorillonite—An Experimental and Theoretical Study

Claro Ignacio Sainz-Díaz [1,*], Nelly L. Jorge [2], Jorge M. Romero [2], André Grand [3] and Alfonso Hernández-Laguna [1,*]

[1] Instituto Andaluz de Ciencias de la Tierra (IACT-CSIC), Consejo Superior de Investigaciones Científicas, Av. de las Palmeras 4, 18100 Armilla, Granada, Spain

[2] Laboratorio de Investigaciones en Tecnología del Medio Ambiente, Área de Química Física, Facultad de Ciencias Exactas y Naturales y Agrimensura, Universidad del Nordeste, Corrientes 3400, Argentina; lidianj@exa.unne.edu.ar (N.L.J.); ing.jorgemromero@gmail.com (J.M.R.)

[3] Université Grenoble Alpes, Commissariat à l'Energie Atomique, Centre National de la Recherche Scientifique, Institute for Nanoscience and Cryogenics, Systèmes Moléculaires et NanoMatériaux pour l'Énergie et la Santé, F-38000 Grenoble, France; andre.grand8@wanadoo.fr

* Correspondence: ci.sainz@csic.es (C.I.S.-D.); a.h.laguna@csic.es (A.H.-L.)

Abstract: Many properties of 2,4-dichlorophenoxyacetic acid (2,4-D) depend on its molecular environment, such as whether it is an isolated molecule, a dimer, or in a crystalline state. The molecular geometry, conformational analysis, and vibrational spectrum of 2,4-D were theoretically calculated using Density Functional Theory (DFT) methods. A new slightly more stable conformer was found, which is different to those previously reported. The most stable conformer shows a dimer by means of hydrogen bonds between the carboxylic groups of both molecules, which agrees with the experimental results. The crystal structure of 2,4-D was also calculated with 3D periodical boundary conditions at the DFT level. From the theoretical IR spectra, a vibrational analysis of this molecular species was accomplished, and the bands were reassigned. ^{1}H and ^{13}C NMR in the dissolution and solid states, respectively, showed intramolecular hydrogen bonds between carboxylic acid groups. The dimer is more stable than the isolated molecule. All these results indicated that the dimer can also exist in the solid state, which could explain the low solubility of this compound. In addition, the intercalation of 2,4-D into the confined interlayer space of montmorillonite was also calculated, and it was found that the adsorption is energetically favourable. This result was experimentally confirmed. These findings predicted that these natural clay minerals, which are found in the environment, can be excellent adsorbents for the 2,4-D pollutant.

Keywords: 2,4-dichlorophenoxyacetic acid; DFT; intercalation in clay minerals; dimer; IR assignment; NMR; crystal structure; hydrogen bonds

1. Introduction

Aiming to augment productivity, intensive agricultural exploitations strongly increase the application of fertilizers and pesticides (herbicides, fungicides, insecticides, and others). Most of these chemicals can infiltrate and leach through water, soil profile, and strata subsoil, contaminating soils, surficial water, and groundwater [1,2]. Herbicides derived from phenoxy-acetic acids have important chemical effects and non-desirable pharmacodynamics and pharmacokinetic effects on mammalian species [3–6]. Phenoxy-acid derivatives are systemic herbicides, which are absorbed by the leaves and/or the roots, that are transported by the sap to the whole body of the plants; these reach the internal tissues, which finally kill the plants. This type of herbicide is considered to be one of the first selective herbicides because it kills many broadleaved plants while causing little damage to the narrow leaves. These derivatives and their metabolites are the most important pollutants of drinking water as they have a very low maximum concentration limit for drinking water. These compounds enter into the formulation of many pesticide preparations and are hazardous to human beings.

The herbicidal activity (HA) of phenoxy-acetic acid is greatly influenced by the substituents on the aromatic group. When a halogen atom—either fluorine or chlorine—is substituted in the aromatic moiety, the HA generally increases. The C-Cl bond is especially stabilized by conjugation with the π electron cloud of the aromatic phenyl group. The position of this halogen also plays a crucial role in relation to the HA [7]. When a hydrogen atom in an aromatic group is substituted by an aliphatic group, the HA does not significantly increase [8]. To increase the activity, the molecule must possess one -COOH group or a group for reacting with plant tissues [8,9]. Therefore, the type and position of substituents are an important issue when looking for a high HA. In this way, a theoretical search about the structural properties of phenoxy acetic compounds would be very useful as a means to start to understand the HA and the toxic properties at the molecular level.

The herbicide 2,4-dichlorophenoxyacetic acid (2,4-D) (Figure 1) widely fulfils the previous HA requirements. Therefore, 2,4-D is a very popular plant growth regulator and is applied as an herbicide for general and broadleaved weed control, especially in the rice ecosystem. This compound has a maximum concentration limit in drinking water of 0.7 μg/L [10], and it is found with a maximum concentration of 0.012 μg/L in underground water sources [11,12] and 1.22 μg/L in river water [11,13] in European Union countries.

In a previous article, the hydrolysis reaction of the 2,4-D was studied [14] by the kinetics of the disappearance of the 2,4-D at particular values of pH and temperature, and a mechanism of the hydrolysis reaction was proposed with a computational model at the gas phase. Theoretical vibrational frequencies were compared with experimental frequencies [15–17]. Karthikeyan and Saravanan [18] determined the Raman spectra and calculated the phenoxy-acetic acid and other Cl-derivatives at the Hartree–Fock level, finding the chlorine substitution effect at the molecule. Badawi [16] calculated eight conformers, where the most stables are energetically very close. In this work, we found one new slightly more stable but different conformer with respect to Badawi's conformers [16]. This molecule can be either an isolated molecule or dimer, due to the strong interactions between the carboxylic groups. Molinelli et al. [19] found that the dimerization rate of 2,4-D is one order of magnitude higher than the interaction with polymers. Previous studies on the crystal structure of 2,4-D [20] by X-ray diffraction experiments showed that, in the solid phase, two molecules are linked together by hydrogen bonds between their carboxylic groups, producing a centre-symmetrical cyclic structure dimer. However, it not well known how the 2,4-D molecules are distributed in the atmospheric aerosols and in the soils and waste waters. A higher knowledge of the molecular interactions of 2,4-D is interesting to investigate the interaction with other soil components like mineral surfaces. Moreover, the molecular environment of this pollutant is critical for the risk evaluation of toxicity and carcinogenicity tests [21] in the interactions with receptors at the molecular level. The use of low-cost adsorbents, like clay minerals, is becoming recognized as a successful method of prevention and remediation environmental strategies. Aquino et al. studied the adsorption of 2,4-D on the Goethite surface [22]. Clay minerals were reported to display excellent adsorbing properties with low prices and easy access to ubiquitous deposits [23].

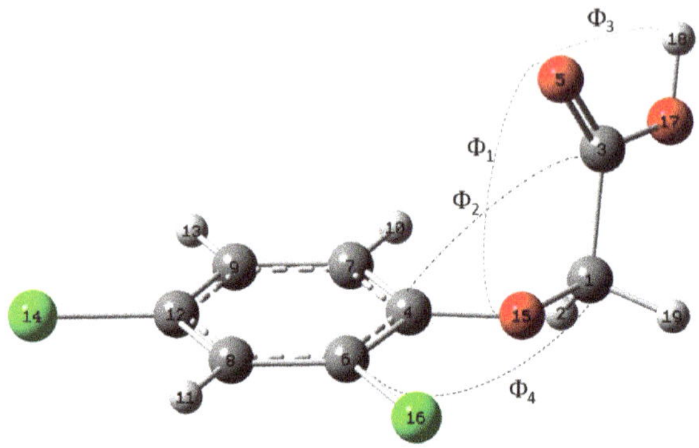

(**a**)

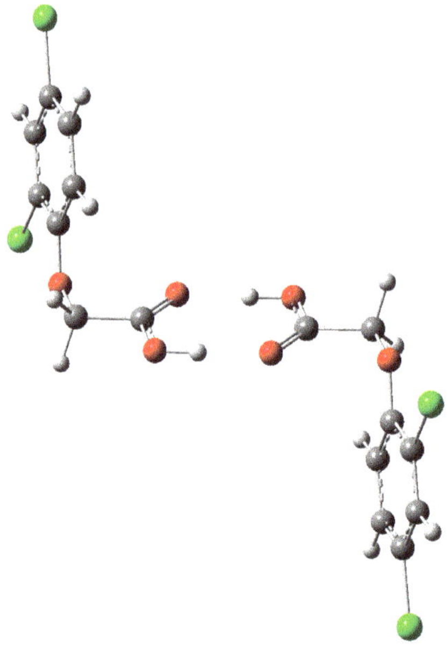

(**b**)

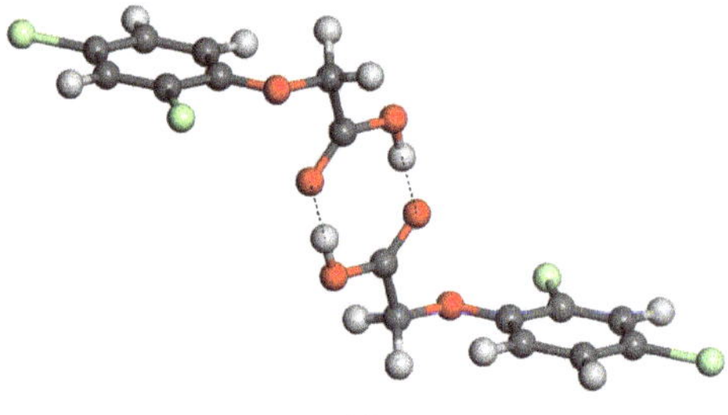

(**c**)

Figure 1. Optimized molecular structure of the most stable conformer I (**a**). Dimer-I of 2,4-dichlorophenoxyacetic acid optimized with B3LYP/6-311++G(d,p) (**b**) and with CASTEP (**c**). The C, H, O, and Cl atoms are in grey, white, red, and green colours, respectively. Dashed lines mean hydrogen bond. This criterium is extended for the rest of this work.

In this work, the acidic form of the molecule, one dimer in the gas phase, and the crystal structure of 2,4-D were studied at the DFT level, and the vibrational spectrum was calculated, studying the dimer effect on the molecular structure and frequencies. These calculations helped the experimental vibrational spectrum to be assigned. Moreover, the ^{1}H and ^{13}C NMR spectra of the compound in the dissolution and solid states, respectively, were also recorded in order to understand the intermolecular situation of 2,4-D. In addition, the intercalation of 2,4-D into the interlayer space of montmorillonite was calculated and determined as an energetically favourable process. This study was compared with additional experimental works of the adsorption and desorption of 2,4-D on Argentinian clay-rich soils.

This paper is organized as follows: (i) the Results and Discussion section starts with geometry optimization and a conformational analysis for the molecule, dimer, and crystal structures of 2,4-D; (ii) next, the intercalation of 2,4-D into montmorillonite is examined; (iii) this process is further studied experimentally by exploring the adsorption of 2,4-D on clay-rich soils; (iv) a complete and comparative analysis of the vibrational spectra between the isolated molecule, dimer, and crystal structures is accomplished along with frequency assignments; (v) another section focuses on the NMR methods; and (vi) the work ends with some concluding remarks.

2. Results and Discussion

2.1. *Structure Calculations*

2.1.1. Molecule

A conformational analysis was performed for a comprehensive exploration of Potential Energy Surface (PES). Combining several dihedral angles (Figure 1a), nine conformers of the 2,4-D acid molecule were found. All of them were fully optimized and did not show any imaginary frequency. Our results were similar to previous work [16]. However, we found a new conformer that is the most stable (conformer I). In our conformational analysis, the second more stable conformer was II, which is similar to previous reports [16]. The energy differences between both conformers are very low, and both can exist at room temperature. However, an important torsional difference can be observed: the torsional angle ϕ_4 is 180° in our minimal conformer I corresponding to a *trans* conformation, whereas this angle is close to 90° in conformer II (Table S1). The calculated geometry of conformer I is in accordance with the X-ray experimental data provided by Smith et al. [20]. The herbicide exists predominantly in non-planar conformations at room temperature. The alkylic chain is perpendicular to the aromatic ring plane, the carbonyl bond is in the same plane as the alkoxy O atom, and the hydroxyl bond is coplanar to the carbonyl group. Our full optimized conformer I (Figure 1a) yielded good results, as can be seen from the RMS values of the differences between the experimental and calculated geometrical parameters (Table S1): the smallest RMS values are between the experimental structure and conformer I, especially the RMS of bond lengths and bond angles. Nevertheless, the RMS of the dihedral angles increases to 5.5°. The origin of this discrepancy is likely due to the different phases in which the experiment and calculations were performed. Intramolecular interactions and steric effects play a competitive role in the conformational stability in the ground state. Considering the phenyl moiety of the molecule, the conjugation effects should be important for stabilizing planar structures, whereas the steric forces would stabilize non-planar structures. The steric forces should be responsible for the stability of the I.

The dipole interactions are also important for determining the relative stability of conformations, so conformer I shows a high value of μ (4.22 D), indicating larger stabilization in polar solvents, like water, the solvent applied to weeds. The same conformer was obtained with the CASTEP program, which have important methodological and electronic differences with GAUSSIAN09.

2.1.2. Dimer

2,4-D can build cyclic dimers by hydrogen bond interactions of the carboxylic acid groups (Figure 1b,c). When choosing the two most stable conformers of 2,4-D, two possible dimers can be considered, D-I (with conformer I) and D-II (with conformer II) (Figure S1). The optimization of both dimers indicated that D-I is 1.02 kcal/mol more stable than D-II. The dichlorophenyl groups are in a disposition approximately perpendicular to the cyclic hydrogen bond plane of the carboxylic groups. Calculated geometric parameters are compared with the experimental data of the 2,4-D crystal [20] (Table S2). Once again, the RMS values between the experimental structure and the dimer D-I are lower than that of D-II (Table S2). Generally, the calculated bond lengths for the dimer agree with the experimental values of the crystal, though slight differences are observed in a few bond angles and dihedral angles. Again, this is probably due to the difference in phase. Nevertheless, one can be confident with the results of these calculations. Similar structure D-I was obtained with CASTEP.

Table 1 shows the total energy ($^{\&}E$) of the molecule and dimer D-I, including the ZPE ($^{\#}E$) corrections. In addition, the BSSE (*E) correction can also be observed for the dimer. From these values, the energy of the interaction between both molecules is −15.0 kcal/mol (considering ZPE). The steric forces also play an important role in stabilizing the ground-state structures of non-planar conformers. The nearly null dipole moments of the dimer indicate that the stabilization is also coming from the fitting of the dipolar moments of the molecules at dimer formation.

Table 1. The total energy [Gaussian] values ($^{\&}E$), corrected for ZPE ($^{\#}E$) and corrected for BSSE (*E) of the optimized 2,4-D molecule and cyclic dimer.

Energy (Ha)	Molecule	Dimer-I
$^{\&}E$	−1454.7351857	−2909.496049
$^{\#}E$	−1454.6069160	−2909.237675
*E		−2909.494342

2.1.3. Crystal Structure

The above results get us to examine the crystal structure of 2,4-D. The asymmetric unit of the crystal lattice of 2,4-D was transformed to a P1 symmetry unit cell with two 2,4-D molecules (Z = 2) and was calculated with CASTEP and periodical boundary conditions. The crystal cell was fully optimized, including atomic positions and cell parameters, resulting in the following cell parameters: a = 7.19 Å, b = 7.86 Å, c = 8.95 Å, α = 90.6°, β = 104.0°, and γ = 110.7°, which are quite close to the experimental lattice cell parameters of the crystal structure (a = 7.12 Å, b = 7.84 Å, c = 9.0 Å, α = 90.7°, β = 104.6°, γ = 110.5°) (Table 2) [20]. This optimization was performed under different computational conditions, with ultrasoft pseudopotentials at 490 and 630 eV energy cutoffs and with norm-conserving pseudopotentials, and in all cases, the cell parameters were close to the experimental values (Table 3), being the closest to the experimental parameters those calculated with 630 eV and ultrasoft pseudopotentials. The carboxylic groups formed dimers with strong hydrogen bonds, d(CO···HOC) = 1.558 Å. The conformation of the molecule was similar to that found above for the isolated molecule at the gas phase. The conformation of the intra-crystalline dimers was similar that in the gas-phase dimer (Figure 2). This dimerization was not described previously. The carbonyl O atoms and the H-C atoms showed electrostatic interactions, d(C=O···H-C) = 2.425 Å. Intermolecular motifs were found with hydrogen bonds between C-H bonds and carbonyl and ether O atoms, forming alternating $R^2_2(8)$ and $R^2_2(6)$ rings [24].

Table 2. Cell parameters of the experimental and calculated (CASTEP) crystal structure of 2,4-D (distances in Å and angles in °).

Parameters	*a*	*b*	*c*	α	β	γ
experimental	7.12	7.84	9.0	90.7	104.6	110.5
Method A [a]	7.31	7.99	9.03	90.2	104.6	111.6
Method B [b]	7.19	7.86	8.95	90.6	104.0	110.7
Method C [c]	7.20	7.87	8.97	90.6	103.9	110.7

[a] With ultrasoft pseudopotentials and E cutoff = 490 eV. [b] Ultrasoft pseudopotentials and E cutoff = 630 eV. [c] Norm-conserving pseudopotentials and E cutoff of 990 eV.

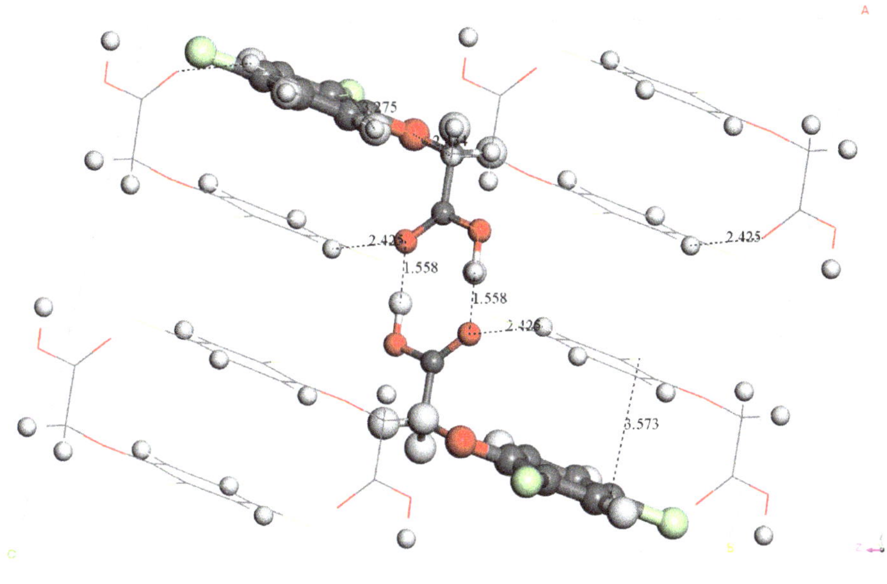

(**a**)

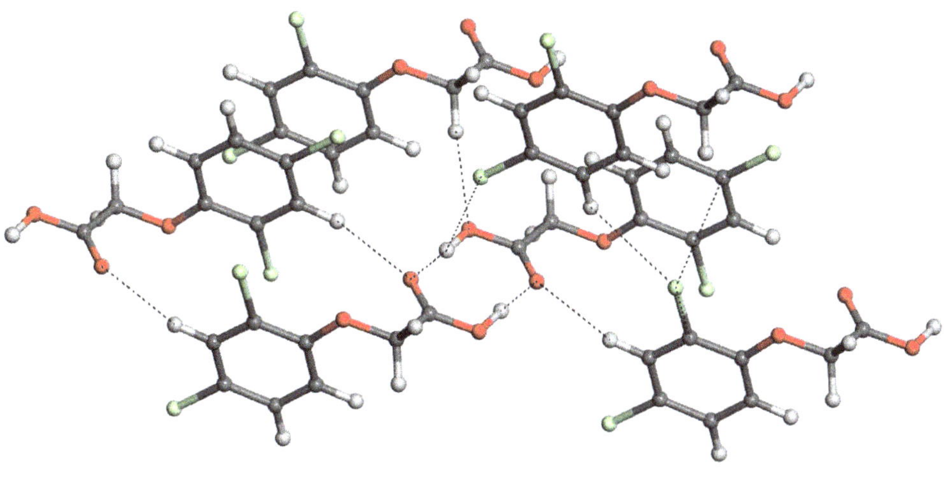

(**b**)

Figure 2. *Cont.*

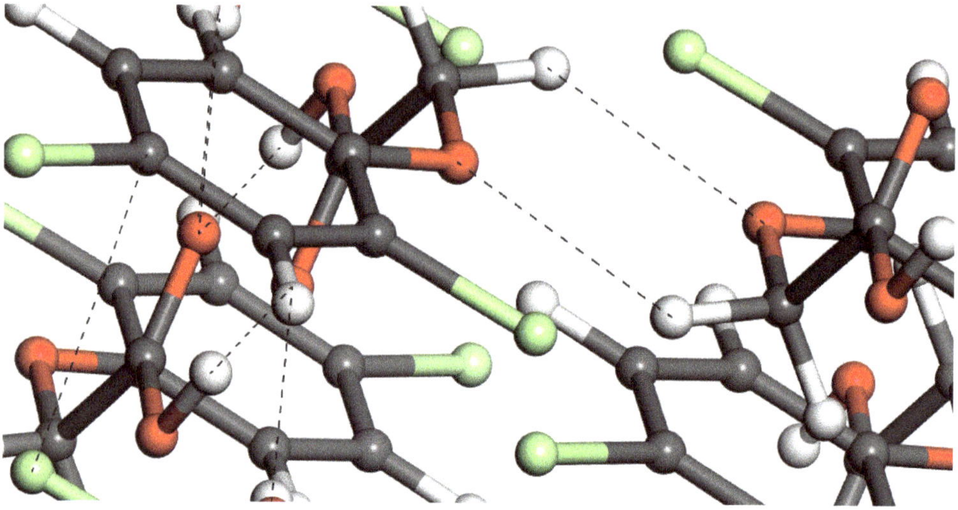

(c)

Figure 2. Optimized (CASTEP) crystal structure of 2,4-D showing different intermolecular interactions. View from the (010) plane the dimer structure is highlighted with balls showing a $R^2{}_2(8)$ ring (**a**). A general view (**b**). A zoom view highlighting the $R^2{}_2(6)$ ring (**c**).

Table 3. The fitting parameters of the adsorption isotherm models.

Freundlich			Langmuir			Temkin		
K_{fdes} (L g^{-1})	$1/n_{des}$	R^2	K_L (L g^{-1})	q_m (m g^{-1})	R^2	K_T (L g^{-1})	B (mg g^{-1})	R^2
0.019	0.526	0.988	1.856	0.028	0.946	20.02	0.006	0.954

The cohesion energy E_{coh} was calculated in Equation (1) as follows:

$$E_{coh} = E_{crys} - 2\,E_{mol} \quad (1)$$

where E_{crys} is the energy of the optimized crystal structure, and E_{mol} is the energy of one 2,4-D molecule optimized in a 3D periodical cubic box with a constant volume of 12 × 12 × 12 Å^3, at the same computational level of the crystal. The E_{coh} was −65.7 kcal/mol.

A dimer of 2,4-D was extracted from this optimized crystal structure, and it was included in a 3D periodical box and optimized. The structure of this dimer is close to the conformation of the above gas phase dimer. The carboxylic groups are joined by both hydrogen bonds with *d*(C=O···HOC) = 1.593 Å, being longer and weaker than in the crystal lattice. The *E* of the hydrogen bonds of this dimer, with respect to the isolated molecules, was −15.8 kcal/mol, which is consistent with the previous values calculated above from isolated molecules. This indicates that the hydrogen bonds between the carboxylic groups is one of the main cohesion forces; however, additional intermolecular interactions contribute to the higher packing energy, that is, E_{coh}, for instance, electrostatic interactions between Cl and O atoms with H atoms, and all aromatic rings are in a parallel disposition with π-π interactions in an inter-ring distance of 3.579 Å (Figure 2). This interaction is enhanced by the presence of Cl atoms with their π electrons. Moreover, the C-Cl···π, C-H···π, C-H···O, and C-Cl···C interactions participated in the stabilization of packing in the crystal lattice, whereas the aromatic rings were on the opposite side of the isolated dimer.

2.1.4. Intercalation of 2,4-D into Montmorillonite

An isolated Na salt of 2,4-D was placed in the centre of a 3D periodical box of $20 \times 20 \times 20$ Å^3 (2,4-DNa) and optimized. The alkoxy chain adopted a perpendicular orientation with respect to the aromatic ring, and the d(CO···Na) distances were 2.22–2.24 Å. Similarly, a Ca salt of 2,4-D was created and optimized, $(2,4\text{-D})_2Ca$ (Figure 3a,b). Two possible configurations were considered with respect to the aromatic rings: (i) both rings at the opposite side (Figure 3a, *anti*) and (ii) both ring at the same sides (Figure 3b, *syn*). In both complexes, the carboxylic groups coordinated with Ca^{2+} by twisting 90° of each other, and both aromatic rings were not parallel with d(O···Ca) distances of 2.28–2.33 Å (Figure 3). The form *anti* was 0.58 kcal/mol more stable than the *syn* one.

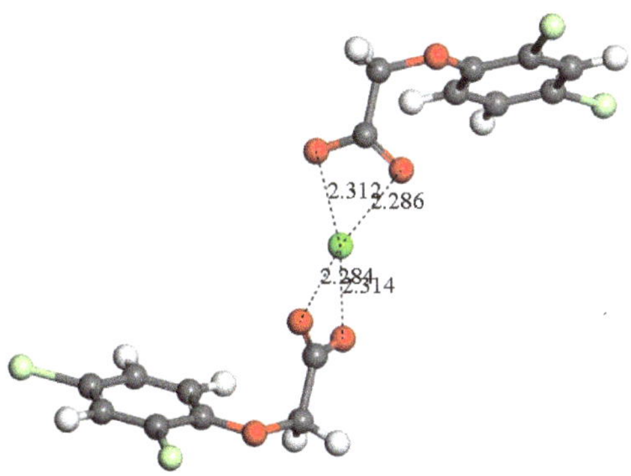

(**a**)

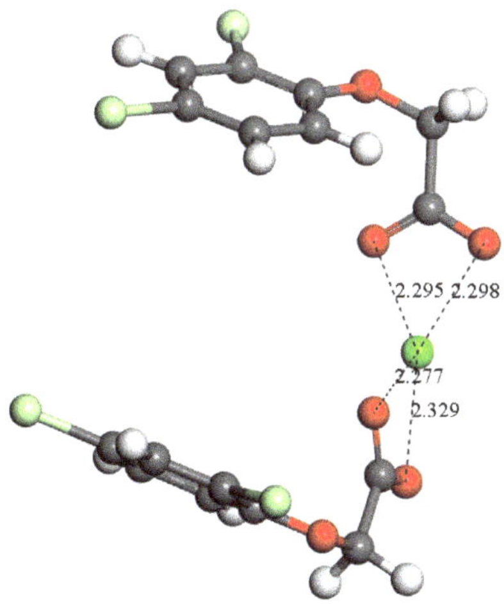

(**b**)

Figure 3. *Cont.*

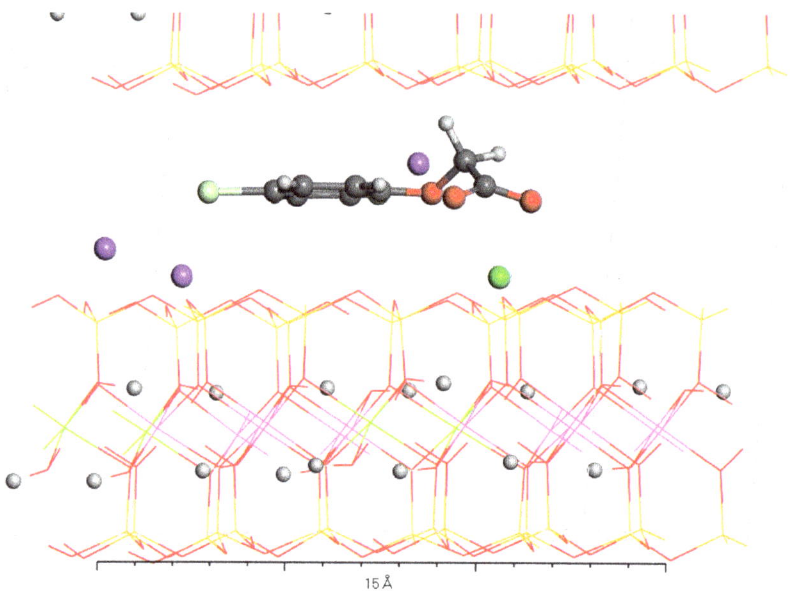

(**c**)

Figure 3. Optimized structure of the $(2,4\text{-}D)_2Ca$ salt, *anti* (**a**) and *syn* (**b**) configuration, and 2,4-D intercalated into montmorillonite (**c**). The H, Si, Al, Mg, Na, Ca, O, C, and Cl atoms are in white, yellow, pink, clear-green, fuchsia, green, red, grey, and pale-green colours, respectively. The H atoms and interlayer atoms are depicted as balls.

A montmorillonite (MNT) supercell with four interlayer Na^+ cations ($MNTNa_4$) was optimized at a constant volume with the $Dmol^3$ method. From this structure, another model of MNT supercell was created with two Na^+ and one Ca^{2+} cations ($MNTNa_2Ca$) and optimized. From $MNTNa_4$, a $2,4\text{-}DCa^+$ monocation was placed in the centre of the interlayer space of MNT exchanged by one Na^+ cation ($MNTNa_3Ca$-2,4-D) and optimized with $Dmol^3$ (Figure 3c). The aromatic ring was parallel to the mineral surface, the (001) plane. The organic molecule was more planar than out of MNT, where the alkoxy group was not perpendicular to the aromatic ring. The carboxylic O atoms are coordinating the Ca^{2+} cation with d(O···Ca) = 2.27–2.35 Å, and one of the carboxylic O atoms is also coordinating one Na^+ cation with d(O···Na) = 2.29 Å, d(O···Ca) = 2.35 Å. The methylene H atoms are oriented to the basal mineral surface O atoms (Figure 3c).

Then, considering the first and last steps of the intercalation process, the energy of this process can be calculated from Equation (2) as follows:

$$MNTNa_4 + (2,4\text{-}D)_2Ca = MNTNa_3Ca\text{-}2,4\text{-}D + 2,4\text{-}DNa \quad (2)$$

And the intercalation energy will be calculated from Equation (3) as follows:

$$E_{intercal} = E_{MNTNa3Ca\text{-}2,4\text{-}D} + E_{2,4\text{-}DNa} - E_{MNTNa4} - E_{(2,4\text{-}D)2Ca} \quad (3)$$

The intercalation energy of 2,4-D in MNT is −21.87 kcal/mol, similar to the adsorption energy of 2,4-D on the Fe oxide surface reported previously [22].

Depending on the experimental conditions, this process can also be considered by calculating Equation (4) as follows:

$$MNTNa_2Ca + 2{,}4\text{-}DNa = MNTNa_3Ca\text{-}2{,}4\text{-}D \tag{4}$$

And the intercalation energy will be calculated from Equation (4) as follows:

$$E_{\text{intercal}} = E_{\text{MNTNa3Ca-2,4-D}} - E_{\text{2,4-DNa}} - E_{\text{MNTNa2Ca}} \tag{5}$$

The intercalation energy for this process is −93.34 kcal/mol. Both processes indicate that the intercalation of 2,4-D in MNT is energetically favourable.

Applying the COSMO continuum solvation model to simulate an aqueous effect, the intercalation energy in an aqueous medium following Equation (4) is −49.57 kcal/mol. All these results predict that MNT and similar clay minerals can be strong candidates for trapping the 2,4-D pollutant and its residues in soils and in the whole Earth environment.

This process was experimentally studied by exploring the adsorption of 2,4-D on clay-rich soils. This process was performed at pH = 6.7, where the 2,4-D is as an anion. Moreover, we included the presence of Ca^{2+} in the solutions in order to reproduce a similar medium from our above intercalation calculations. The adsorption of 2,4-D increases with the initial concentration of herbicide (Figure 4). The adsorption isotherm curve is not linear, but rather type L. This indicates a physical and multilayer adsorption. With low-concentration solutions, the slope of this curve is very high, showing a high affinity of the herbicide and the adsorbent.

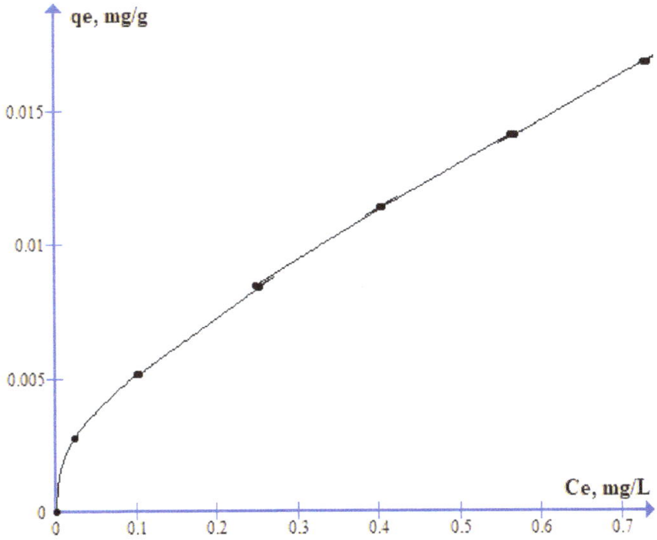

Figure 4. Adsorption isotherm of 2,4-D on clay-rich soil.

This isotherm was fitted to the Langmuir, Temkin, and Freundlich models for comparison purposes. The model with the best fitting is Freundlich (Table 3), indicating a multilayer adsorption. Therefore, our experimental results corroborated our theoretical calculations predicting the high adsorption capacity of clay minerals for intercalating 2,4-D.

2.2. Vibrational Analysis: IR Frequencies

Conformer I of the 2,4-D molecule had C_1 symmetry, and as expected, they presented (19 × 3–6) 51 normal vibrations modes. The structure of the dimer possessed a C_s symmetry and (38 × 3–6) 108 normal mode vibrations, with all of them being IR active. The frequencies of the isolated molecule, dimer, and crystal structure were calculated and assigned with CASTEP and compared with the experimental data (Table 4). The main differences between theoretical and experimental frequency values come from the lack of non-bonding intermolecular interactions in the isolated molecule, whereas most of the experimental spectra come from the crystal structure. Currently, most of experimental FT-IR analysis are performed on solid samples, and many theoretical works on molecules frequencies are compared with experimental values obtained on solid samples, not in the gas phase, where the intermolecular interactions are different. However, in this work, we showed the frequencies of 2,4-D in different states (Table 4), as an isolated molecule, as a dimer, and as a solid crystal phase and compared them with the experimental values obtained in the gas phase, non-polar dissolution, and solid state in contrast to previous works.

Table 4. Observed and calculated wavenumbers (cm^{-1}) and assignments for molecule, dimer, and crystal structures of 2,4-D.

Experimental			Molecule [b]	Dimer [b]	Crystal [b]	Assignments [c]
Gas [a]	dissolution	solid				
3580	3251 [d] (molecule)		3636			ν(OH)
3080		3075 [e]	3141 [f], 3136 [g], 3120 [h]	3140 [f], 3136 [g], 3121 [h]	3157 [f], 3138 [g], 3126 [h]	ν(CH) ring
2940		2979 [e]	3019	3023	3028	$\nu(CH_2)_{as}$ OCH_2
		2953 [e]	2955	2961	2980	$\nu(CH_2)_s$, OCH_2
	2495 [d] (dimer)	2700–2550 [i]		2782_s, 2628_{as}	2691_s, 2576_{as}	$\nu(OH)_s$, with H bond with C=O.
1820, 1760	1739 [d]	1733 [e],1735 [i], 1667 [j]	1777	1681	1660_s, 1611_{as}	ν(C=O)
1580		1595, 1580 [e]	1571, 1554	1569, 1555	1572,1560	ν(C=C) + δ(CH) ring
		1486–1479 [e]	1455, 1375	1373	1458, 1379	δ(CH) (ring)
				1431	1466, 1441	δ(OH)
		1480 [i]				$\delta(CH_2)_s + \delta(OH)$
1480		1449 [e], 1435 [k]	1419	1412	1420	$\delta(CH_2)_s$
1420			1288, 1253			δ(CH) (ring) + δ(OH) + $\delta(CH_2)$
		1431 [e]			1420	$\delta(CH_2)_{twist}$ OCH_2
1250		1393–1298 [e]	1355_s, 1224_{as}	1339_s, $(1274, 1226, 1212)_{as}$	1357_s, $(1282, 1238, 1212)_{as}$	$\delta(CH_2)_{wagging}$
1320		1235 [e], 1234 [i]	1234, 1136, 1086, 1032	1135, 1089	1246, 1144, 1092	δ(CH) (ring)
1280		1145 [e]	1104			$\delta(OH) + \delta(CH_2)$
		1105 [e], 1089 [k]	1068	1073	1069	ν(C-O) + δ(CH) ring + δ(OH)
			994	993	999	$\gamma(CH_2)$ OCH_2
				1037	1028	γ(OH)
880, 760		873–796 [e]	856, 779	908, 858, 777	926, 861, 779	γ(CH) ring
850		838 [e]			885	$\nu(C\text{-}CH_2)$

[a] Experimental data from gas phase were not assigned but extracted from the spectrum [25]. [b] Calculations based on DFT/PBE-plane waves. [c] *s*: symmetrical; *as*: asymmetrical. [d] Mainly a dimer in CCl_4 dissolution [19]. [e] Experimental data extracted from solid state spectrum in KBr [26] and interpreted in Ref. [16]. [f] Aromatic CH bond in ortho to both Cl substituents. [g] CH bond in meta position with respect to the alkoxy substituent. [h] CH bond in ortho position with respect to the alkoxy group. [i] From Ref. [21] in solid state extracted from spectrum. [j] From Ref. [27] in solid state. [k] From Ref. [28].

2.2.1. Region 4000–1900 cm^{-1}

This region is specific to the ν(OH) and ν(CH) stretching vibration modes. The calculated frequency for the ν(OH) stretching in the 2,4-D molecule was 3636 cm^{-1}, which is related with the experimental value of 2,4-D in the gas phase at around 3580 cm^{-1} [25]. However, the range of frequency decreased to 3251 when the 2,4-D was in non-polar dissolution due to various possible intermolecular interactions [19]. These interactions were confirmed with the band at 2495 cm^{-1} as the result of the dimerization process of 2,4-D [19]. Our calculations corroborated these assignments with bands at 2782–2628 cm^{-1} in the dimer and 2691–2576 cm^{-1} in the crystal structure, where hydrogen bonds were formed between the carboxylic H and O atoms. In the crystal structure, the intermolecular interactions altered the frequencies of some vibration modes. The numerous bands and the broadening of some of them in the frequency range of 3200–2500 cm^{-1} of the IR spectrum of 2,4-D [26–28] hint at the clustering of units in forming dimers. In other words, the formation of clusters was very likely. So, a bathochromic shift was consequently produced with respect to the isolated molecule frequencies.

In the ν(CH) mode bands, our calculations can distinguish the different H atoms of the benzenic ring. The H atom between both Cl atoms appeared at higher frequencies (3157–3140 cm^{-1}) than the H atom in the meta position with respect to the alkoxy group (3138–3136 cm^{-1}), probably due to the electric field of both Cl atoms. In the crystal structure, this band appeared at higher frequencies due to the additional electric field of the vicinal carbonyl groups in the crystal lattice (3157 cm^{-1}). The lowest frequency of this mode corresponded to the H atom in the ortho position with respect to the alkoxy group (3120–3126 cm^{-1}) (Table 4). These vibration modes showed similar frequencies in isolated molecule, dimer, and crystal states since these bonds did not participate in the intermolecular interactions.

The ν(C–H) modes of the CH_2 groups were assigned to the bands at 3028–3019 and 2980–2955 cm^{-1} to antisymmetric and symmetric modes, respectively. In the crystal structure, these modes appeared at higher frequencies due to interactions with the O atoms of the vicinal molecules.

2.2.2. Region 1800–1200 cm^{-1}

In this region, the ν(C=O) mode appeared at lower frequencies in the crystal structure (1660–1611 cm^{-1}) than the isolated molecule (1777 cm^{-1}) due to the hydrogen bonds between this group and the OH groups of the vicinal molecules in the crystal lattice. The same was observed in the dimer (1681 cm^{-1}). In this dimer formation, an eight-atom cycle was formed, and the stretching mode also implied a deformation of this cycle. This effect can be experimentally observed in this mode, which appears at 1810–1760 cm^{-1} in the gas phase, at 1739 cm^{-1} as a dimer in dissolution, and at 1733–1667 cm^{-1} in the solid state. The calculated vibrations of both groups are coupled in the crystal lattice, distinguishing the symmetrical and anti-symmetrical modes (Table 4).

In the molecule, the C=C stretching modes were at 1572–1554 cm^{-1}, which were coupled to the other vibration modes, such as to the ring deformations in the plane and δ(CH) in-plane bending. No significant differences were observed between the different phases. The δ(OH) in-plane mode appeared at higher frequencies in the crystal structure than the dimer, and this one was higher than in the isolated molecule, following the sequence: 1466–1441 > 1431 > 1288–1253 cm^{-1}.

2.3. *Raman Spectrum*

The Raman spectrum of the crystal structure of 2,4-D optimized with DFT was calculated (Figure 5). Most of the bands are consistent with the experimental spectrum [29], with the only difference being in the intensity of the band at 2570 cm^{-1}, probably due to different conditions being used in the experiment and calculations and the small theoretical model not allowing disorders.

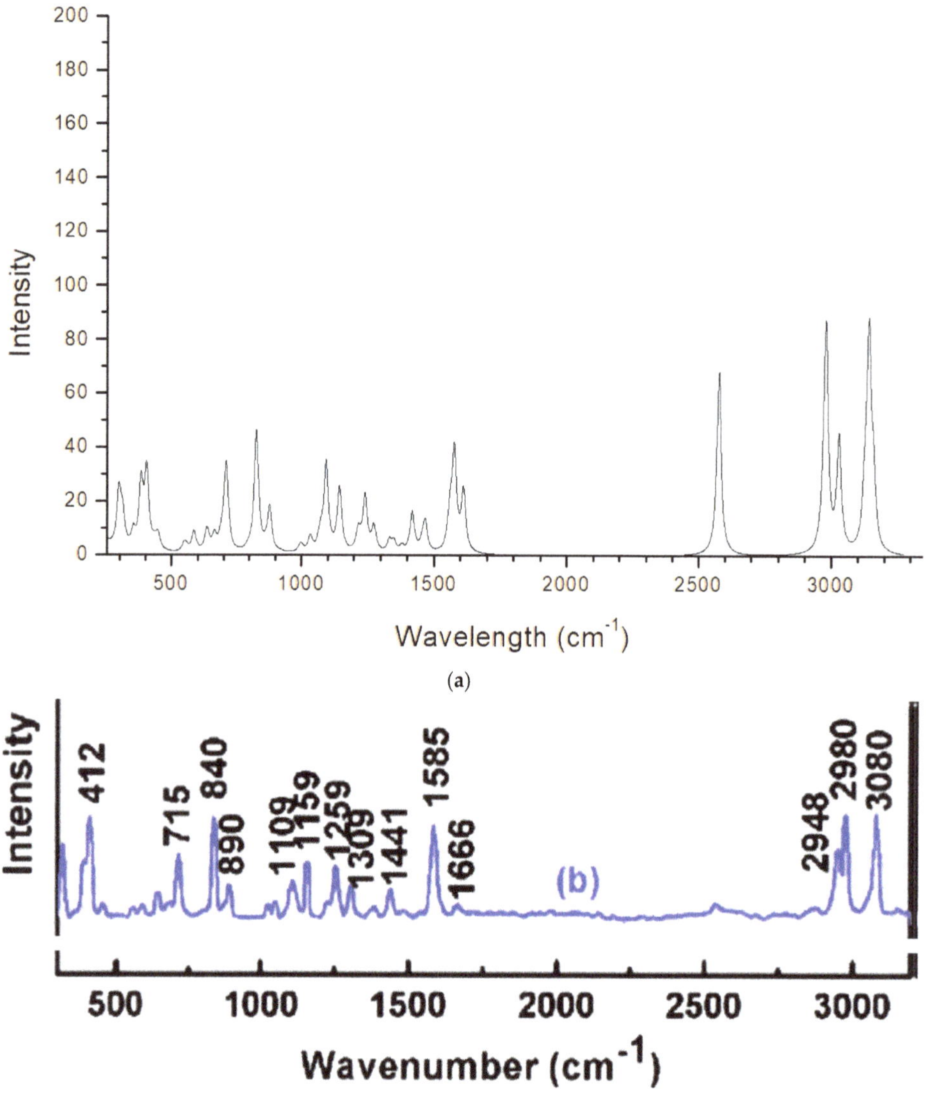

Figure 5. Raman spectrum calculated (CASTEP) for the optimized 2,4-D crystal structure (**a**) and experimental from Ref. [29] (**b**).

2.4. Experimental NMR Analysis

In Figure 6 and Table 5, the ^{1}H chemical shifts (δ) are depicted as being consistent with previous results [21,27]. These shifts were assigned to the H atoms of 2,4-D based on multiplicity and chemical shifts. The chemical shifts from 7.09 to 7.47 ppm can be assigned to the phenyl H atoms (Figure 6b). The H(11) (numbering from Figure 1) appeared at 7.47 ppm due the deshielding effect of both Cl substituents at β position. The δ of H(13) appeared at 7.29 ppm due to the partial deshielding effect of one Cl atom. This proton showed a doublet of doublets due to a double coupling, one with the H(11) at meta position with *J* = 2.5 Hz, and other with H(10) at ortho position with higher coupling constant (*J*). The proton H(10) appeared at 7.09 ppm as a doublet with a coupling with the vicinal H atom at ortho position. The highest intensity signal at 4.87 ppm was assigned to CH_2 (Figure 6a). This δ value could come from the deshielding from the ether O(15) atom and the carboxylic group. Finally, the broad and very low intensity chemical shift at 9.8 ppm can be assigned to the acidic H atom.

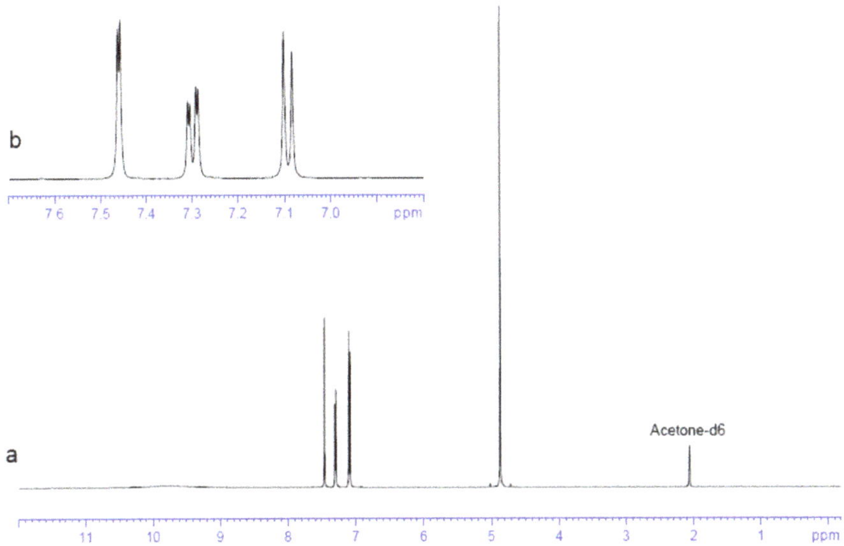

Figure 6. ^{1}H NMR spectrum of 2,4-D in solution (**a**) and from 7.0 to 7.6 ppm (**b**).

Table 5. ^{1}H and ^{13}C NMR chemical shifts of 2,4-D in acetone-6D solution.

C number (Figure 1)	δ ^{1}H (ppm)	δ ^{13}C (ppm)
3 (C=O)		168.7, 170.5 [b]
1 (CH_2)	4.87, 4.83 [a]	65.3, 68.6 [b]
4 (O-C_{arom})		152.9, 154.2 [b]
6 (Cl-C_{arom})		123.3, 122.2 [b]
8 (H-C_{arom})	7.47, 7.48 [b]	129.6, 129.2 [b]
12 (Cl-C_{arom})		125.8, 127.9 [b]
9 (H-C_{arom})	7.29, 7.27 [b]	127.7, 127.1 [b]
7 (H-C_{arom})	7.09, 6.86 [b]	114.8, 115.6 [b]

[a] Experimental data [21], where the aromatic H were not identified (7.58–7.07 ppm). [b] From Ref. [27].

The ^{13}C CPMAS spectra of 2,4-D in dissolution (Figure 7a) and solid state (Figure 7b,c) were recorded, and the chemical shifts are presented in Table 5. The C atoms (1), (2), (5), (7), and (8) were assigned using chemical shifts and proton–carbon correlations on the HSQC 2D map. The three quaternary carbons (C(4), C(6), and C(12)) are discriminated by long range couplings on Heteronuclear Multiple-Bond Correlation (HMBC) experiment. The carboxylic C atom was found at the largest chemical shift (168.7 ppm), and the methyl C atom at the lowest δ (65.3 ppm). All aromatic C atoms appeared in the range of 114.8–152.9 ppm, yielding C(4) with the highest δ, due to the deshielding effect of the substituent (–O-CH_2-COOH). C(8) also appeared with a high δ because of the deshielding of the two Cl atoms in β.

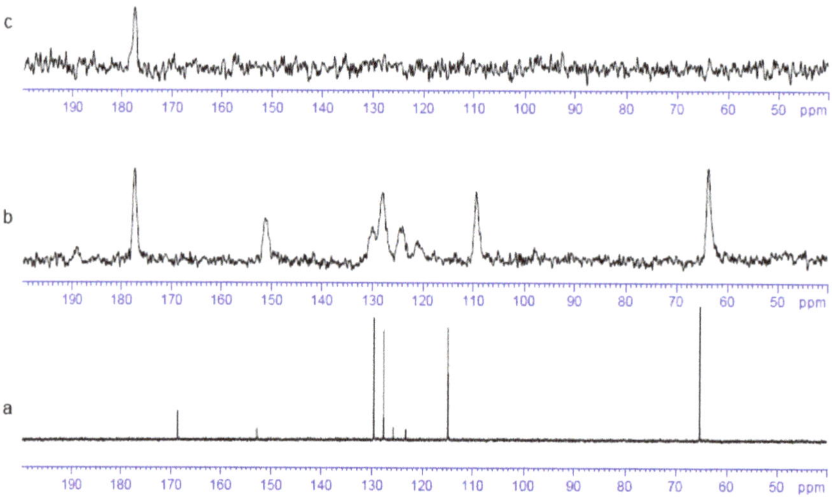

Figure 7. ^{13}C NMR spectra in solution: liquid phase (**a**); crystal phase (**b**); and high-power ^{13}C decoupling (HPDEC) of the crystal phase of 2,4-D (**c**).

In solid state, the ^{13}C NMR spectrum is similar to that in dissolution with only small differences in intensities and chemical shifts. The broadening infrastructure of the signals was due to different local environments in the crystal lattice. The chemical shift of the aromatic C atom C(7) (in meta position with with respect to the Cl atoms) and the CH_2 group in solid state was slightly lower than in dissolution due to the different local environments in each case. Similar differences were observed between the dimer in solid state and the isolated dimer where the aromatic rings have different orientations and π-π interactions between the aromatic rings in a crystal state (Figures 1 and 2). The δ of the carboxylic C atom in the solid state is higher than in dissolution due to the effects of the intermolecular hydrogen bonds, indicating that they are stronger than in dissolution. Considering the similarity of spectra at both phases (Figure 7a,b), the 2,4-D molecule forms dimers in the crystal lattice. When the HPDEC spectrum was recorded, all chemical shifts disappeared, with the exception of the carboxylic C chemical shift, indicating that the relaxation time of this band strongly increased. This acceleration can be the result of a double proton transfer into the dimeric form. This led to a very rapid proton exchange between both monomers as predicted using the NMR spectra. The carbonyl group of the solvent (acetone) can likely solvate the carboxylic groups, decreasing the strength of the intermolecular H bonds of the dimer in the solution. This can explain the chemical shift difference in the carboxylic C atoms between the solid state and when in the solution.

Moreover, the high dissociation energy of the dimer form of 2,4-D would indicate the solvation could not be strong enough to destroy both hydrogen bonds in the dimer.

3. Methods

3.1. Computational Details

The molecular geometry at the gas phase was optimized starting from the geometry of the molecule extracted from the crystal structure [20]. All the calculations were performed within DFT. The isolated molecule and the dimer structure were optimized by using the hybrid functional B3LYP [30,31] and the 6-311++G(d,p) basis set [32–34]. For the vibrational studies of the monomer and dimer of 2,4-D, we established calculations for the geometries and harmonic frequencies in cartesian coordinates of the most stable conformers to confirm the true minimum energy critical point. The systems studied in this work are neutral molecules. The basis set superposition errors (BSSEs) in the total energies of the cyclic dimer were corrected using the counterpoise method [35]. All computations in the gas phase were carried out with the GAUSSIAN09 package [36]. The geometries were represented with the GAUSSVIEW program [37].

The crystal structure was taken from experimental crystallographic studies [20,24] (CCDC 695525). This structure was transformed to a P1 symmetry form, and the unit cell was optimized at the DFT level by using the CASTEP code with 3D periodical boundary conditions based on plane waves [38,39]. The generalized gradient approximation (GGA), Perdew–Burke–Ernzerhof (PBE) [40] correlation exchange functional, and ultrasoft pseudopotentials were used [41]; the Tkatchenko corrections for long-range dispersion interactions [42] and cutoff energies of 490 and 630 eV were also used. For the spectroscopic calculations, the structure was optimized with norm-conserving pseudopotentials with the Koelling–Harmon relativistic treatment using a cutoff energy of 990 eV and $2 \times 2 \times 2$ k points of the reciprocal Brillouin space. The vibrational analysis was calculated by atomic finite displacements with harmonic approximation. The same procedure was applied to the isolated dimer and molecule embedded in 3D periodical boxes for comparison purposes. The Raman spectrum of the crystal structure was simulated for a laser light of 514.5 nm at 10 K with a smoothing Lorentzian smearing of 20 cm^{-1}.

An MNT crystal structure model was chosen with a chemical composition of the unit cell as $NaSi_8(Al_3Mg)O_{20}(OH)_4$ [43]. The octahedral Mg^{2+} cations were placed considering previous studies of cation ordering in smectites [44]. For the intercalation calculations, a supercell of $2 \times 2 \times 1$ was created with 3D periodical boundary conditions, yielding $Na_4Si_{32}(Al_{12}Mg_4)O_{80}(OH)_{16}$.

The intercalation of 2,4-D into montmorillonite was performed with $Dmol^3$ code [45] based on LCAO (Linear Combination of Atomic Orbitals) with a double-ξ basis set with polarization functions and using GGA and PBE correlation exchange functional, applying periodical boundary conditions. The Tkatchenko method was also used for dispersion corrections. DFT-based semi-core pseudopotentials (DSPPs), including relativistic correction into the core [46], were also used. The frequencies of the main vibration modes were calculated from the Hessian matrix obtained from the harmonic approximation with finite atomic displacements [47]. The effect of a solvent environment (water) was simulated by using the COSMO continuum solvation model [48] with a dielectric constant of 78.54.

3.2. ^1H and ^{13}C NMR Spectra

The ^1H and ^{13}C NMR spectra of analytical standards of 2,4-D (Sigma-Aldrich, St. Louis, MO, USA) were performed by using a Bruker Advance III spectrometer (Billerica, MA, USA), working at 500 MHz for protons. ^1H spectrum was recorded in the dissolution of acetone-D6, and the ^{13}C spectra was recorded in acetone dissolution and solid state by using 2D-NMR Correlation Spectroscopy (COSY), Heteronuclear Multiple-Bond Correlation (HMBC), and Heteronuclear Single-Quantum Coherence Spectroscopy (HSQC). The spectra were performed at Cross-Polarization and Magic Angle Spinning (CPMAS) and high-power ^{13}C decoupling (HPDEC) under magic angle spinning at a speed of 11 KHz.

3.3. Experimental Sorption of 2,4-D on Clay-Rich Soils

Clay-rich soil samples taken from Mercedes, an Argentinian location, was chosen for our sorption experiments. These samples were homogenized, dried, and sieved at a mesh diameter of 2 mm. 2,4-D was purchased from Sigma-Aldrich (USA).

The batch equilibrium method was used in our adsorption essays. The adsorption isotherms were performed mixing 1 g of soil with 30 mL of aqueous solutions of 2,4-D at 0.10, 0.25, 0.50, 0.75, 1.0, and 1.255 mg L^{-1}. All aqueous solutions had $CaCl_2$ at 0.01 M and pH = 6.72. Blank samples without soil were used simultaneously under the same conditions in order to check the loss of 2,4-D during contact time. The mixtures were stirred for 24 h at 298 K, then centrifugated for 15 min at 3000 rpm, and the supernatant was analysed by HPLC. The averaged results from the triplicate experiments were considered. The experimental data were fitted to the Freundlich, Langmuir, and Temkin models of the adsorption isotherm [49] using the least-squares method.

4. Conclusions

In the present work, a structural analysis of the isolated molecule and dimer of 2,4-D acid was carried out at the DFT/B3LYP level of theory, employing a 6-311++G(d,p) basis set, finding one particularly stable conformer for the isolated 2,4-D molecule and one dimer for this conformer. Moreover, the crystal structure of 2,4-D was also calculated at the DFT level, which agrees with the experimental crystal data. The theoretical vibrational spectra of the most stable conformer of the isolated molecule, dimer, and crystal structure are also calculated at different theoretical levels. Experimental and theoretical spectra were matched for exhaustive harmonic vibrational analysis, achieving a complete assignment of the bands. ^{1}H and ^{13}C NMR spectra were experimentally determined. For the isolated molecule, theoretical calculations predicted that non-planar conformation is predominant, which has a high dipolar moment, that is preferred in polar solvents. This molecule forms a dimer, which is present in the crystal structure, reproducing the X-ray diffraction data. From the experiments in crystals, only one conformer appeared, which agreed with our calculated conformer. The broadening of some vibrational bands of the IR spectrum strongly suggests the existence of molecular clustering in the crystal. Our calculations pointed out that the intermolecular hydrogen bonds between the carboxylic groups in crystal structure and dimer are strong. The ^{13}C NMR results showed that the dimer structure exists in the solid crystal structure. The high cohesion energy of the crystal structure can justify the stability of the dimer.

Moreover, our calculations showed that the intercalation of 2,4-D into the confined interlayer space of montmorillonite was energetically favourable. This result was experimentally confirmed with the isotherm of the adsorption and desorption of 2,4-D on clay-rich soils. Hence, we predict that this clay mineral can be a good medium for trapping the 2,4-D pollutant.

Supplementary Materials: The following supporting information can be downloaded at: https://www.mdpi.com/article/10.3390/molecules30020367/s1, Figure S1: Conformers of the dimer of 2,4-dichlorophenoxyacetic acid; Table S1: Geometrical parameters of the optimized structures for the main 2,4-D conformers and compared with experimental values; Table S2: Geometric features of the optimized structures for the main 2,4-D dimer conformers and compared with experimental values.

Author Contributions: A.H.-L.: Conceptualization, investigation, writing—editing and review, N.L.J.: Conceptualization, investigation, writing—editing and review; J.M.R.: Investigation; C.I.S.-D.:

Conceptualization, investigation, writing—editing and review; A.G.: Conceptualization, investigation, NMR spectra. All authors have read and agreed to the published version of the manuscript.

Funding: This research was funded by Andalusian project P18-RT-3786, the Spanish MINECO project FIS2016-77692-C2-2-P, and Spanish Science and Innovation Ministry (project PID2022-137603OB-I00).

Institutional Review Board Statement: Not applicable.

Informed Consent Statement: Not applicable.

Data Availability Statement: The original contributions presented in this study are included in the article, and further inquiries can be directed to the corresponding authors.

Acknowledgments: The authors are thankful to the Computational Centres of the University of Granada, CESGA, and CSIC for computational facilities.

Conflicts of Interest: The authors declare no conflicts of interest.

References

1. Eriksen, S.H.; Jensen, B.; Schneider, I.; Kaasgaard, S.; Olsen, J. Uptake of phenoxyacetic acid by Penicilliumchrysogenum. *Appl. Microbiol. Biotech.* **1995**, *42*, 945–950. [CrossRef] [PubMed]
2. Santagostino, A.; Leone, M.P.; Maci, R.; Casale, A.; Marabini, L. Effects of Phenoxyacetic Acid Herbicides on Chicken Embryo Liver Drug Metabolizing Enzymes. *Pharmacol. Toxicol.* **1991**, *68*, 110–114. [CrossRef]
3. Arnold, E.K.; Beasley, V.R. The pharmacokinetics of chlorinated phenoxy acid herbicides: A literature review. *Vet. Hum. Toxicol.* **1989**, *31*, 121–125. [PubMed]
4. Gehring, P.J.; Betso, J.E. Phenoxy Acids: Effects and Fate in Mammals. In *Ecological Bulletins 27, Chlorinated Phenoxy Acids and Their Dioxins*; Ramel, C., Ed.; Oikos Editorial Office: Stockholm, Sweden, 1978; pp. 122–133.
5. Munro, I.C.; Carlo, G.L.; Orr, J.C.; Sund, K.G.; Wilson, R.M.; Kennelpohl, E.; Lunch, B.S.; Jablinske, M.; Lee, N.L. A comprehensive, integrated review and evaluation of the scientific evidence relating to the safety of the herbicide 2,4-D. *J. Am. Coll. Toxicol.* **1992**, *11*, 559–664. [CrossRef]
6. Cheah, U.B.; Ooi, G.G. Potential approaches towards degradation of 2,4-D. In Proceedings of the National Seminar and Workshop on Rice Field Weed Management, Penang, Malaysia, 7–9 June 1988; pp. 237–243.
7. Turker, L. AM1 Treatment of Some Phenoxyacetic Acid Herbicides. *Turk. J. Biol.* **2000**, *24*, 291–298.
8. Gruzdyev, G.S.; Zinchenko, V.A.; Kalinin, V.A.; Slovtsov, R.I. *The Chemical Protection of Plants*; Mir Publishers: Moscow, Russia, 1983.
9. Cremlyn, R. *Pesticides: Preparation and Mode of Action*; John Wiley: New York, NY, USA, 1979.
10. United States Environmental Protection Agency. National Primary Drinking Water Regulation. 2015. Available online: https://www.epa.gov/ground-water-and-drinking-water/national-primary-drinking-water-regulations (accessed on 20 May 2020).
11. Muszynski, P.; Brodowska, M.S.; Paszko, T. Occurrence and transformation of phenoxy acids in aquatic environment and photochemical methods of their removal: A review. *Environ. Sci. Pollut. Res.* **2019**, *27*, 1276–1293. [CrossRef]
12. Loos, R.; Locoro, G.; Comero, S.; Contini, S.; Schewesig, D.; Werres, F.; Balsaa, P.; Gans, O.; Weiss, S.; Blaha, L.; et al. Pan–European survey on the occurrence of selected polar organic persistent pollutants in ground water. *Water Res.* **2010**, *44*, 4115–4126. [CrossRef]
13. Loos, R.; Gawlik, B.M.; Locoro, G.; Rimaviciute, E.; Contini, S.; Bidoqlio, G. EU-wide survey of polar organic persistent pollutants in European river waters. *Environ. Pollut.* **2009**, *157*, 561–568. [CrossRef] [PubMed]
14. Romero, J.M.; Jorge, N.L.; Grand, A.; Hernández-Laguna, A. Hydrolysis reaction of 2,4-dichlorophenoxyacetic acid. A kinetic and computational study. *Chem. Phys. Lett.* **2015**, *639*, 57–62. [CrossRef]
15. Andino, M.G.; Profeta, M.I.; Romero, J.M.; Jorge, N.L.; Castro, E.A. Theoretical Studies on the Structure and Spectroscopic Properties of 2,4-D (2,4-Diclorofenoxiacetic Acid). In *Methodologies and Applications for Chemoinformatics and Chemical Engineering*; IGI Global: Hershey, PA, USA, 2013; pp. 180–190.
16. Badawi, H.M. Molecular structure and vibrational assignments of 2,4-dichlorophenoxyacetic acid herbicide. *Spectrochim. Acta Part A Mol. Biomol. Spectrosc.* **2010**, *77*, 24–27. [CrossRef]
17. Sundaraganesan, N.; Meganathan, C.; Karthikeyan, B. FT-IR, FT-Raman spectra and quantum chemical calculations of some chloro substituted phenoxy acetic acids. *Spectrochim. Acta Part A Mol. Biomol. Spectrosc.* **2008**, *70*, 430–438. [CrossRef]

18. Karthikeyan, B.; Saravanan, B. Raman spectral analysis of phenoxyacetic acid and some chloro substituted phenoxyacetic acids. *Spectrochim. Acta Part A Mol. Biomol. Spectrosc.* **2006**, *63*, 619–623. [CrossRef] [PubMed]
19. Molinelli, A.; O'Mahony, J.; Nolan, K.; Smyth, M.R.; Jakusch, M.; Mizaikoff, B. Analyzing the Mechanisms of Selectivity in Biomimetic Self-Assemblies via IR and NMR Spectroscopy of Prepolymerization Solutions and Molecular Dynamics Simulations. *Anal. Chem.* **2005**, *77*, 5196–5204. [CrossRef]
20. Smith, G.; Kennard, C.H.L.; White, A.H. Herbicides. Part I. Crystal structure of 2,4-D (2,4-dichlorophenoxyacetic acid). *J. Chem. Soc. Perkin Trans.* **1976**, *2*, 791–792. [CrossRef]
21. Mehmood, Z.; Williamson, M.P.; Kelly, D.E.; Kelly, S.L. Human cytochrome P450 3A4 is involved in the biotransformation of the herbicide 2,4-dichlorophenoxyacetic acid. *Environ. Toxicol. Pharmacol.* **1996**, *2*, 397–401. [CrossRef] [PubMed]
22. Aquino, A.J.A.; Tunega, D.; Haberhauer, G.; Gerzabek, M.H.; Lischka, H. Quantum Chemical Adsorption Studies on the (110) Surface of the Mineral Goethite. *J. Phys. Chem. C* **2007**, *111*, 877–885. [CrossRef]
23. Meunier, A. *Clays*; Springer: Berlin/Heidelberg, Germany, 2005.
24. Sieron, L.; Kobylecka, J.; Turek, A. Crystal Packing and Supramolecular Motifs in Four Phenoxyalkanoic Acid Herbicides—Low-Temperature Redeterminations. *Org. Chem. Int.* **2011**, *4*, 608165. [CrossRef]
25. NIST Standard Reference Data Program Collection. U.S. Secretary of Commerce, USA. 2018. Available online: https://www.nist.gov/srd (accessed on 20 May 2020).
26. SDBSWeb. National Institute of Advanced Industrial Science and Technology. 2010. Available online: https://sdbs.db.aist.go.jp/ (accessed on 20 May 2020).
27. Naz, N.; Sirajuddin, M.; Harder, A.; Mustansar-Abbas, S.; Ali, S.; Wadood, A.; Ghufran, M.; Rehman, G.; Mirza, B. Synthesis, characterization, biological screenings and molecular docking study of Organotin(IV) derivatives of 2,4-dichlorophenoxyacetic acid. *J. Mol. Struct.* **2019**, *1179*, 662–671. [CrossRef]
28. Trivedi, N.S.; Kharkar, R.A.; Mandavgane, S.A. Uttilization of cotton plant ash and char for removal 4-dichlorophenoxyacetic acid. *Resour. Effic. Technol.* **2016**, *2*, S39–S46.
29. Jia, J.L.; Jin, X.Y.; Liu, Q.L.; Liang, W.L.; Lin, M.S.; Xu, H.H. Preparation, Characterization and Intracellular Imaging of 2,4-Dichlorophenoxyacetic Acid Conjugated Gold Nanorods. *J. Nanosci. Nanotechnol.* **2016**, *16*, 4936–4942. [CrossRef] [PubMed]
30. Becke, A.D. Density-functional thermochemistry. III. The role of exact exchange. *J. Chem. Phys.* **1993**, *98*, 5648–5652. [CrossRef]
31. Lee, C.; Yang, W.; Parr, R.G. Development of the Colle-Salvetti correlation-energy formula into a functional of the electron density. *Phys. Rev. B* **1988**, *37*, 785–789. [CrossRef] [PubMed]
32. Vosko, S.H.; Wilk, L.; Nusair, M. Accurate spin dependent electron liquid correlation energies for local spin density calculation: A critical analysis. *Can. J. Phys.* **1980**, *58*, 1200–1211. [CrossRef]
33. McLean, A.D.; Chandler, G.S. Contracted Gaussian basis sets for molecular calculations. I. Second row atoms, Z = 11−18. *J. Chem. Phys.* **1980**, *72*, 5639–5648. [CrossRef]
34. Krishnan, R.; Binkley, J.S.; Seeger, R.; Pople, J.A. Self-consistent molecular orbital methods. XX. A basis set for correlated wave functions. *J. Chem. Phys.* **1980**, *72*, 650–654. [CrossRef]
35. Boys, S.F.; Bernardi, F. The calculation of small molecular interactions by the differences of separate total energies. Some procedures with reduced errors. *Mol. Phys.* **1970**, *19*, 553–566. [CrossRef]
36. Frisch, M.J.; Truck, G.W.; Schlegel, H.B.; Scuseria, G.E.; Robb, M.A.; Cheeseman, J.R.; Scalmani, G.; Barone, V.; Mennucci, B.; Petersson, G.A.; et al. *Gaussian 09, Revision B.01*; Gaussian, Inc.: Wallingford, CT, USA, 2009.
37. Frisch, A.; Nielson, A.B.; Holder, A.J. *GAUSSVIEW User Manual*; Gaussian. Inc.: Pittsburgh, PA, USA, 2000.
38. Clark, S.J.; Segall, M.D.; Pickard, C.J.; Hasnip, P.J.; Probert, M.J.; Refson, K.; Payne, M.C. First principles methods using CASTEP. *Zeits. Kristall.* **2005**, *220*, 567–570. [CrossRef]
39. Segall, M.D.; Lindan, P.J.D.; Probert, M.J.; Pickard, C.J.; Hasnip, P.J.; Clark, S.J.; Payne, M.C. First-principles simulation: Ideas, illustrations and the CASTEP code. *J. Phys. Condens. Matter* **2002**, *14*, 2717–2744. [CrossRef]
40. Perdew, J.P.; Burke, K.; Ernzerhof, M. Generalized gradient approximation made simple. *Phys. Rev. Lett.* **1996**, *77*, 3865–3868. [CrossRef]
41. Vanderbilt, D. Soft Self-Consistent Pseudopotentials in a Generalized Eigenvalue Formalism. *Phys. Rev. B* **1990**, *41*, 7892–7895. [CrossRef]
42. Tkatchenko, A.; Scheffler, M. Accurate Molecular Van Der Waals Interactions from Ground-State Electron Density and Free-Atom Reference Data. *Phys. Rev. Lett.* **2009**, *102*, 073005. [CrossRef]
43. Sainz-Díaz, C.I.; Palin, E.J.; Hernández-Laguna, A.; Dove, M.T. Octahedral cation ordering of illite and smectite. Theoretical exchange potential determination and Monte Carlo simulations. *Phys. Chem. Miner.* **2003**, *30*, 382–392. [CrossRef]

44. Ortega-Castro, J.; Hernández-Haro, N.; Dove, M.T.; Hernández-Laguna, A.; Sainz-Díaz, C.I. Density functional theory and Monte Carlo study of octahedral cation ordering of Al/Fe/Mg cations in dioctahedral 2: 1 phyllosilicates. *Am. Mineral.* **2010**, *95*, 209–220. [CrossRef]
45. Baker, J.; Kessi, A.; Delley, B.K. The generation and use of delocalized internal coordinates in geometry optimization. *J. Chem. Phys.* **1996**, *105*, 192–212. [CrossRef]
46. Delley, B.K. Hardness conserving semilocal pseudopotentials. *Phys. Rev. B* **2002**, *66*, 155125. [CrossRef]
47. Biovia. *Materials Studio v2020*; Dassault Inc.: Waltham, MA, USA, 2020.
48. Delley, B. The conductor-like screening model for polymers and surfaces. *Mol. Simul.* **2006**, *32*, 117–123. [CrossRef]
49. Jorge, N.L.; Garrafa, M.V.; Romero, J.M.; Jorge, M.J.; Jorge, L.C.; Delfino, M.R.; Meruvia-Rojas, Y.V.; Hernández-Laguna, A.; Sainz-Díaz, C.I. Adsorption of Ciprofloxacin on Clay Minerals in Argentinian Santa Rosa-Corrientes Soils. *Molecules* **2024**, *29*, 1760. [CrossRef]

molecules

MDPI

Article

Mechanistic Pathways in Cyanide-Mediated Benzoin Condensation: A Comprehensive Electron Localisation Function (ELF) and Catastrophe Theory Analysis of the Umpolung Reaction

Michal Michalski [1] and Slawomir Berski [2,*]

[1] Centre of New Technologies, University of Warsaw, 02-097 Warsaw, Poland; m.michalski@cent.uw.edu.pl
[2] Faculty of Chemistry, University of Wroclaw, 50-383 Wroclaw, Poland
* Correspondence: slawomir.berski@uwr.edu.pl

Abstract: This research investigates the mechanism of the cyanide-type umpolung reaction in benzoin condensation using topological analysis of ELF and catastrophe theory. The study achieves a comprehensive understanding of the evolution of chemical bonds and non-bonding electron density in the reaction of benzaldehyde and cyanide ions. The results reveal that the reaction proceeds through five transition state structures, with the formation of Lapworth's cyanohydrin being the rate-determining step. The study characterises topological catastrophes in the evolution of the ELF field and provides a detailed description of the evolution of electron density in the mechanism of the reaction. An in-depth analysis of ELF catastrophes confirms the well-established Lapworth mechanism.

Keywords: umpolung reaction; benzoin condensation; BET; ELF

1. Introduction

In 1903, Lapworth provided a pioneering account of the mechanism underlying the cyanide-catalysed benzoin condensation [1], which has since been widely studied in the field of organic chemistry [2–6]. This reaction belongs to the class of umpolung-type reactions, where the polarity of the functional group is reversed due to its chemical modification [7]. The proposed mechanism involves the formation of a cyanohydrin intermediate, which serves as a crucial metastable state in facilitating the reaction. The benzoin condensation is a dimerisation reaction involving two aldehyde molecules, where one aldehyde donates a hydrogen atom and the other accepts it. Notably, this reaction proceeds without generating a small byproduct molecule. Wiberg in his pioneering investigation observed inconsistencies with the classic Lapworth mechanism when using a 66% ethanol–water solvent, suggesting deviations from the expected mechanism [8,9]. Breslow and colleagues examined the influence of salts on the hydrophobic acceleration and anti-hydrophobic effects, finding that the reaction rate in dimethyl sulfoxide (DMSO) is approximately 50 times greater than in water without additives and roughly 10,000 times greater than in ethanol [10,11]. While protonic solvents like methanol, ethanol, and water are commonly employed, DMSO stands out as an aprotic solvent that does not promote proton transfer or intermediate formation, potentially explaining the significantly enhanced reaction rate observed in this medium [12]. This highlights the need for further investigation to gain a deeper understanding of the umpolung mechanism.

In this work, a theoretical study of the Lapworth mechanism of cyanide-catalysed benzoin condensation in the gas phase was conducted using Density Functional Theory (DFT) and Bonding Evolution Theory (BET) [13] methods. The BET is based on the topological analysis of the electron localisation function (ELF) developed by Silvi and Savin [14,15] and on Thom's catastrophe theory [16]. This method has been employed in previous studies to analyse the mechanisms of organic reactions [17–21] and offers an attractive alternative to Hilbert space descriptions. The electronic structure of the reaction can be characterised through ELF topology, with ELF regions, or basins, classified into core basins (C(X)) and valence basins (V(A,B,...)), each associated with a local maximum or attractor in the ELF field. Two primary types of valence basins can be identified: monosynaptic basins, V(A), representing lone pairs in Lewis structures, and disynaptic basins, V(A,B), which denote covalent bonds between atoms A and B [22,23]. As the reaction progresses, structural changes in the ELF field can be observed, with attractors and critical points (CPs) undergoing abrupt transitions called catastrophes. Three main types of topological catastrophes—fold, cusp, and elliptic umbilic—describe unique CP transformations. To qualitatively track the ELF field's evolution, Hausdorff distances between consecutive points along the intrinsic reaction coordinate (IRC) path can be calculated [25**?**] to capture the ELF field's stepwise progression. In this study, we aim to answer the following questions: (1) What is the order of bond breaking and formation from the BET perspective? (2) What is the ELF-topological depiction of polarity reversal on the carbon atom, the core of the "umpolung" type reaction? (3) Does the representation of the reaction mechanism based on ELF topological analysis and catastrophe theory align with the classical Lewis structure representation of the mechanism? We believe the presented study provides a deeper understanding of the mechanism of the umpolung reaction.

2. Results and Discussion

In this work, we present a theoretical study of the cyanide-catalysed benzoin condensation, which belongs to the class of umpolung reactions. The study is divided into two parts: (I) an energy analysis of stationary points on PES and (II) an analysis of the bond-breaking and bond-creating events using BET and Hausdorff distances. In the first part, we employ DFT to determine thermodynamics for the reaction. In the second part, we use the topological analysis of ELF to investigate the bond-breaking and bond-creating events during the reaction. We use the Hausdorff distances to quantitatively evaluate the changes in the topology of the electron density during the reaction. The proposed approach of combining energy analysis and ELF analysis allows for a comprehensive examination of the reaction mechanism, providing insights into the changes in the electronic structure and the topology of the electron density during the reaction.

2.1. The Energetic of the Reaction and Geometrical Analysis

The mechanism of the cyanide-catalysed benzoin condensation was formally studied and presented in Figure 1. An analysis of PES revealed that the reaction proceeds through five transition state structures in the gas phase at 298.15 K in the absence of solvent assistance. The Gibbs free energy barriers were depicted in Figure 2, with the corresponding transition state structures shown in Figure 3, and the energy profiles illustrated in Figure 4.

In the first step (TS1), the nucleophile cyanide ion attacks the carbon atom of the C=O group in benzaldehyde. As a result, an electron pair from the C1=O1 bond is transferred to the oxygen atom, leading to the formation of a negatively charged oxygen atom. The second step (TS2) is associated with the migration of a hydrogen atom to the O1 atom, resulting from the breaking of the C1-H1 bond. During this step, an electron pair from the O1 atom is transferred toward the migrating H1 atom, which now bears a positive charge. This step also results in a change in the hybridisation of the C1 atom from sp^3 to sp^2, as well as a reversal of the polarity of the carbonyl group. The product of this step, known in the literature as the Lapworth's cyanohydrin intermediate [12,26], serves as the key intermediate in the subsequent steps of the reaction. In the third step (TS3), due to the negative charge on the C1 atom, Lapworth's intermediate reacts with benzaldehyde in a second nucleophilic addition, leading to the formation of a new C1-C15 bond. An electron pair from the C15=O2 group is transferred to the oxygen atom, resulting in the formation of a negatively charged oxygen atom. The fourth step (TS4) involves the transfer of a single electron pair from the O2 atom to the H1 atom, resulting in a migration of the H1 proton and the formation of a new O-H bond. This step marks the formation of the final product, benzoin. Lastly, in the fifth step (TS5), a single electron pair from the O1 atom is moved towards the C=O bond, while another electron pair from the C1-C8 bond is moved to the carbon atom in the -CN group. This results in the release of the cyanide ion and the formation of benzoin as the final product.

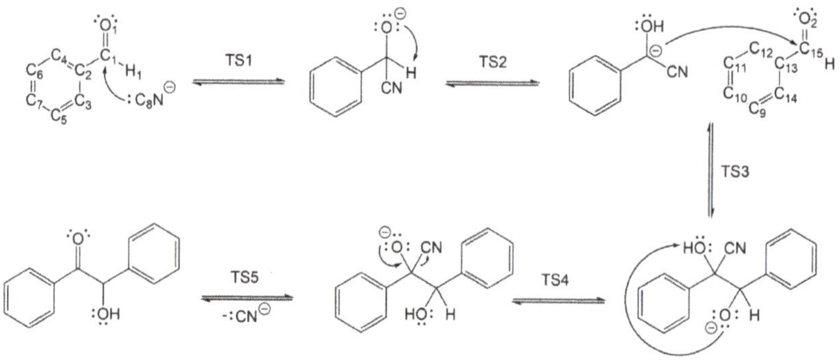

Figure 1. The Lapworth mechanism of the cyanide-catalysed benzoin condensation.

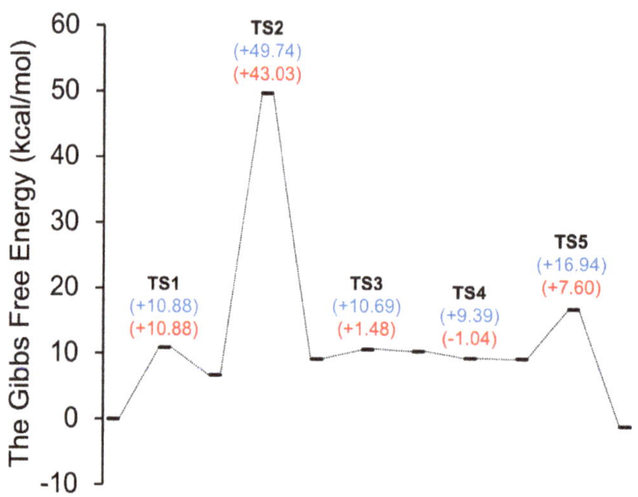

Figure 2. The Gibbs free energy of activation barriers and relative Gibbs free energies for umpolung transition states (TSs), with blue values representing relative energies and red values indicating activation energies from the reagent side, $\Delta G_a(I)$.

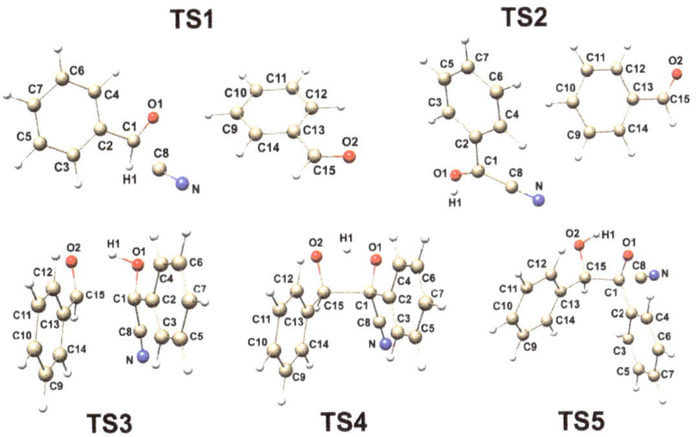

Figure 3. Optimised transition state (TS) structures for all steps of the umpolung reaction at the DFT(B3LYP)/6-311++G(d,p) computational level.

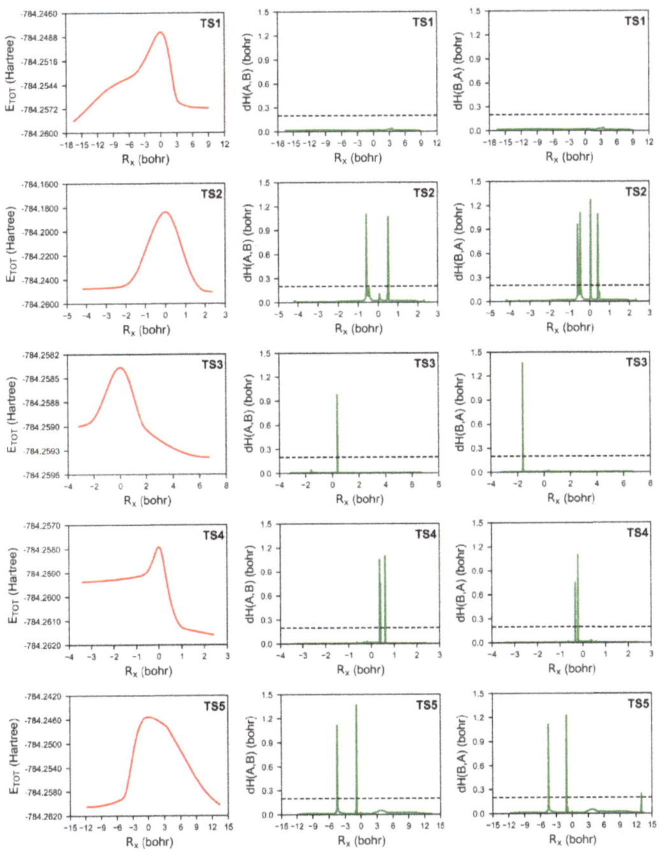

Figure 4. The evolution of the total energy (red lines), E_{TOT}, and both Hausdorff distances (green lines), dH(A,B) and dH(B,A), for the umpolung reaction as a function of the IRC coordinate. The peaks above the dashed line (dH = 0.2) indicate significant topological changes in the ELF map.

The Gibbs free energies of activation, $\Delta G_a(\mathrm{I}, \mathrm{II})$, associated with the transition states TS1-TS5 were calculated using the B3LYP electron density functional and are presented in Table 1. The values of $\Delta G_a(\mathrm{I}, \mathrm{II})$ were determined for both the reactant side, $\Delta G_a(\mathrm{I})$, and the product side, $\Delta G_a(\mathrm{II})$, of each transition state.

Table 1. The values (kcal/mol) of the Gibbs free energy of activation, $\Delta G_a(\mathrm{I}, \mathrm{II})$, and the Gibbs free energy of reaction, ΔG_r, calculated at DFT(B3LYP)/6-311++G(d,p) level.

Energy	TS1	TS2	TS3	TS4	TS5
$\Delta G_a(\mathrm{I})$	10.88	43.03	1.48	−1.04	7.60
$\Delta G_a(\mathrm{II})$	4.17	40.53	0.26	0.05	17.77
ΔG_r	6.71	2.50	1.22	−1.09	−10.17

The first step (TS1) is characterised by the nucleophilic attack of the cyanide ion on the carbon atom of the carbonyl group in benzaldehyde. The imaginary frequency of TS1 was calculated to be 218.860i cm^{-1}, and its vibrational vector corresponds to the formation of the C1-C8 bond. The distance of the C1-C8 bond in the pre-reaction and post-reaction complexes was found to be 4.260 and 1.531 Å, respectively. The activation energy associated with this step, $\Delta G_a(\mathrm{I})$, was calculated to be 10.88 kcal/mol. This value is 6.71 kcal/mol higher than the corresponding value for the product. The ΔG_r was found to be positive in the gas phase, indicating that the reaction is thermodynamically not spontaneous.

The second step (TS2) is characterised by the transfer of a proton between the C1 and O1 atoms, resulting in the formation of the Lapworth's cyanohydrin intermediate. This step is associated with a change in the hybridisation of the C1 atom from sp^3 to sp^2 and is accompanied by an elongation of the C1-O1 bond length from 1.328 Å to 1.422 Å, which is within the typical range of single C-O interactions [27]. The $\Delta G_a(\mathrm{I})$, $\Delta G_a(\mathrm{II})$, and ΔG_r associated with this step were calculated to be 43.03, 40.53, and 2.50 kcal/mol, respectively. These values indicate that the energy barriers for this step are significantly higher than those for TS1, making it the rate-determining step of the reaction. It is noteworthy that this step has the highest energy activation barrier among all the steps, making it the most energetically difficult step in the reaction mechanism. This problem was investigated using DFT calculations with the CPCM solvation model (DMSO, water, ethanol, methanol) [12]. No significant impact on the barrier height was observed when considering solvent effects. However, a proton transfer system assisted by methanol was simulated, showing a decrease in the barrier height by approximately 25 kcal/mol. We presume that the step leading to the formation of the cyanohydrin intermediate may proceed via a solvent-assisted proton shift transition state in protonic solvent [12].

The reaction of the Lapworth's intermediate with benzaldehyde leads to the formation of a C1-C15 bond. The C1-C15 bond length in the pre-reaction, transition state (TS3), and post-reaction complexes are 2.480, 2.091, and 1.760 Å, respectively. The imaginary frequency of TS3 is associated with the formation of the C1-C15 bond. The activation energy of the step, as calculated using the B3LYP method, is relatively low at 1.48 kcal/mol when considering the energy barrier from the side of the reagents. However, the energy barrier from the side of the product is even lower at 0.26 kcal/mol. Thus, the Gibbs free energy of reaction is positive, indicating that this process is not spontaneous.

In the fourth step (TS4), a proton transfer occurs between the O1 and H1 atoms. The associated activation energy, as calculated using B3LYP, is −1.04 kcal/mol. This value is unphysical and negative, which is indicative of a barrier-less reaction. Analysis of PES reveals that the C-O distance between the O1 and H1 atoms elongates from 1.186 to 1.242 Å during this step. The negative sign of the activation energy is attributed to the fact that the zero-point vibrational energy of the pre-reaction complex is larger than that of the transition state, 143.627 kcal/mol and 141.667 kcal/mol, respectively. To further verify these results, the activation energies were also calculated using additional DFT functionals and basis sets, as presented in Table 2. The results obtained using these alternative methods also indicated a negative activation energy, consistent with the literature on barrier-less reactions [28–30].

In the final step (TS5), the mechanism of benzoin condensation proceeds via the elimination of the -CN group. The vibrational frequency of TS5 has been determined to be 129.984i cm^{-1}, corresponding to the breaking of the C1-C8 bond. The $\Delta G_a(I)$ and ΔG_r associated with this step have been calculated to be 7.60 and 17.77 kcal/mol, respectively. These values indicate that at room temperature and in the gas phase, this process is thermodynamically spontaneous.

Table 2. The values (kcal/mol) of the Gibbs free energy of activation, $\Delta G_a(I, II)$, and the Gibbs free energy of reaction, ΔG_r, for the fourth umpolung reaction step.

Energy	B3LYP	M06-2X	BHHLYP	wB97XD	PBE0
			6-31+G(d,p)		
$\Delta G_a(I)$	−1.16	−1.15	−1.02	−1.36	−1.08
$\Delta G_a(II)$	−0.31	0.15	1.32	0.55	−0.40
ΔG_r	−0.85	−1.30	−2.34	−1.91	−0.68
			6-311++G(d,p)		
$\Delta G_a(I)$	−1.04	−1.10	−0.71	−1.01	−1.20
$\Delta G_a(II)$	0.05	0.68	1.72	0.77	−0.27
ΔG_r	−1.09	−1.78	−2.43	−1.78	−0.93
			6-311++G(2df,p)		
$\Delta G_a(I)$	−0.93	−1.14	−0.86	−1.16	−0.86
$\Delta G_a(II)$	−0.30	0.39	1.38	0.60	1.38
ΔG_r	−0.63	0.75	−2.24	−1.76	−2.24

The overall ΔG_r value for the umpolung reaction is −0.83 kcal/mol, indicating that the reaction is thermodynamically spontaneous under the conditions considered. This overall reaction character aligns with the energetic profile, as each step described earlier contributes to the overall process. It is important to note that these calculations were performed in the gas phase and thus do not account for potential solvent effects. The solvent can play a crucial role in stabilising reaction intermediates and transition states, which may further reduce the activation barriers and enhance the thermodynamic favorability of the reaction.

2.2. *The Mechanism of the Reaction—BET Analysis*

The Hausdorff distances for TS1 were found to exhibit a flat baseline with no strong peaks (as seen in Figure 4). This indicates that no catastrophes related to the evolution of the ELF field attractors were observed. However, at a value of $R_x = -0.137$ Bohr, a change in V(C8) basin synapticity to V(C1,C8) was observed, signalling the formation of a C1-C8 bond with a population of 2.58e. The V(C1,C8) attractor was found to be localised in the position of V(C8) without any additional changes in the ELF field. A topological analysis of the ELF for the post-reaction complex revealed a flux of electron density from V(O1,C1) to V_1(O1), which is in good agreement with the elongation of the C1-O1 bond length. The C1-C8 and C8-N bonds were found to be described by V(C1,C8) and V(C8,N) basins with populations of 2.24e and 4.33e, respectively.

The formation of the Lapworth's cyanohydrin is characterised by six ELF field catastrophes. This conclusion can also be derived from the six peaks in the Hausdorff distance where dH > 0.2. The first peak corresponds to an annihilation of V_1(C8,N) and V_2(C8,N) into a single V(C8,N) basin. Following this, the second and third peaks correspond to the annihilation and recreation of the V_2(O1) basin. These reorganisations of the valence electron density do not have a simple chemical interpretation. The fourth peak is related to the breaking of the C1-H1 bond. This leads to the formation of a proton surrounded by electron density, known as a dressed proton, in the form of the V(H1) basin and the non-bonding V(C1) basin in the vicinity of the C1 atom. The fifth and sixth peaks correspond to the creation of the V_3(O1) attractor in the valence shell of the O1 atom and the formation of the O1-H1 bond from V_3(O1) and V(H1) basins. A topological analysis of the ELF for the post-reaction complex revealed a flux of electron density from the V_1(O1) and V_2(O1) basins to the V(O1,H1) basin. The V(C1) basin is now described by two $V_{1,2}$(C1) basins with populations of 0.55e and 0.45e, respectively. These results align well with the APT charge data, as shown in Figure 5. In the metastable Lapworth's cyanohydrin formation, the C1 atom initially exhibits a positive charge, which shifts to negative values as the proton transfer reaction progresses. This charge inversion on the C1 atom is indicative of electron density redistribution that stabilises the reaction intermediate. Conversely, the H1 atom starts with a negative partial charge, which reverses to a positive charge during the course of the reaction. This reversal signifies a transition in the electron density environment around the H1 atom, where it gains a partial positive character as it approaches the dressed proton state.

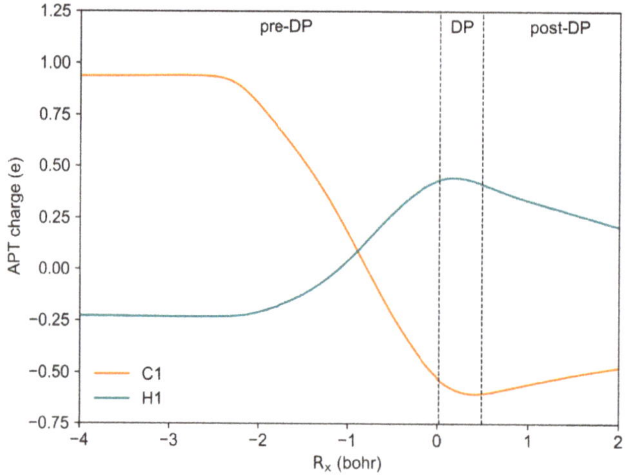

Figure 5. The evolution of APT charge of C1 and H1 atoms during the proton transfer in Lapworth's cyanohydrin formation. The "DP" refers to the dressed proton state.

The formation of the C1-C15 bond in the TS3 is characterised by the occurrence of two ELF catastrophes. This is in good agreement with the calculated Hausdorff distances. The first peak, observed at $R_x = -1.620$ Bohr, corresponds to the ELF topological catastrophe, where the V(C15) basin is created in association with the (3,−1) CP. The second peak corresponds to the formation of the C1-C15 bond, as evidenced by the annihilation of the V(C15) and V(C1) basins, along with the corresponding (3,−1) CP. This results in the creation of the V(C1,C15) basin, which is populated with a density of 1.08e. Notably, the population of the V(C1,C15) basin reaches a final value of 1.76e at the last step of the IRC path.

The proton transfer between the O1 and O2 atoms has been observed in TS4. The analysis of the Hausdorff distances, as depicted in Figure 4, reveals the presence of five peaks associated with five ELF catastrophes. The first peak, observed at $R_x = -0.330$ Bohr, corresponds to the annihilation of the V(H1,O1) basin and results in the creation of the dressed proton V(H1) and monosynaptic V(O1) basins. The basin population of V(H1) after the catastrophe is 0.43e. The second and third peaks do not have a straightforward chemical interpretation. The V_3(O2) basin is created at $R_x = -0.210$ Bohr, and the V_2(C8,N) basin is annihilated and merges with V_1(C8,N) to form a single V(C8,N) basin with a population of 4.35e. These ELF changes are also reflected by changes in the number of (3,−1) CPs. In the fourth catastrophe, the V_3(O2) and V(H1) basins are annihilated, resulting in the creation of the V(H1,O2) basin. At this point, the O2-H1 bond is represented by the V(H1,O2) basin with a population of 1.59e. The final peak corresponds to the annihilation of the third V_3(O1) basin.

The elimination of the cyanide group via TS5 is characterised by four ELF field catastrophes in the final stages. The first peak, observed in both dH(A,B) and dH(B,A), corresponds to a displacement of the V(O1) attractor into a new region without any significant ELF changes. The second peak results in the annihilation of the V(C1,C8) basin and the creation of the V(C1) and V(C8) basins. This is followed by an annihilation of the V(C1) basin. In the final catastrophe, the V(C8,N) basin is split into two $V_{1,2}$(C8,N) basins with an overall population of 3.53e.

In summary, the analysis of IRC path has revealed multiple ELF catastrophes at various stages of the chemical reactions studied. The Hausdorff distances and basin populations obtained from the analysis were found to be consistent with the proposed Lapworth mechanism. The ELF topology of the reaction was found to be in good agreement with the mechanistic understanding of the umpolung reaction.

3. Materials and Methods

The calculations for this study were performed in the gas phase, at 0 K, using the DFT method with the B3LYP electron density functional [31,32] as implemented in the Gaussian16 (G16) program (version C.01) [33]. The electron wave functions were described using the Pople basis set, 6-311++G(d,p) [34,35]. The optimisation of molecular geometries was performed using the Berny algorithm and obtained structures were verified by the calculation of infrared spectra to confirm its character on the potential energy hypersurface (PES). The presence of one imaginary frequency for transition states (TSs) and zero for reagents and products indicates that the stationary points correspond to the expected states.

The reaction path was calculated using IRC [36,37] implemented in G16. The initial Hessian matrix was calculated and recalculated every five steps. The correct localisation of TSs on PES was verified through an analysis of IRC paths and the subsequent optimisation of molecular structures to energy minima. These steps ensure that the results obtained are consistent with the expected reaction mechanism. The Gibbs free energies of activation, $\Delta G_a(\mathrm{I})$, and $\Delta G_a(\mathrm{II})$, have been calculated as the difference between the sum of electronic and thermal free energies between TS and reagents and products, respectively. The Gibbs free energy of reaction, ΔG_r, has been calculated as the difference between the sum of electronic and thermal free energies between products and reagents. The $\Delta G_a(\mathrm{I,II})$ and ΔG_r values were calculated for a temperature of 298.15 K and pressure of 1 atm.

The evolution of chemical bonds was studied by comparing the topology of ELF for electronic structures of molecules localised at all points on IRC paths. Due to the flat character of PES, the gradient magnitude for the damped velocity Verlet (DVV) stopping criteria was turned off using the IOp(1/108 = −1) keyword in G16. Single-point calculations at the DFT(B3LYP)/6-311++G(d,p) computational level were performed to obtain a set of molecular orbitals for each molecular structure.

The topological analysis of ELF was performed using TopMod09 and MultiWFN programs [38,39]. A parallelepiped grid of points with a step size of 0.05 Bohr was used. The wave functions for the minima on PES and each structure on the IRC path were obtained by single-point DFT calculations at the DFT(B3LYP)/6-311++G(d,p) computational level. The analysis of catastrophes of the ELF field was performed using the MultiWFN and AyudaTop programs [39,40].

The Hausdorff distances [**?**], dH(A,B) and dH(B,A), were calculated for two sets of coordinates for the attractors of the ELF field. The A set contains the coordinates of all attractors for the electronic structure of molecule A on the IRC path preceding point B. The use of Hausdorff distances in the topological analysis of ELF in the context of reaction mechanisms has been previously studied by us and has been found to be an effective tool to investigate topological catastrophes in the evolution of chemical bonds [25]. This approach allows for a quantitative evaluation of the changes in the topology of the electron density during the reaction, providing a deeper understanding of the reaction mechanism.

4. Conclusions

The mechanism of the cyanide-type umpolung reaction in benzoin condensation has been investigated in detail by utilising the modern topological analysis of ELF and catastrophe theory. By applying this approach, a comprehensive understanding of the evolution of chemical bonds and non-bonding electron density in the reaction of benzaldehyde and cyanide ions was achieved. The use of the topological analysis of ELF and BET proved to be an effective method for characterising the mechanisms of this type of chemical reaction. The results may be summarised as follows:

1. The umpolung mechanism in the gas phase and without solvent assistance proceeds through five transition state structures. The formation of Lapworth's cyanohydrin is not spontaneous, as evidenced by the positive ΔG_r value. Among these steps, the formation of the Lapworth's cyanohydrin has the highest $\Delta G_a(I)$ value, making it the rate-determining step. Additionally, it was observed that the proton transfer in TS4 does not occur simultaneously with C-C bond formation and has a barrier-less character, as indicated by the negative $\Delta G_a(I)$ value.
2. The evolution of the electron localisation function (ELF) field in the studied umpolung mechanism is characterised by 17 distinct catastrophes. While some changes within the ELF field lack straightforward chemical interpretation, the application of topological analysis alongside catastrophe theory has enabled a detailed mapping of electron density evolution. In particular, a comprehensive analysis of ELF catastrophes in the cyanide-type umpolung reaction reveals that the formation of the C1-C8 bond occurs through a shift in the synapticity of the V(C8) basin to a V(C1,C8) basin at a distance of 2.029 Å. This was followed by proton transfer, resulting in the formation of Lapworth's cyanohydrin and the appearance of additional electron density near the C1 atom, as indicated by the V(C1) basin. The formation of the C1-C15 bond in TS3 is characterised by two distinct ELF catastrophes: one describing the creation of non-bonded electron density near the C15 atom and the other describing the formation of the C1-C15 bond through the annihilation of the V(C1) and V(C15) basins. The proton transfer between the O1 and O2 atoms is found to involve the creation of a dressed-atom-type V(H1) basin. Finally, the dissociation of the C1-C8 bond and the elimination of the cyanide group occur through the annihilation of the disynaptic V(C1,C8) basin and the creation of monosynaptic V(C1) and V(C8) basins.

3. The applications of topological analysis of ELF and catastrophe theory confirm the well-established Lapworth mechanism for this reaction.

Author Contributions: Conceptualisation, M.M.; methodology, M.M.; formal analysis, M.M.; investigation, M.M.; writing—original draft preparation, M.M.; writing—review and editing, S.B.; visualisation, M.M.; supervision, S.B. All authors have read and agreed to the published version of the manuscript.

Funding: This research received no external funding.

Institutional Review Board Statement: Not applicable.

Informed Consent Statement: Not applicable.

Data Availability Statement: The raw data supporting the conclusions of this article will be made available by the authors on request.

Acknowledgments: The authors are grateful to the Wroclaw Centre for Networking and Supercomputing for the generous allocation of computer time.

Conflicts of Interest: The authors declare no conflicts of interest.

References

1. Lapworth, A. XCVI.—Reactions involving the addition of hydrogen cyanide to carbon compounds. *J. Chem. Soc. Trans.* **1903**, *83*, 995–1005. [CrossRef]
2. Bugaut, X.; Glorius, F. Organocatalytic umpolung: N-heterocyclic carbenes and beyond. *Chem. Soc. Rev.* **2012**, *41*, 3511–3522. [CrossRef] [PubMed]
3. Juhl, M.; Lee, J.W. Umpolung Reactivity of Aldehydes toward Carbon Dioxide. *Angew. Chem. Int. Ed.* **2018**, *57*, 12318–12322. [CrossRef] [PubMed]
4. Wang, S.; Cheng, B.Y.; Sršen, M.; König, B. Umpolung Difunctionalization of Carbonyls via Visible-Light Photoredox Catalytic Radical-Carbanion Relay. *J. Am. Chem. Soc.* **2020**, *142*, 7524–7531. [CrossRef] [PubMed]
5. Chen, Y.H.; Duan, M.; Lin, S.L.; Liu, Y.W.; Cheng, J.K.; Xiang, S.H.; Yu, P.; Houk, K.N.; Tan, B. Organocatalytic aromatization-promoted umpolung reaction of imines. *Nat. Chem.* **2024**, *16*, 408–416. [CrossRef]
6. Bai, D.; Guo, X.; Wang, X.; Xu, W.; Cheng, R.; Wei, D.; Lan, Y.; Chang, J. Umpolung reactivity of strained C–C σ-bonds without transition-metal catalysis. *Nat. Commun.* **2024**, *15*, 2833. [CrossRef]
7. Shi, Q.; Liu, W.H. Reactivity Umpolung of Amides in Organic Synthesis. *Asian J. Org. Chem.* **2023**, *12*, e202300473. [CrossRef]
8. Wiberg, K.B. The Deuterium Isotope Effect of Some Ionic Reactions of Benzaldehyde. *J. Am. Chem. Soc.* **1954**, *76*, 5371–5375. [CrossRef]
9. Wiberg, K.B. The Deuterium Isotope Effect of the Methanolysis of Some Organometallic Compounds. *J. Am. Chem. Soc.* **1955**, *77*, 5987–5990. [CrossRef]
10. Kool, E.T.; Breslow, R. Dichotomous salt effects in the hydrophobic acceleration of the benzoin condensation. *J. Am. Chem. Soc.* **1988**, *110*, 1596–1597. [CrossRef]
11. Breslow, R. Hydrophobic effects on simple organic reactions in water. *Accounts Chem. Res.* **1991**, *24*, 159–164. [CrossRef]
12. He, Y.; Xue, Y. Mechanism Insight into the Cyanide-Catalyzed Benzoin Condensation: A Density Functional Theory Study. *J. Phys. Chem. A* **2010**, *114*, 9222–9230. [CrossRef] [PubMed]
13. Krokidis, X.; Noury, S.; Silvi, B. Characterization of Elementary Chemical Processes by Catastrophe Theory. *J. Phys. Chem. A* **1997**, *101*, 7277–7282. [CrossRef]
14. Becke, A.D.; Edgecombe, K.E. A simple measure of electron localization in atomic and molecular systems. *J. Chem. Phys.* **1990**, *92*, 5397–5403. [CrossRef]
15. Savin, A.; Becke, A.D.; Flad, J.; Nesper, R.; Preuss, H.; von Schnering, H.G. A New Look at Electron Localization. *Angew. Chem. Int. Ed. Engl.* **1991**, *30*, 409–412. [CrossRef]
16. Thom, R. *Structural Stability and Morphogenesis*; Taylor & Francis Group LLC: London, UK, 1972.
17. Polo, V.; Andres, J.; Berski, S.; Domingo, L.R.; Silvi, B. Understanding Reaction Mechanisms in Organic Chemistry from Catastrophe Theory Applied to the Electron Localization Function Topology. *J. Phys. Chem. A* **2008**, *112*, 7128–7136. [CrossRef]

18. Berski, S.; Andrés, J.; Silvi, B.; Domingo, L.R. The Joint Use of Catastrophe Theory and Electron Localization Function to Characterize Molecular Mechanisms. A Density Functional Study of the Diels-Alder Reaction between Ethylene and 1,3-Butadiene. *J. Phys. Chem. A* **2003**, *107*, 6014–6024. [CrossRef]
19. Berski, S.; Andrés, J.; Silvi, B.; Domingo, L.R. New Findings on the Diels-Alder Reactions. An Analysis Based on the Bonding Evolution Theory. *J. Phys. Chem. A* **2006**, *110*, 13939–13947. [CrossRef]
20. Polo, V.; Andres, J.; Castillo, R.; Berski, S.; Silvi, B. Understanding the Molecular Mechanism of the 1,3-Dipolar Cycloaddition between Fulminic Acid and Acetylene in Terms of the Electron Localization Function and Catastrophe Theory. *Chem.— Eur. J.* **2004**, *10*, 5165–5172. [CrossRef]
21. Andrés, J.; Berski, S.; Silvi, B. Curly arrows meet electron density transfers in chemical reaction mechanisms: From electron localization function (ELF) analysis to valence-shell electron-pair repulsion (VSEPR) inspired interpretation. *Chem. Commun.* **2016**, *52*, 8183–8195. [CrossRef]
22. Silvi, B.; Savin, A. Classification of chemical bonds based on topological analysis of electron localization functions. *Nature* **1994**, *371*, 683–686. [CrossRef]
23. Silvi, B. The synaptic order: A key concept to understand multicenter bonding. *J. Mol. Struct.* **2002**, *614*, 3–10. [CrossRef]
24. Hausdorff, F. *Grundzüge der Mengenlehre*; Göschens Lehrbücherei/Gruppe I: Reine und Angewandte Mathematik Series; Von Veit: Leipzig, Germany, 1914.
25. Michalski, M.; Berski, S. Understanding the molecular mechanism of the chlorine atom transfer between ammonia and hypochlorous acid with electron localisation function (ELF). *Mol. Phys.* **2021**, *119*, e1961035. [CrossRef]
26. Han, G.; Liu, X.; Cao, Z.; Sun, Y. Photocatalytic Pinacol C–C Coupling and Jet Fuel Precursor Production on ZnIn2S4 Nanosheets. *ACS Catal.* **2020**, *10*, 9346–9355. [CrossRef]
27. Bruno, I.J.; Cole, J.C.; Edgington, P.R.; Kessler, M.; Macrae, C.F.; McCabe, P.; Pearson, J.; Taylor, R. New software for searching the Cambridge Structural Database and visualizing crystal structures. *Acta Crystallogr. Sect. B* **2002**, *58*, 389–397. [CrossRef]
28. Revell, L.E.; Williamson, B.E. Why Are Some Reactions Slower at Higher Temperatures? *J. Chem. Educ.* **2013**, *90*, 1024–1027. [CrossRef]
29. Bursch, M.; Mewes, J.M.; Hansen, A.; Grimme, S. Best-Practice DFT Protocols for Basic Molecular Computational Chemistry. *Angew. Chem. Int. Ed.* **2022**, *61*, e202205735. [CrossRef]
30. Shukla, P.B.; Mishra, P.; Baruah, T.; Zope, R.R.; Jackson, K.A.; Johnson, J.K. How Do Self-Interaction Errors Associated with Stretched Bonds Affect Barrier Height Predictions? *J. Phys. Chem. A* **2023**, *127*, 1750–1759. [CrossRef]
31. Becke, A.D. Density-functional thermochemistry. III. The role of exact exchange. *J. Chem. Phys.* **1993**, *98*, 5648–5652. [CrossRef]
32. Lee, C.; Yang, W.; Parr, R.G. Development of the Colle-Salvetti correlation-energy formula into a functional of the electron density. *Phys. Rev. B* **1988**, *37*, 785–789. [CrossRef]
33. Frisch, M.J.; Trucks, G.W.; Schlegel, H.B.; Scuseria, G.E.; Robb, M.A.; Cheeseman, J.R.; Scalmani, G.; Barone, V.; Petersson, G.A.; Nakatsuji, H.; et al. *Gaussian~16 Revision C.01*; Gaussian Inc.: Wallingford, CT, USA, 2016.
34. Krishnan, R.; Binkley, J.S.; Seeger, R.; Pople, J.A. Self-consistent molecular orbital methods. XX. A basis set for correlated wave functions. *J. Chem. Phys.* **1980**, *72*, 650–654. [CrossRef]
35. Frisch, M.J.; Pople, J.A.; Binkley, J.S. Self-consistent molecular orbital methods 25. Supplementary functions for Gaussian basis sets. *J. Chem. Phys.* **1984**, *80*, 3265–3269. [CrossRef]
36. Fukui, K. Formulation of the reaction coordinate. *J. Phys. Chem.* **1970**, *74*, 4161–4163. [CrossRef]
37. Fukui, K. The path of chemical reactions—The IRC approach. *Accounts Chem. Res.* **1981**, *14*, 363–368. [CrossRef]
38. Noury, S.; Krokidis, X.; Fuster, F.; Silvi, B. Computational tools for the electron localization function topological analysis. *Comput. Chem.* **1999**, *23*, 597–604. [CrossRef]
39. Lu, T.; Chen, F. Multiwfn: A multifunctional wavefunction analyzer. *J. Comput. Chem.* **2012**, *33*, 580–592. [CrossRef]
40. Michalski, M. AyudaTop. 2024. Available online: https://github.com/Parecido/AyudaTop (accessed on 6 July 2024).

molecules

MDPI

Article

Computation of the pK_a Values of Gallic Acid and Its Anionic Forms in Aqueous Solution: A Self-Similar Transformation Approach for Accurate Proton Hydration Free Energy Estimation

Marcin Molski

Department of Quantum Chemistry, Faculty of Chemistry, Adam Mickiewicz University of Poznań, ul. Uniwersytetu Poznańskiego 8, 61-614 Poznań, Poland; mamolski@amu.edu.pl

Abstract: The Gibbs free energies of gallic acid (GA) and its anionic forms in aqueous solution were computed utilizing density functional theory (DFT) at the LSDA, M062X, B3LYP/QZVP levels, in conjunction with the SMD solvation model. The pK_a values corresponding to the four-step deprotonation of GA were determined through a non-linear self-similar transformation expressed as, pK_a = a·pK_a(the)^c which establishes a link between theoretical and experimental pK_a values. This approach replaces the previously employed linear relationship, pK_a = a·pK_a(the) + b. The proposed model demonstrates high accuracy in reproducing the experimental pK_a1 = 4.16 ± 0.02, pK_a2 = 8.55 ± 0.01, pK_a3 =11.40 ± 0.10, pK_a4 =12.8 ± 0.40 values of GA, with a standard error (SE) of 0.045 and a mean absolute error (MAE) of 0.019 in pK_a unit. Furthermore, it facilitates the precise determination of the Gibbs free energy of the proton hydration, yielding $\Delta G(H^+)_{aq}$ = 259.4272(75) [kcal mol^{-1}]. This result conforms acceptably with the experimental value of $\Delta G(H^+)_{aq}$ = −259.5 [kcal mol^{-1}].

Keywords: DFT method; gallic acid; multi-step deprotonation; proton hydration; Gibbs energy; pK_a descriptor; selectivity; reactivity

1. Introduction

Deprotonation represents a fundamental chemical reaction that occurs across both biological [1,2] and abiotic [2–4] systems, thereby representing a significant research area focused on molecular reactivity and selectivity [3,4]. Monocarboxylic acids [5,6], along with their combinations with phenols and polyphenols [7,8], are particularly compelling subjects of study due to the presence of multiple reactive carboxyl and hydroxyl groups that engage actively in the deprotonation process. Experimental and theoretical analyses of protonation strive to assess the reactivity of each active group, identifying the sequential order of proton removal from the base molecule and evaluating the relative susceptibility of each group to deprotonation [3,4,9]. The chemical reactivity of these compounds is largely influenced by the electrophilic nature of the carboxylic moiety, as well as the resonance stabilization that occurs upon proton dissociation. These factors jointly contribute to both the acidity and predominant chemical reactivity of the carboxylic and phenolic groups. They can be numerically (quantitatively) characterized by the parameter pK_a, which stands for the negative base-10 logarithm of an acid dissociation constant in a specific solvent, usually of a hydrophilic nature facilitating the proton detachment. Experimentally and theoretically determining the pK_a values of the compounds, especially in aqueous solution, is essential for a comprehensive understanding of numerous chemical processes and provides insight into the deprotonation state of a molecule within a specific solvent environment. Theoretical approaches to calculating pK_a parameters have garnered significant interest, particularly for applications involving molecules that have not yet been synthesized, those for which experimental determination of pK_a is challenging, and larger molecules where local environmental factors alter the intrinsic pK_a values. An example of the latter includes certain amino acids incorporated within polypeptide chains [2]. Achieving chemical accuracy in pK_a calculations is inherently challenging, as an error of 1.36 [kcal mol^{-1}] in the change in free energy associated with deprotonation in a solvent corresponds to a deviation of 1 pK_a unit [2]. In this regard, gallic acid (GA, Figure 1) is among the most extensively studied compounds [3,4,9] with respect to selectivity and functional group activity, owing to its four active hydrogens in carboxyl moiety and hydroxyl groups attached to the benzene ring.

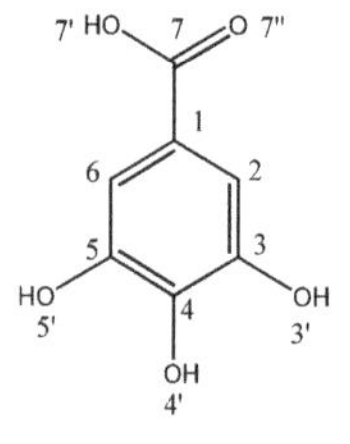

Figure 1. Gallic acid, 3,4,5-trihydroxybenzoic acid.

Research has demonstrated [3,4] that the deprotonation sequence in GA in water solution follows the order $O_{7'}$-H, $O_{4'}$-H, $O_{3'}$-H, $O_{5'}$-H, necessitating the use of distinct deprotonation descriptors: pK_a1($O_{7'}$-H), pK_a2($O_{4'}$-H), pK_a3($O_{3'}$-H), pK_a4($O_{5'}$-H) associated with the four-step deprotonation of GA presented in Figure 2.

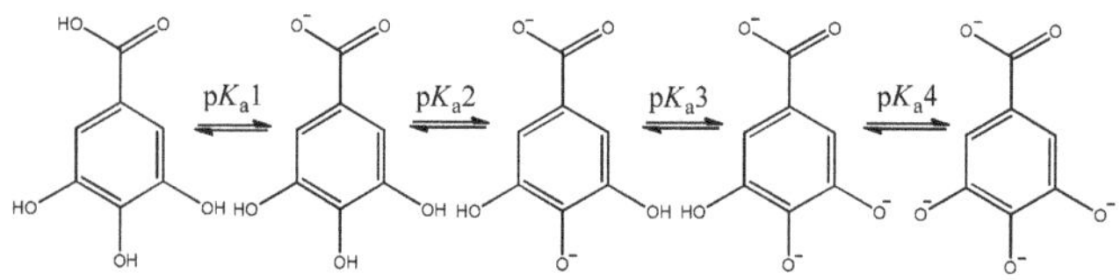

Figure 2. Four-step deprotonation of gallic acid and related pK_a parameters.

Despite the broad utility of the pK_a descriptor, the acquisition of accurate and unique experimental data for GA remains challenging. They are commonly determined through ultraviolet–visible (UV-Vis) and Raman spectroscopy [10–13], or various analytical techniques [14]. For GA, the following experimental pK_a1,2,3,4 values were reported with the standard errors if available: Set I (4.50, 7.05, 8.75, 10.25) [11], Set II (4.4 ± 0.1, 8.8 ± 0.1, 10.0 ± 0.1, 11.4 ± 0.1) [12], Set III (4.44, 8.54, 10.05, 11.30) [14], Set IV (4.16 ± 0.02, 8.55 ± 0.01, 11.40 ± 0.10, 12.8 ± 0.40) [13]. The availability of experimental pK_a values, measured with varying degrees of precision, facilitates a comparison with the theoretical results generated by different models and methods employed in the calculations. A straightforward approach to the theoretical simulation of solvent effects involves the application of various solvation models, with particular emphasis on the Conductor-like Polarizable Continuum Model (C-PCM) [15], the Integral Equation Formalism Polarizable Continuum Model (IEF-PCM) [16], and the universal Solvation Model Density (SMD) based on solute electron density [17]. They are widely employed to calculate the theoretical pK_a values of chemical compounds in solution through the relationship show below [18–21]:

$$pK_a = \frac{G(R^-)_{sol} - G(RH)_{sol} + G(H^+)_{sol} - RTln(V)}{RTln(10)} = \frac{\Delta G + G(H^+)_{sol} - 1.8942}{1.3642} \quad (1)$$

in which R is the gas constant, T = 298.15 [K] is the temperature (which is used in all calculations), and RT·ln(V) = 1.8942 [kcal mol^{-1}] stands for the correction [21]

$$G^{1\ [mol]} = G^{1\ [atm]} - 1.8942\ [kcal\ mol^{-1}] \quad (2)$$

for the reference state (V = 24.46 [L], T = 298.15 [K]), from 1 [atm] to 1 [mol], whereas ΔG in [kcal mol^{-1}] denotes the Gibbs free energy of the deprotonation according to the reaction

$$RH \xrightleftharpoons{solvent} R^- + H^+$$

The Gibbs free energy of the solvated proton $G(H^+)_{sol}$, is related to the solvation energy of the proton $\Delta G(H^+)_{sol}$ by the relation [21]

$$G(H^+)_{sol} = \Delta G(H^+)_{sol} + G(H^+)_{gas} \quad (3)$$

in which $G(H^+)_{gas}$ = −6.2883 [kcal mol^{-1}] [22] is the free energy of a proton in the gas phase under a pressure of 1 [atm]. From (1) and (2) it follows that Equation (1) is useful evaluating pK_a using the calculated (theoretical) $G(H^+)_{sol}{}^{atm}$ value, whereas $G(H^+)_{sol}{}^{mol}$ = $G(H^+)_{sol}{}^{atm}$ −1.8942 [kcal mol^{-1}] can be compared to the experimental data. To estimate the Gibbs free energies of the solvated species involved in the deprotonation process, quantum-chemical computational methods at different levels of theory have been employed [5,6]. The calculations performed for the water medium revealed that the application of Equations (1)–(3) in the calculation of pK_a presents three primary challenges:

(i) The accuracy of pK_a reproduction is contingent upon the computational method employed to determine G(RH) and G(R^-); it varies within an MAE (mean absolute error) range of 0.51–2.86 pK_a for monoprotic acids, depending on the method used in the calculations [5].
(ii) The value of $\Delta G(H^+)_{aq}$, as determined by experimental [23], theoretical [24–30], and mixed [30] approaches, ranges from −244.9 to −266.7 [kcal mol^{-1}], which significantly influences the predicted pK_a values.
(iii) Equation (1) was effectively utilized to reproduce pK_a parameters for monoprotic [5–7] compounds; however, its application to polyprotic molecules undergoing multi-step proton detachment yielded [4] results that deviated significantly from the experimental ones.

To address the first challenge, calculations are conducted using a range of available semi-empirical and ab initio methods, incorporating various basis sets and solvation models to identify those that most accurately reproduce experimental pK_a values [5,6]. Additionally, a linear similarity relation [5,31–33] such as

$$pK_a = a \cdot pK_a(the) + b \quad (4)$$

is employed to enhance the accuracy of pK_a reproduction by adjusting the a and b parameters to the set of experimental pK_a = pK_a(exp) data and theoretical pK_a(the) values. This approach improves the accuracy of pK_a prediction from MAE = 0.51–2.86 to

MAE = 0.30–0.67 in pK_a units for 22 monoprotic acids analyzed [5]. In the case of thiols, the MAE = 0.57–0.96 [32], whereas for phenols, carboxylic acids, and amines, the MAE values are lower than 0.7 pK_a units [31]. However, this level of accuracy still remains higher than the experimental error range of 0.01–0.1 in pK_a units obtained, e.g., for the multi-step deprotonation of GA [12,13]. To concurrently address the first and second challenges, Dutra et al. [6] proposed an approach wherein the average $\langle G(H^+)_{sol} \rangle$ of the $G(H^+)_{sol}$ values could be calculated using the equation below

$$G(H^+)_{sol} = 1.3642 \cdot pK_a(exp) + G(RH)_{sol} - G(R^-)_{sol} + 1.8942 \quad (5)$$

depending on the experimental pK_a(exp) and the theoretical $G(RH)_{sol}$ and $G(H^-)_{sol}$ values obtained using diverse calculation methods for a specific class of compounds with similar chemical structures. This average value is subsequently incorporated into Equation (1) to form Equation (6):

$$pK_a = \frac{G(R^-)_{sol} - G(RH)_{sol} + \left\langle G(H^+)_{sol} \right\rangle - 1.8942}{1.3642} \quad (6)$$

which can be used to calculate the pK_a values for all compounds within the studied class. Calculations performed using this scheme for 22 monocarboxylic acids in an aqueous medium revealed [6] that the combination of density functional theory (DFT) with the local spin density approximation (LSDA) functional provided superior performance in nearly all calculations, achieving accuracy levels comparable to those obtained using the G4CEP (Gaussian 4 Compact Effective Pseudopotential) method employed by Silva and Custodio [5]. In this study, the calculations conducted for 22 monoprotic acids using the G4CEP method, in conjunction with (i) the SMD solvation model, (ii) SMD +1 H_2O (with one explicit water molecule), and (iii) the SMD + 1 H_2O with a linear correction based on Equation (4) and experimental pK_a(exp) values, yielded mean absolute errors of 0.83, 0.51, 0.30 in pK_a units, respectively. A comparative analysis with the results obtained by DFT at the B3LYP and BMK theory levels revealed significant errors reaching up to 2 pK_a units. In contrast, the largest deviations observed for the G4CEP method rarely exceeded 1 pK_a unit [5].

Based on the results reported above, the primary objectives of the present study were fourfold:

(i) To employ a modified version of the method proposed by Dutra et al. [6] to determine the pK_a1,2,3,4 parameters for the four-stage deprotonation of GA, within the bounds of experimental errors of 0.01–0.40 pK_a units.

(ii) To calculate the Gibbs free energy of the hydrated proton $G(H^+)_{aq}$ and proton hydration $\Delta G(H^+)_{aq}$ by fitting this parameter via the improved version of Equation (4) to the experimental pK_a values reported in [11–14].

(iii) To select the experimental data that most accurately represent the pK_a values for GA within the framework of the proposed approach.

(iv) To demonstrate that the accurate reproduction of the pK_a of GA, within the range of experimental errors, can be achieved using the DFT method and LSDA/QZVP level of theory combined with the SMD solvation model, without the explicit inclusion of a water molecule.

2. Results and Discussion

To achieve the primary objectives of this work, the geometries of the neutral GA molecule and its four anionic forms (Figure 2) were optimized in the water medium employing the DFT method at the different theory levels (LSDA, M062X, B3LYP), and the basis set by QZVP and the SMD solvation model and the reasons for their application are set out in the Materials and Methods section. The optimized geometries of the neutral GA molecule and its anionic forms generated at the LSDA/QZVP theory level are presented in Figure 3, while Tables 1 and 2 report the corresponding Gibbs free energy values and their differences used in the determination of the pK_a1,2,3,4 parameters.

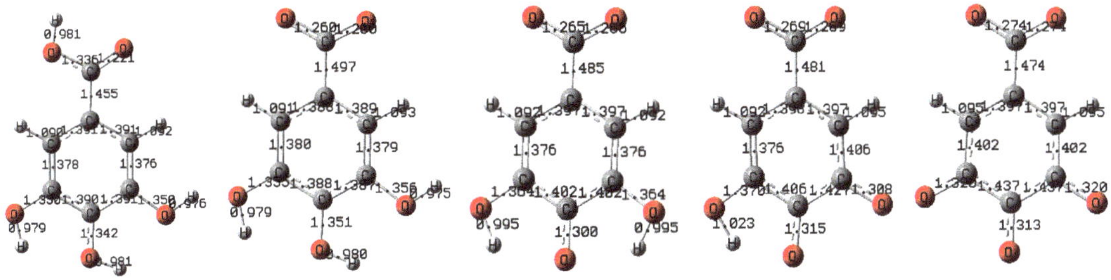

Figure 3. The optimized geometries of neutral GA molecule and its anionic forms GA^{-N} N = 1,2,3,4. The calculations were performed in a water medium at the LSDA/QZVP theory level, using the SMD solvation model. The values of the bond lengths are presented.

Table 1. The differences ΔG_N [kcal mol^{-1}] in Gibbs free energies of neutral GA^0 molecule and its anionic forms GA^{-N} N = 1,2,3,4 including the zero-point energy corrections from Table S1. The calculations were performed in the water medium at the LSDA, M062X, and B3LYP/QZVP theory levels, using the SMD solvation model.

	ΔG_N [kcal mol^{-1}]	LSDA	M062X	B3LYP
ΔG_1	$G(GA^{-1})_{aq}$–$G(GA^0)_{aq}$	267.6773	270.6210	272.7539
ΔG_2	$G(GA^{-2})_{aq}$–$G(GA^{-1})_{aq}$	270.8450	279.5730	280.5099
ΔG_3	$G(GA^{-3})_{aq}$–$G(GA^{-2})_{aq}$	283.2258	289.4136	290.8582
ΔG_4	$G(GA^{-4})_{aq}$–$G(GA^{-3})_{aq}$	295.9617	297.0059	298.7058

Table 2. The values of the parameters a, c (dimensionless), and $G(H^+)_{aq}$ ([kcal mol^{-1}]) fitted via Equation (7) to the four sets of experimental pK_a for GA reported in [11–14]. The indicators of goodness of the fit/standard error (SE) of estimation and the coefficient of determination, R^2, are presented.

	LSDA	M062X	B3LYP	LSDA	M062X	B3LYP
Parameter		**Set I**			**Set II**	
$G(H^+)_{aq}$	−265.47(32)	−62.1(3.2)	−266.6(2.5)	−265.775(20)	−268.12(65)	−270.51(44)
a	5.87(59)	2.02(80)	2.84(94)	7.87(64)	5.36(48)	5.90(96)
c	0.177(38)	0.51(11)	0.41(10)	0.114(33)	0.247(66)	0.216(61)
SE	0.3861	0.2031	0.2630	0.4959	0.4020	0.4383
R^2	0.9918	0.9977	0.9962	0.9910	0.9941	0.9930
Parameter		**Set III**			**Set IV**	
$G(H^+)_{aq}$	−265.764(26)	−267.78(51)	−270.29(38)	−265.7155(75)	−266.81(45)	−269.63(15)
a	7.59(83)	4.91(57)	5.48(60)	7.271(57)	3.60(34)	4.33(16)
c	0.126(23)	0.275(41)	0.240(40)	0.1858(31)	0.420(31)	0.364(13)
SE	0.3327	0.2261	0.2666	0.0454	0.1453	0.0766
R^2	0.9959	0.9981	0.9973	1.0000	0.9995	0.9994

In this study, we examined two potential scenarios (Figure 4) for the deprotonation of GA, involving the dianion A, which possesses a higher energy, at $G(GA^{-2}) = -642.539452$ [Ha], than its conformer B, at $G(GA^{-2}) = -642.544071$ [Ha]. The energy disparity arises from a conformational change that contributes additional energy 2.90 [kcal mol^{-1}]. The rotation barrier for an A $\rightarrow$ B transition is equal to 2.93 [kcal mol^{-1}], whereas for B $\rightarrow$ A it is 5.61 [kcal mol^{-1}]. These results were calculated using the free energy of the transition state $[AB]^{\ddagger}$ $G^{\ddagger} = -642.534128$ [Ha]. Despite the energetic difference between dianions A and B, the reaction pathways for both molecules are isoenergetic as $\Delta G_2 + \Delta G_3 = \Delta G'_2 + \Delta G'_3$. However, incorporating the low-energy dianion B into the calculations yields $\Delta G'_2$ and $\Delta G'_3$ values that fail to accurately reproduce pK_a2 and pK_a3 descriptors. For this reason, all calculations were carried out using scheme A.

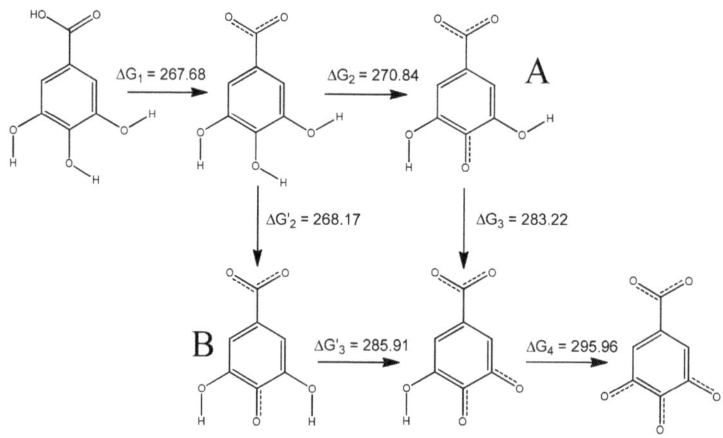

Figure 4. Potential scenarios for the deprotonation of GA, involving dianions A and B. The values (in [kcal mol^{-1}]) of Gibbs free energies for the A reaction pathway are taken from Table S1 and Table 1, whereas $\Delta G'_2 = 268.1668$ and $\Delta G'_3 = 285.9096$ [kcal mol^{-1}]. Both processes are isoenergetic as $\Delta G_2 + \Delta G_3 = \Delta G'_2 + \Delta G'_3$.

The calculations performed at the LSDA/QZVP theory level using the SMD solvation model indicated that the method based on Equations (5) and (6) yields an unsatisfactory reproduction of the experimental pK_a values of GA. For example, set IV provides the mean value $\langle G(H^+)_{aq} \rangle = -263(5)$ [kcal mol^{-1}], exhibiting a substantial standard error, and produces pK_a = 1.77, 4.09, 13.16, 22.50. Attempts to refine these results using Equation (4) also failed to generate satisfactory agreement with the experimental pK_a (exp) values of 4.16, 8.55, 11.40, and 12.80. For the adjusted parameters a = 0.362 (125) and b = 5.47 (1.65) obtained with a standard error (SE) of 2.0479 of estimation and a coefficient of determination (R^2) of 0.8077, the corrected pK_a values were equal to 6.11, 6.95, 10.24, 13.62. This discrepancy suggests that the relationship between the theoretical and experimental pK_a values is more complex than the linear relation (4) utilized so far. Accordingly, in this study, the linear similarity relation (4) was replaced with a non-linear self-similar formula like that shown below

$$\begin{aligned} pK_aN &= a \cdot pK_a(\text{the})^c = a\left[\frac{\Delta G_N + G(H^+)_{aq} - 1.8942}{1.3642}\right]^c \\ \Delta G_N &= G(GA^{-N})_{aq} - G(GA^{-(N-1)})_{aq} \; N = 1,2,3,4 \end{aligned} \tag{7}$$

which simplifies to the linear relation (4) with b = 0 when c approaches 1. The parameters a, c, and $G(H^+)_{aq}$ in Equation (7) fitted to the four sets of available experimental GA data have subsequently been employed to reproduce the pK_a1,2,3,4 parameters characterizing the four-step deprotonation of GA. The results of these calculations are reported in Tables 2 and 3.

Table 3. The values of pK_a1,2,3,4 reproduced by Equation (7) with parameters from Tables 1 and 2 evaluated for the four sets of experimental pK_a data reported in [11–14]. The mean absolute error (MAE) of the data reproduction is presented.

	pK_a (exp)		pK_a		pK_a (exp)		pK_a	
Parameter	**Set I**	**LSDA**	**M062X**	**B3LYP**	**Set II**	**Set I**	**LSDA**	**M062X**
pK_a1	4.50	4.52	4.52	4.52	4.4(1)	4.61	4.40	4.40
pK_a2	7.05	6.92	6.95	6.93	8.8(1)	8.67	8.67	8.67
pK_a3	8.75	9.06	8.91	8.96	10.0(1)	10.40	10.32	10.35
pK_a4	10.25	10.05	10.17	10.14	11.4(1)	11.13	11.20	11.17
MAE		0.165	0.090	0.115		0.253	0.163	0.178
Parameter	**Set III**	**LSDA**	**M062X**	**B3LYP**	**Set IV**	**LSDA**	**M062X**	**B3LYP**
pK_a1	4.44	4.55	4.44	4.44	4.16(2)	4.161	4.155	4.168
pK_a2	8.54	8.45	8.46	8.46	8.55(1)	8.536	8.607	8.600
pK_a3	10.05	10.32	10.23	10.26	11.40(10)	11.44	11.28	11.37
pK_a4	11.30	11.12	11.19	11.16	12.80(40)	12.78	12.87	12.87
MAE		0.163	0.118	0.108		0.019	0.063	0.060

The results obtained reveal that the approach proposed enables both the precise determination of the Gibbs free energy of the hydrated proton $G(H^+)_{aq} = -265.7155(75)$ [kcal mol^{-1}] and the accurate reproduction of the experimental pK_a values by taking advantage of the LSDA/QZVP theory level and set IV of the datapoints [13]. The results fall within the range of experimental errors, with MAE = 0.019 in pK_a units. As experimental data are typically represented by the Gibbs free energy of the proton hydration, this parameter can also be determined through a fitting procedure utilizing Equations (3) and (7). The findings presented in Table 4 demonstrate that parameters $\Delta G(H^+)_{aq} = -259.4272(75)$ [kcal mol^{-1}], a = 7.271(57), and c = 0.1858(31) reproduce set IV [13] of pK_a with R^2 = 1.0000 and SE = 0.0454.

Table 4. The values of the parameters a, c (dimensionless), and $\Delta G(H^+)_{aq}$ ([kcal mol^{-1}]) fitted to the four sets of experimental pK_a for GA via Equation (7) including Equation (3) and ΔG_N calculated at the LSDA/QZVP theory level using the SMD solvation model and a water medium. The indicators of goodness of the fit/standard error (SE) of estimation and the coefficient of determination (R^2) are presented. The mean absolute error (MAE) and the normalized mean absolute error (NMAE defined in Equation (8)) of the reproduced pK_aN parameters are presented.

Parameter	Set I	Set II	Set III	Set IV
$\Delta G(H^+)_{aq}$	−259.19(32)	−259.486(20)	−259.476(26)	−259.4272(75)
a	5.87(59)	7.871(39)	7.59(43)	7.271(57)
c	0.177(38)	0.114(33)	0.126(23)	0.1858(31)
R^2	0.9918	0.9910	0.9959	1.0000
SE	0.3861	0.4959	0.3327	0.0454
pK_a1	4.51	4.40	4.44	4.16
pK_a2	6.92	8.67	8.45	8.54
pK_a3	9.08	10.40	10.32	11.44
pK_a4	10.05	11.13	11.12	12.78
MAE	0.160	0.199	0.134	0.019
NMAE		1.992		0.463

The $\Delta G(H^+)_{aq}$ =−259.4272(75) calculated using the LSDA/QZVP theory level and set IV is consistent with the experimental value of −259.5 [kcal mol^{-1}] obtained by Lim et al. [23]. This value corresponds to the average −4.5 [V] of five measurements of the standard hydrogen potential (−4.43, −4.43, −4.44, −4.48, −4.73 [V]), which allows for the determination of a range of the experimental values of $\Delta G(H^+)_{aq}$ from −254 to −261 [kcal mol^{-1}] [23]. The reproduction of the experimental pK_a values IV for GA using relationship (7) proved to be more accurate than that obtained using linear Equation (4). The parameters $<\Delta G(H^+)_{aq}>$ = −258.6(7.4) [kcal mol^{-1}], a = 0.36(13), and b = 5.4(1.7) reproduce set IV of pK_a with R^2 = 0.8079 and SE = 0.7118, yielding corrected pK_a values of 5.96, 6.25, 7.38, 8.55. These results clearly indicate that, for the multivariate deprotonation of GA, the linear transformation (4) should be replaced by the self-similar formula given in Equation (7). Furthermore, the plots of the relationship linking pK_aN and ΔG_N (Figure 5) generated for sets I, II, III, and IV highlight the strongly non-linear nature of this relation. The graphs of the pK_aN as function of ΔG_N qualitatively have the same form for ΔG_N calculated at the LSDA, M062X, and B3LYP/QZVP theory levels.

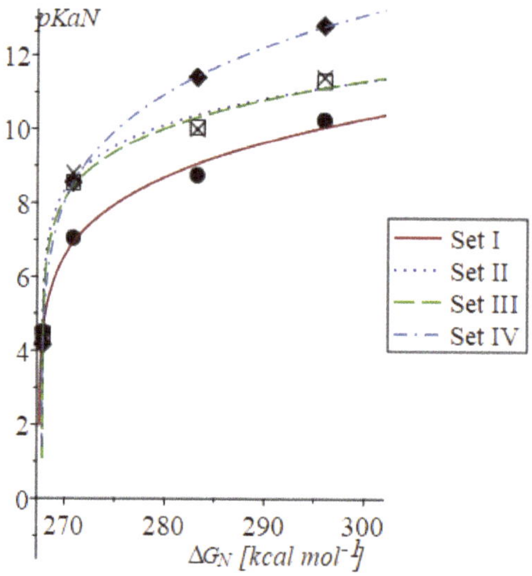

Figure 5. Plots of the self-similar relationship (7) with ΔG_N calculated at the LSDA/QZVP theory level, using the SMD model (water medium) and the four experimental datasets represented by the graphical symbols I—•; II—□; III—x; and IV—♦.

In works [5,6], it was demonstrated that the accuracy of reproducing the experimental pK_a values for monoprotic acids can be enhanced through the implementation of an SMD solvation model that incorporates an additional explicitly included water molecule. Given that the model proposed in this work effectively reproduces set IV, achieving a maximum R^2 of 1.0000, and a satisfactory MAE = 0.019, NMAE = 0.463 in pK_a units, the inclusion of the water molecule should not affect the accuracy of the calculations. To substantiate this hypothesis, the calculations were performed utilizing the scheme proposed by Kelly et al. [21], which generates a = 6.15(12), c = 0.2648(79), and $\Delta G(H^+)_{aq}$ = −260.166(44) [kcal mol^{-1}] from ΔG_1 = 266.4521, ΔG_2 = 268.6051, ΔG_3 = 278.0447, and ΔG_4 = 299.3631 [kcal mol^{-1}], values that reproduce the pK_a of set IV with SE = 0.073, MAE = 0.030, and NMAE = 0.7651 in pK_a units. These results exhibit a lower degree of accuracy compared to those obtained without the inclusion of the water molecule in the calculations at the LSDA/QZVP theory level. To investigate the impact of dispersion effects on the pK_a and $\Delta G(H^+)_{aq}$ parameters, calculations were performed employing the QZVP basis set, the SMD solvation model, and (i) the functional developed by Head-Gordon and collaborators [34], which incorporates Grimme's dispersion model D2 [34], and the B3LYP functional combined with the D3 dispersion model. Such a combination has been recommended in [35]. The computational results, summarized in Table S2, indicate that dispersion effects have a negligible influence on the deprotonation of GA. It is noteworthy, however, that the value of the parameter $\Delta G(H^+)_{aq} = -264.10(23)$ [kcal mol^{-1}] is consistent with $\Delta G(H^+)_{aq} = -264.29$ [kcal mol^{-1}] reported by Zhan and Dixon [27]. To evaluate the impact of basis sets on the accuracy of the reproduction of the pK_a and $\Delta G(H^+)_{aq}$ parameters for GA, calculations were conducted at the 6311++G(d,p), aug-cc-pVQZ, QZVP/LSDA theory levels in conjunction with the SMD solvation model. The results obtained are presented in Table 5. The findings indicate that employing the QZVP basis set is not essential for accurately calculating the pK_a parameters in the scheme proposed, as the less extensive 6−311++G(d,p) basis set also yields results consistent with the experimental data. However, an analysis of the $\Delta G(H^+)_{aq}$ parameter values reveals that the use of larger, more comprehensive basis sets is necessary to achieve result that align closely with the experimental one of $\Delta G(H^+)_{aq} = -259.5$ [kcal mol^{-1}] reported by Lim et al. [23].

Table 5. The effect of basis set on the theoretical reproduction of proton hydration energy $\Delta G(H^+)_{aq}$ and pK_aN parameter values. The calculations were performed in the water medium at the LSDA/QZVP theory levels, using the SMD solvation model. The experimental value of $\Delta G(H^+)_{aq} = -259.5$ [kcal mol^{-1}] was reported by Lim et al. [23].

	6311++G(d,p)	aug-cc-pVQZ	QZVP
$\Delta G(H^+)_{aq}$	−257.1671(51)	−259.2358(72)	−259.4272(75)
a	7.156(34)	7.268(54)	7.271(57)
c	0.1934(19)	0.1869(29)	0.1858(31)
R^2	1.0000	1.0000	1.0000
SE	0.0265	0.0430	0.0454
pK_a1	4.159	4.160	4.161
pK_a2	8.542	8.537	8.536
pK_a3	11.423	11.435	11.436
pK_a4	12.788	12.778	12.776
MAE	0.011	0.017	0.019
NMAE	0.268	0.427	0.463

3. Materials and Methods

For monocarboxylic acids, the best reproduction of the pK_a values was achieved [5,6] using the composite G4CEP and DFT method in conjunction with the LSDA [36], M06-2X [37], and B3LYP [38,39] functionals, and the SMD solvation model as well as the Dunning's [40] basis sets aug-cc-pVDZ and aug-cc-pVTZ. In light of the findings presented in [5,6], in this work, calculations were conducted utilizing the DFT method implemented in Gaussian vs. 16 software [40] in conjunction with the LSDA, M062X, and B3LYP functionals. To identify a basis set capable of accurately reproducing the Gibbs energy for GA and its anionic forms, preliminary calculations were performed on the energy of GA in an aqueous environment, employing the DFT/LSDA method and the SMD solvation model, utilizing basis sets cc-pVQZ, aug-cc-pVQZ [40], and QZVP [41]. The resulting energy values (in [Ha]: E_1(cc-pVQZ) = −643.496286; E_2(aug-cc-pVQZ) = −643.501176; E_3(QZVP) = −643.511057), and their corresponding differences (in [kcal mol^{-1}]: $\Delta E_{12} = 3.0685$; $\Delta E_{23} = 6.2004$; $\Delta E_{13} = 9.2689$), guided the selection of the quadruple zeta valence polarized (QZVP) basis set [41] for subsequent calculations, as it yielded the lowest energy value for the GA. A similar situation was observed for the M062X and B3LYP functionals; consequently, the DFT method at the LSDA, M062X, and B3LYP/QZVP level of theory, combined with the SMD solvation model, was employed for the primary calculations. The input structures were constructed by taking advantage of the Gauss View-6.1 graphical interface [42], and calculations were carried out using the Gaussian version 16 A software [42,43]. Since the computations for the compounds studied at the LSDA,M062X,B3LYP/QZVP level are time consuming and their convergence depends on the initial geometry input, the optimization was divided into two stages: (i) the determination of an approximate geometry at the LSDA,M062X,B3LYP/cc-pVQZ theory levels in the gas phase, and (ii) the final calculation at the LSDA,M062X,B3LYP/QZVP theory levels in the water medium using the SMD solvation model. Thus, optimization was performed at each stage, with the geometry determined at the lower level serving as the starting point for optimization at the higher level. The optimized structures of the compounds considered in this work are displayed in Figure 3. The parameters a, c, $G(H^+)_{aq}$, and $\Delta G(H^+)_{aq}$ were fitted to the four sets of experimental pK_a values for GA, including set I [4.50, 7.05, 8.75, 10.25] [11], set II [4.4(1), 8.8(1), 10.0(1), 11.4(1)] [12], set III [4.44, 8.54, 10.05, 11.3] [14], and set IV [4.16(2), 8.55(1), 11.4(1), 12.8(4)] [13]. The calculations were performed by taking advantage of Sigma Plot vs. 11 software—the results of calculations together with indicators of goodness of the fit, R^2, coefficient of determination, and SE, standard error of estimation, are presented in Tables 2 and 4. The fitted parameters were utilized to calculate the pK_aN values for GA, as presented in Tables 3 and 4, along with the mean absolute errors (MAEs) and the normalized mean absolute errors (NMAEs) associated with the reproduction of the experimental pK_aN(exp) data, defined as follows

$$\mathrm{MAE} = \frac{1}{4}\sum_{N=1}^{4}|\mathrm{p}K_a\mathrm{N}\text{-}\mathrm{p}K_a\mathrm{N(exp)}| \quad \mathrm{NMAE} = \frac{1}{4}\sum_{N=1}^{4}\frac{|\mathrm{p}K_a\mathrm{N}\text{-}\mathrm{p}K_a\mathrm{N(exp)}|}{u_N} \tag{8}$$

Here, u_N denotes experimental errors in the measurements provided for datasets II (u_N = 0.1) [12] and IV (u_1 = 0.02, u_2 = 0.01, u_3 = 0.10, u_4 = 0.40) [13]. A value of NAME ≤ 1 indicates the reproduction of experimental data within the accuracy of the measurement error.

4. Conclusions

The proposed method for determining the pK_a parameters of GA enables the simultaneous validation of both the method used and the experimental data employed in the calculations. Among the four datasets analyzed, the non-linear model demonstrated the best agreement with dataset IV [13], which exhibited the highest statistical consistency when reproduced using the DFT method at the LSDA/QZVP theory level and the SMD solvation model. The parameters presented in Table 3 reproduce experimental dataset IV with a standard error, SE, of 0.0454 and a coefficient of determination, R^2, of 1.0000, achieving reproduction of the experimental pK_a values [13] for GA within the limits of experimental errors as NMAE = 0.463 pK_a units. These findings confirm the validity of the self-similar Formula (7) in particular and the overall calculation methodology in general. Additionally, the reproduction of the experimental pK_a values for GA using relationship (7) has been proved to be more accurate than that obtained with Equation (4). The results obtained indicate that, for the multivariate deprotonation of GA, the linear transformation (4) should be replaced by the self-similar formula given in Equation (7). The application of Equation (7) in conjunction with the ΔG_N values calculated using the DFT method at the LSDA/QZVP theory level, alongside the SMD solvation model, is sufficient to achieve a satisfactory reproduction of the measured pK_a values of GA. Thus, the utilization of the computationally time-consuming G4CEP method and the SMD model with the inclusion of an additional water molecule, as suggested in works [5,6], is deemed unnecessary for this objective. It is acknowledged that the proposed approach represents a significant simplification of the real deprotonation process due to the complex reactions that occur between GA and water. These reactions result in substantial modifications to the structure of GA [44] and its anionic forms, which in turn have a pronounced effect on the kinetics and thermodynamics of GA deprotonation. The proposed model is not derived from the first principles but constitutes a quasi-phenomenological framework, wherein the semi-empirical parameters a and c encapsulate all *hidden variables* that are not explicitly incorporated into the analysis. A first-principles model should, at the molecular level, account for all possible structures of GA, including polymorphs, rotamers, and tautomers and their anionic forms, as well as their interactions with solvent molecules and the formation of complex polymolecular systems. It will be essential to undertake research in this regard in the near future.

Supplementary Materials: The following supporting information can be downloaded at: https://www.mdpi.com/article/10.3390/molecules30030742/s1, Table S1. The Gibbs and zero-point energies [Ha] of neutral GA^0 molecule and its anionic forms GA^{-N} N = 1,2,3,4 calculated in the water medium at the LSDA, M062X, B3LYP/QZVP theory levels, using the SMD solvation model. Table S2. The effect of dispersion on the theoretical reproduction of proton hydration energy $\Delta G(H^+)_{aq}$ and pK_aN parameter values. The calculations were performed in a water medium at the wB97XD/D2//QZVP, B3LYP/D3/QZVP theory levels, using the SMD solvation model. The theoretical value of $\Delta G(H^+)_{aq} = -264.29$ [kcal mol^{-1}] is reported by Zhan and Dixon [27]. For comparison, the results obtained at the B3LYP/QZVP level of theory without taking the dispersion effect into account are shown.

Funding: This research received no external funding.

Institutional Review Board Statement: Not applicable.

Informed Consent Statement: Not applicable.

Data Availability Statement: All new data created are reported in this work.

Acknowledgments: This research was supported by the computing grant PL0293-01 from Supercomputing and Networking Center (PCSS) in Poznań, Poland. I would like to thank Piotr Kopta from PCSS for technical support and Julian Molski for his constructive comments.

Conflicts of Interest: The author declares no conflicts of interest.

References

1. Gregory, M.T.; Gao, Y.; Cui, Q.; Yang, W. Multiple deprotonation paths of the nucleophile 3′-OH in the DNA synthesis reaction. *Proc. Natl. Acad. Sci. USA* **2021**, *118*, e2103990118. [CrossRef] [PubMed]
2. Alongi, K.S.; Shields, G.S. Theoretical Calculations of Acid Dissociation Constants: A Review Article. *Annu. Rep. Comput. Chem.* **2010**, *6*, 113–138. [CrossRef]
3. Marino, T.; Galano, A.; Russo, N. Radical scavenging ability of gallic acid toward OH and OOH radicals. Reaction mechanism and rate constants from the density functional theory. *J. Phys. Chem. B* **2014**, *118*, 10380–10389. [CrossRef] [PubMed]
4. Molski, M. Theoretical study on the radical scavenging activity of gallic acid. *Heliyon* **2023**, *9*, e12806. [CrossRef]
5. De Souza Silva, C.d.S.; Custodio, R. Assessment of pK_a determination for monocarboxylic acids with an accurate theoretical composite method: G4CEP. *J. Phys. Chem. A* **2019**, *123*, 8314–8320. [CrossRef]
6. Dutra, F.R.; Silva, C.d.S.; Custodio, R. On the accuracy of the direct method to calculate pK_a from electronic structure calculations. *J. Phys. Chem. A* **2021**, *125*, 65–73. [CrossRef]
7. Walton-Raaby, M.; Floen, T.; García-Díez, G.; Mora-Diez, N. Calculating the aqueous pK_a of phenols: Predictions for antioxidants and cannabinoids. *Antioxidants* **2023**, *12*, 1420. [CrossRef]

8. Morency, M.; Néron, S.; Iftimie, R.; Wuest, J.D. Predicting pK_a values of quinols and related aromatic compounds with multiple OH groups. *J. Org. Chem.* **2021**, *86*, 14444–14460. [CrossRef]
9. Badhani, B.; Kakkar, R. Influence of intrinsic and extrinsic factors on the antiradical activity of gallic acid: A theoretical study. *Struct. Chem.* **2018**, *29*, 359–373. [CrossRef]
10. Masoud, M.S.; Ali, A.E.; Haggag, S.S.; Nasr, N.M. Spectroscopic studies on gallic acid and its azo derivatives and their iron(III) complexes. *Spectrochim. Acta A Mol. Biomol. Spectrosc.* **2014**, *120*, 505–511. [CrossRef]
11. Agrawal, M.D.; Bhandari, C.S.; Dixit, M.K.; Sogani, N.C. 3,4,5-Trihydroxybenzoesäure als Chelatbildner, 1. Mitt.: Praseodym. *Monatsh. Chem.* **1976**, *713*, 75–82. [CrossRef]
12. Huguenin, J.; Hamady, S.O.S.; Bourson, P. Monitoring deprotonation of gallic acid by Raman spectroscopy. *J. Raman Spectrosc.* **2015**, *46*, 1062–1066. [CrossRef]
13. Kipton, H.; Powell, J.; Taylor, M.C. Interactions of iron(II) and iron(III) with gallic acid and its homologues: A potentiometric and spectrophotometric study. *Aust. J. Chem.* **1982**, *35*, 739–756. [CrossRef]
14. Loginova, L.F.; Medyntsev, V.V.; Khomutov, B.I. Acidic properties of gallic acid and stability constants of iron complexes. *Zh. Obshch. Him.* **1972**, *42*, 739–742.
15. Cossi, M.; Rega, N.; Scalmani, G.; Barone, V. Energies, structures, and electronic properties of molecules in solution with the C-PCM solvation model. *J. Comput. Chem.* **2003**, *24*, 669–681. [CrossRef]
16. Tomasi, J.; Mennucci, B.; Cancès, E. The IEF version of the PCM solvation method: An overview of a new method addressed to study molecular solutes at the QM ab initio level. *J. Mol. Struct. THEOCHEM* **1999**, *464*, 211–226. [CrossRef]
17. Marenich, A.V.; Cramer, C.J.; Truhlar, D.G. Universal Solvation Model Based on Solute Electron Density and on a Continuum Model of the Solvent Defined by the Bulk Dielectric Constant and Atomic Surface Tensions. *J. Phys. Chem. B* **2009**, *113*, 6378–6396. [CrossRef]
18. Xu, L.; Coote, M.L. Methods To Improve the Calculations of Solvation Model Density Solvation Free Energies and Associated Aqueous pK_a Values: Comparison between Choosing an Optimal Theoretical Level, Solute Cavity Scaling, and Using Explicit Solvent Molecules. *J. Phys. Chem. A* **2019**, *123*, 7430–7438. [CrossRef]
19. Pezzola, S.; Venanzi, M.; Galloni, P.; Conte, V.; Sabuzi, F. Towards the "Eldorado" of pKa Determination: A Reliable and Rapid DFT Model. *Molecules* **2024**, *29*, 1255. [CrossRef]
20. Bryantsev, V.S.; Diallo, M.S.; Goddard, W.A. Calculation of solvation free energies of charged solutes using mixed cluster/continuum models. *J. Phys. Chem. B* **2008**, *112*, 9709–9719. [CrossRef]
21. Kelly, C.P.; Cramer, C.J.; Truhlar, D.G. Adding explicit solvent molecules to continuum solvent calculations for the calculation of aqueous acid dissociation constants. *J. Phys. Chem. A* **2006**, *110*, 2493–2499. [CrossRef] [PubMed]
22. Fifen, J.J.; Dhaouadi, Z.; Nsangou, M. Revision of the thermodynamics of the proton in gas phase. *J. Phys. Chem. A* **2014**, *118*, 11090–11097. [CrossRef]
23. Lim, C.; Bashford, D.; Karplus, M. Absolute pKa calculations with continuum dielectric methods. *J. Phys. Chem.* **1991**, *95*, 5610–5620. [CrossRef]
24. Tissandier, M.D.; Cowen, K.A.; Feng, W.Y.; Gundlach, E.; Cohen, M.H.; Earhart, A.D.; Coe, J.V.; Tuttle, T.R. The Proton's Absolute Aqueous Enthalpy and Gibbs Free Energy of Solvation from Cluster-Ion Solvation Data. *J. Phys. Chem. A* **1998**, *102*, 7787–7794. [CrossRef]
25. Klots, C.E. Solubility of protons in water. *J. Phys. Chem.* **1981**, *85*, 3585–3588. [CrossRef]
26. Tawa, G.J.; Topol, I.A.; Burt, S.K.; Caldwell, R.A.; Rashin, A.A. Calculation of the aqueous solvation free energy of the proton. *J. Chem. Phys.* **1998**, *109*, 4852–4863. [CrossRef]
27. Zhan, C.-G.; Dixon, D.A. Absolute Hydration Free Energy of the Proton from First-Principles Electronic Structure Calculations. *J. Phys. Chem. A* **2001**, *105*, 11534–11540. [CrossRef]
28. Marković, Z.; Tošović, J.; Milenkovic, D.; Markovic, S. Revisiting the solvation enthalpies and free energies of the proton and electron in various solvents. *Comput. Theor. Chem.* **2016**, *1077*, 11–17. [CrossRef]
29. Rimarčík, J.; Lukeš, V.; Klein, E.; Ilčin, M. Study of the solvent effect on the enthalpies of homolytic and heterolytic N–H bond cleavage in p-phenylenediamine and tetracyano-p-phenylenediamine. *J. Mol. Struct. THEOCHEM* **2010**, *952*, 25–30. [CrossRef]
30. Mejías, A.; Lago, S. Calculation of the absolute hydration enthalpy and free energy of H+ and OH−. *J. Chem. Phys.* **2000**, *113*, 7306–7316. [CrossRef]
31. Galano, A.; Pérez-González, A.; Castañeda-Arriaga, R.; Muñoz-Rugeles, L.; Mendoza-Sarmiento, G.; Romero-Silva, A.; Ibarra-Escutia, A.; Rebollar-Zepeda, A.M.; León-Carmona, J.R.; Hernández-Olivares, M.A.; et al. Empirically Fitted Parameters for Calculating pK_a Values with Small Deviations from Experiments Using a Simple Computational Strategy. *J. Chem. Inf. Model.* **2016**, *56*, 1714–1724. [CrossRef] [PubMed]

32. Pérez-González, A.; Castañeda-Arriaga, R.; Verastegui, B.; Carreón-González, M.; Alvarez-Idaboy, J.R.; Galano, A. Estimation of empirically fitted parameters for calculating pK a values of thiols in a fast and reliable way. *Theor. Chem. Acc.* **2018**, *137*, 5. [CrossRef]
33. Miguel, E.L.M.; Silva, P.L.; Pliego, J.R. Theoretical Prediction of pK_a in Methanol: Testing SM8 and SMD Models for Carboxylic Acids, Phenols, and Amines. *J. Phys. Chem. B* **2014**, *118*, 5730–5739. [CrossRef] [PubMed]
34. Chai, J.-D.; Head-Gordon, M. Long-range corrected hybrid density functionals with damped atom–atom dispersion corrections. *Phys. Chem. Chem. Phys.* **2008**, *10*, 6615–6620. [CrossRef]
35. Schröder, H.; Hühnert, J.; Schwabe, T. Evaluation of DFT-D3 dispersion corrections for various structural benchmark sets. *J. Chem. Phys.* **2017**, *146*, 044115. [CrossRef]
36. Parr, R.G.; Yang, W. *Density-Functional Theory of Atoms and Molecules*; Oxford University Press: New York, NY, USA; Oxford, UK, 1989.
37. Zhao, Y.; Truhlar, D.G. The M06 suite of density functionals for main group thermochemistry, thermochemical kinetics, noncovalent interactions, excited states, and transition elements: Two new functionals and systematic testing of four M06-class functionals and 12 other functionals. *Theor. Chem. Acc.* **2008**, *120*, 215–241. [CrossRef]
38. Becke, A.D. Density-functional thermochemistry. III. The role of exact exchange. *J. Chem. Phys.* **1993**, *98*, 5648–5652. [CrossRef]
39. Lee, C.; Yang, W.; Parr, R.G. Development of the Colle-Salvetti correlation-energy formula into a functional of the electron density. *Phys. Rev. B* **1988**, *37*, 785–789. [CrossRef]
40. Dunning, T.H. Gaussian basis sets for use in correlated molecular calculations. I. The atoms boron through neon and hydrogen. *J. Chem. Phys.* **1989**, *90*, 1007–1023. [CrossRef]
41. Weigend, F.; Ahlrichs, R. Balanced basis sets of split valence, triple zeta valence and quadruple zeta valence quality for H to Rn: Design and assessment of accuracy. *Phys. Chem. Chem. Phys.* **2005**, *7*, 3297–3305. [CrossRef]
42. Frisch, M.J.; Trucks, G.W.; Schlegel, H.B.; Scuseria, G.E.; Robb, M.A.; Cheeseman, J.R.; Scalmani, G.; Barone, V.; Petersson, G.A.; Nakatsuji; et al. *Gaussian 16*; Revision A.03; Gaussian, Inc.: Wallingford, CT, USA, 2016.
43. Glendening, E.D.; Reed, A.E.; Carpenter, J.E.; Weinhold, F. NBO Version 3.1. 2017.
44. Braun, D.E.; Bhardwaj, R.M.; Florence, A.J.; Tocher, D.A.; Price, S.L. Complex Polymorphic System of Gallic Acid—Five Monohydrates, Three Anhydrates, and over 20 Solvates. *Cryst. Growth Des.* **2013**, *13*, 19–23. [CrossRef]

molecules

MDPI

Article

Binding Free Energy Analysis of Colicin D, E3 and E8 to Their Respective Cognate Immunity Proteins Using Computational Simulations

Mahesh Koirala [1,2,*] and Clifton K. Fagerquist [1]

[1] Department of Agriculture, Produce Safety & Microbiology, Western Regional Research Center, Agricultural Research Service, U.S., Albany, CA 94710, USA; clifton.fagerquist@usda.gov
[2] Department of Energy, Research Participation Program Administered by the Oak Ridge Institute for Science and Education, U.S., Oak Ridge, TN 37830, USA
* Correspondence: mahesh.koirala@usda.gov

Abstract: Colicins are antimicrobial proteins produced by bacteria for the purpose of destroying neighboring bacteria. Colicin activity is neutralized by a specific cognate immunity protein in order to protect the host. This study investigates the structural and binding mechanisms underlying the interaction of colicin-D, -E3 and -E8 to their respective immunity proteins (ImD, Im3 and Im8) using structure prediction, molecular dynamics (MD) simulations and MM-PBSA approach of free energy calculations. High-confidence colicin-immunity (Col-Im) complex structures predicted using AlphaFold2 were subjected to MD simulations of 150 ns with GROMACS and were analyzed for the binding free energy calculation using gmx_MMPBSA. Results showed that the complex of Col_E3-Im3 exhibited the most favorable binding free energy, driven by strong van der Waals and electrostatic interactions. Col_D-ImD and Col_E8-Im8 also showed the favorable binding. Electrostatics and hydrogen bonding emerged as a key factor driving binding and stability, while polar solvation acted as a destabilizing factor across all systems. These outcomes provide an understanding of the molecular mechanisms of Col-Im systems, with potential applications for developing natural antimicrobials for food safety.

Keywords: colicins; immunity proteins; molecular dynamics simulations; MM-PBSA; protein–protein interactions; binding free energy; AlphaFold2; food safety

1. Introduction

Colicins are plasmid-encoded proteins secreted by pathogenic bacteria to attack competing bacteria that occupy the same local environment [1]. Colicins destroy the DNA, RNA (t-RNA or ribosomal RNA) or outer membrane of neighboring bacteria. For colicin to not attack the bacterial host that produced it, an immunity gene is co-expressed with the colicin gene. The immunity protein strongly binds to its cognate colicin until it is secreted from the host cell. Upon attachment of the colicin/immunity complex to the outer membrane receptor of a neighboring bacterial cell, the complex undergoes rearrangement such that the immunity protein detaches from its colicin partner. Although there is an associated metabolic burden to the bacterial host for plasmid carriage, plasmid genes can provide a survival benefit to the host, especially in environments where bacteria compete for limited resources. Colicins effectively act as natural antibiotics deployed by bacteria against bacterial competitors [1].

The only defense against colicin is the cognate immunity protein, whose plasmid-encoded genes must be expressed by the bacterium under attack. A plasmid-encoded immunity gene may be expressed in excess (or large excess) beyond that of the 1:1 stoichiometric ratio of colicin-to-immunity in order to have sufficient immunity protein in the event of an external colicin attack, i.e., a basal expression of the immunity gene [2]. However, colicin and their cognate immunity genes are under SOS regulatory control, which means their genes are repressed by a DNA-binding protein (LexA) that binds to a DNA inverted repeat (SOS box), blocking the expression of genes downstream [3]. The SOS response is triggered by DNA damage: single-stranded DNA, UV radiation or DNA-damaging antibiotics. RecA is the master regulator of the SOS response, and its initiation results in the self-cleavage of LexA, thus unblocking the expression of colicin and immunity genes [3]. It is essential that a colicin nuclease, once translated, is inactivated so that it does not destroy host DNA or RNA. In consequence, the immunity protein must recognize and bind quickly and tightly to its cognate colicin. While the colicin and immunity genes are co-expressed, it is essential to have, at least, a slight excess of the immunity protein so that the colicin is quickly inactivated while inside the host cell.

Colicin-bearing plasmids (D, E3 and E8) were found to be quite prevalent amongst environmental isolates of pathogenic Shiga toxin-producing *E. coli* (STEC) collected from agricultural regions and previously analyzed by our group [4–7]. As such, there is a need to better understand these three distinct colicin nucleases, whose targets are tRNA, rRNA, and DNA, respectively. E3 and E8 attach to the same outer membrane receptor BtuB, whereas D gains entry by attachment to FepA [1]. In the current study, we have calculated the binding energies of immunity proteins: ImD, Im3 and Im8 to their respective colicins: D, E3 and E8 to better understand the interactions between colicins and their cognate immunity proteins.

1.1. Protein–Protein Interactions

Protein–protein interactions (PPIs) are essential to every biological process and act as the basis for signal transduction, cellular organization, immune responses, and metabolic regulation. Proteins interact, forming the complexes associated in a precise manner, which are critical for the proper functioning of the cells and organisms. Several diseases, including cancer, neurogenerative disorders, and infection, occur due to irregularities in these interactions, making PPIs a major aspect of biomedical research [8–12]. The structural complexity of PPIs represents challenges as well as opportunities in drug discovery. Many PPIs have historically been classified as "undruggable" due to the large, flat binding surfaces that lack clear pockets for small-molecule drugs. However, recent advances in high-throughput screening, computational modeling, and fragment-based drug discovery have enabled scientists to target these previously subtle interactions [10]. Moreover, PPIs are increasingly recognized as potential therapeutic targets in diseases where protein interactions drive pathological processes, such as the p53-MDM2 interaction in cancer [13,14]. PPIs also play a crucial role in microbial competition, which is of great interest in the area of food production and safety. Bacterial proteins like colicins obstruct the growth of competing microorganisms and influence microbial ecology. Colicins and their corresponding immunity proteins are known to contribute to bacterial fitness, leading to harnessing and improving food safety by targeting harmful pathogens in foodborne bacteria [1]. Understanding these interactions is essential for developing novel biocontrol approaches to boost food safety and avoid adulteration in agricultural systems.

Recent developments in computational techniques and resources, including molecular dynamics (MD) simulations, machine learning (ML) algorithms, and protein structure prediction using artificial intelligence (AI), have modernized and advanced the study of PPIs [15]. Tools like AlphaFold have dramatically improved the accuracy of protein structure prediction, while MD simulations allow for the investigation of PPI dynamics and binding free energy calculations in unprecedented detail [16]. These computational tools, alongside experimental approaches like X-ray crystallography, cryo-electron microscopy, nuclear magnetic resonance (NMR), and native state mass spectrometry, provide a comprehensive view of PPI networks and their role in cellular function [17–20].

1.2. Binding Free Energy

Binding free energy is a critical thermodynamic quantity that provides insights into the stability and affinity of molecular complexes, including protein–protein, protein–ligand and protein–DNA/RNA interactions. It is crucial for understanding molecular recognition, biological interactions, and complex formation, which are crucial in drug design, protein engineering, and biomolecular studies. Molecular dynamics (MD) simulations combined with free energy calculation methods have proven to be computationally effective tools for estimating these energies [21–23].

The binding free energy (ΔG_{bind}) is represented as the free energy difference between the bound complex and the unbound components. It can be stated mathematically as:

$$\Delta G_{bind} = G_{complex} - G_{prot1} - G_{prot2} \tag{1}$$

where $G_{complex}$ is the free energy of the bound complex, G_{prot1} and G_{prot2} are the free energies of the unbound proteins separately. Various computational approaches exist to estimate binding free energy. Among these, Molecular Mechanics Poisson-Boltzmann Surface Area (MM-PBSA) and Molecular Mechanics Generalized Born Surface Area (MM-GBSA) methods are widely used due to their balance of computational efficiency and accuracy.

1.3. Molecular Mechanics Poisson-Boltzmann Surface Area (MM-PBSA)

In this method, the binding free energy is approximated by summing the gas-phase molecular mechanics (MM) energy, solvation-free energies, and an entropic term. The binding free energy is calculated as:

$$\Delta G_{bind} = \Delta E_{MM} + \Delta G_{solv} - T\Delta S \tag{2}$$

In which

$$\Delta \mathrm{E}_{\mathrm{MM}} = \Delta E_{int} + \Delta E_{elec} + \Delta \mathrm{E}_{vdw} \tag{3}$$

$$\Delta G_{solv} = \Delta G_{polar} + \Delta G_{non-polar} \tag{4}$$

where $\Delta \mathrm{E}_{\mathrm{MM}}$ is the difference in molecular mechanics energy (van der Waals and electrostatic interactions). ΔG_{solv} is the solvation-free energy through polar contributions estimated using the Poisson-Boltzmann (PB) equation and nonpolar contributions valued based on surface area (SA). $T\Delta S$ denotes the entropic involvement, typically estimated using normal mode analysis, quasi-harmonic analysis, or Schlitter's formula [24].

1.4. Molecular Mechanics Generalized Born Surface Area (MM-GBSA)

This method is like MM-PBSA but uses the Generalized Born (GB) model to evaluate solvation-free energy instead of the PB equation. MM-GBSA follows the same general formula as MM-PBSA, with the solvation term (ΔG_{solv}) computed using the GB model. These approaches use the representative conformational snapshots of the molecular system obtained from molecular dynamics simulations. With the average of the generated snapshots, MM-PBSA and MM-GBSA provide approximate estimates of the binding free energy by integrating the dynamic behavior of the molecules [21,25]. Additionally, the entropy term ($-T\Delta S$) is essential for estimating the overall binding free energy. Accurate entropy calculations can be computationally challenging and more complex. Different approaches, such as normal mode analysis, Schlitter's entropy [24], and quasi-harmonic approximation, are frequently used to estimate this term [26].

In the current study, we focus on the interaction between bacterial protein complexes, specifically colicin D, E3 and E8 and their immunity protein cognates. These protein complexes are crucial to microbial competition, bacterial survival and regulation within bacterial communities, directly affecting the safety of food production environments. We use molecular dynamics simulations and free energy calculations, providing insights into the structural and energetic mechanisms guiding the stability and specificity of these protein complexes. The findings may provide a significant contribution to a better understanding of PPIs and open new paths for food safety research.

2. Results

2.1. Structure of Colicin/Immunity Complexes

The structural details of the three colicin/immunity protein complexes colicin D-immunity D (Col_D-ImD), colicin E3-immunity E3 (Col_E3-Im3), and colicin E8-immunity E8 (Col_E8-Im8) predicted from AlphaFold2 (Figures 1a, 2a and 3a) reveal both similarities and unique features that contribute to their binding mechanisms and the neutralization of colicin toxicity. It is found that all the colicins (Col_D, Col_E3, and Col_E8) have extended helical structures that play a vital role in defining their architecture. The smaller immunity proteins (ImD, Im3, and Im8) bind tightly to critical regions of the colicins, temporarily disabling their toxic effects. This binding is a key structural feature that ensures the immunity proteins can effectively inhibit the enzymatic function (ribonuclease or DNase) of the colicins, allowing the host to be protected. The electrostatic interactions between the colicins and their cognate immunity protein are shown (Figures 1b, 2b and 3b), with negatively charged regions shown in red color and positively charged regions shown in blue color. In Col_D-ImD, ImD has distinct positively and negatively charged regions that lie within the charge complementary of Col_D. This kind of electrostatic interaction is essential for guiding the tight binding between the two proteins, maintaining the stability of the complex. Similarly, in Col_E3-Im3, Im3 shows tight electrostatic interaction with Col_E3 near the ribonuclease domain. The charge distribution on Im3 is critical for stabilizing the binding interaction that temporarily disables Col_E3 toxicity. A closer view of the electrostatic potential highlights the charge distribution at the binding interface of Col_E8-Im8, where Im8 wraps around the DNase domain of Col_E8. This electrostatic mapping between two regions of negative and positive charges further stabilizes their interaction. Hence, all three complexes separately show electrostatic complementarity, playing a central role in driving the binding interactions.

Another important feature of these interactions is the number of hydrogen bonds between the colicins and their corresponding immunity proteins (Table 1). Col_D-ImD represents the robust intermolecular interactions with 13 initial hydrogen bonds. Where Col_E3-Im3 presents 11 hydrogen bonds in the native state, Col_E8-Im8 forms only 10 hydrogen bonds. This reflects the relatively weaker interaction in Col_E8-Im8 compared to Col_E3-Im3. This helps to underscore the importance of hydrogen bonding in determining the binding strength of the colicin/immunity complexes. The differences in charge distribution and bond formation confirm that each immunity protein can only bind to and neutralize its respective colicin, providing a clear example of how electrostatic potential drives protein–protein interactions in these systems. Understanding these structural and electrostatic features provides valuable insight into the protective mechanisms that immunity proteins use to safeguard against colicin toxicity.

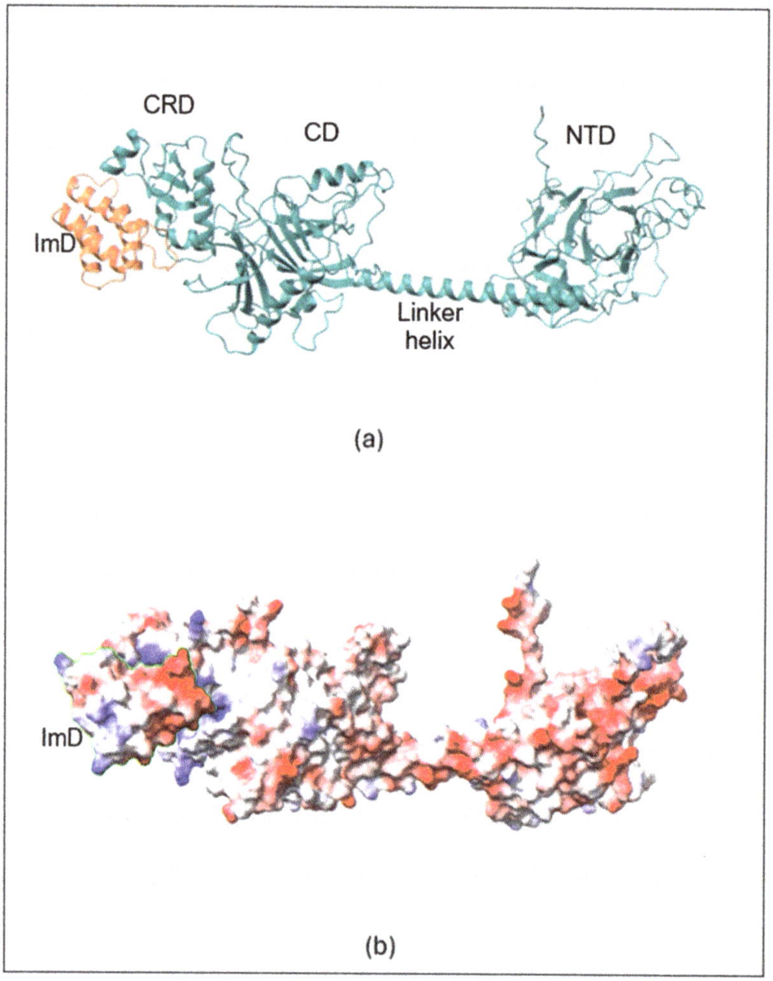

Figure 1. (**a**) AlphaFold2 predicted structure of Col_D-ImD. Central domain (CD), N-terminal domain (NTD), C-terminal RNase domain (CRD). (**b**) Electrostatic mapping on the complex with ImD outlined by the green line.

Table 1. Intermolecular hydrogen bonds and buried (interfacial) water molecules.

Protein Complex	H-Bonds Before Simulation	Average H-Bonds After Simulation	Average Number of Buried Water Molecules
Col_D-ImD	13	15.83	20.15
Col_E3-Im3	11	13.56	23.55
Col_E8-Im8	10	8.50	11.81

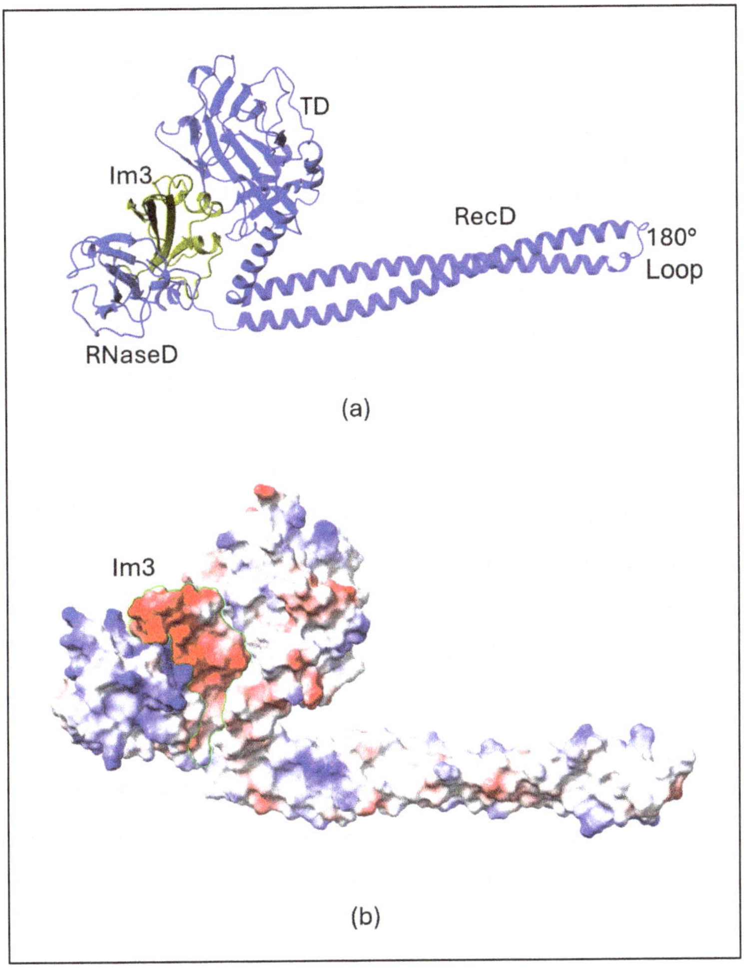

Figure 2. (**a**) AlphaFold2 predicted structure of Col_E3-Im3. Receptor domain (RecD), translocation domain (TD) and RNase domain (RNaseD). (**b**) Electrostatic mapping on the complex with Im3 outlined by the green line.

2.2. MD Simulations

MD simulations provide comprehensive insights into how a protein interacts with its biological partner, offering a holistic understanding crucial for guiding protein–protein interactions. The root-mean-square deviation (RMSD) analysis of the colicin/immunity complexes over the last 50 ns of the simulation shows robust stability and convergence for all complexes, as shown in Figure 4. Col_D-ImD shows the structural stability and consistent equilibrium dynamics with RMSD varying between 0.8 and 1.2 nm across both independent runs. With RMSD values ranging from 0.6 to 1.2 nm, Col_E3-Im3 shows moderate fluctuations, highlighting the stable dynamics and comparable trends. Meanwhile, in Col_E8-Im8, RMSD values vary between 0.6 and 1.0 nm, relating slightly more pronounced variations but suggesting stability with dynamic movement throughout the simulation period. These results demonstrate that all three complexes maintain overall structural stability during the analyzed timeframe. The RMSD cluster distribution analysis for Colicin

D, E3, and E8 complexes (Figure 5) provides insights into their conformational stability over 100–150 ns of the simulations. The clustering analysis revealed that each complex primarily occupies a few dominant conformational states, with the largest cluster consistently capturing most of the trajectory frames. Colicin D-ImD exhibited a well-converged distribution with minimal variation between runs, whereas Colicin E3-Im3 showed noticeable differences between Run 1 and Run 2, indicating a broader sampling of conformations. In contrast, Colicin E8-Im8 displayed a more compact cluster distribution, highlighting stable interaction interfaces, confirming that the simulations adequately captured the relevant conformational space and supporting the robustness of the chosen simulation time.

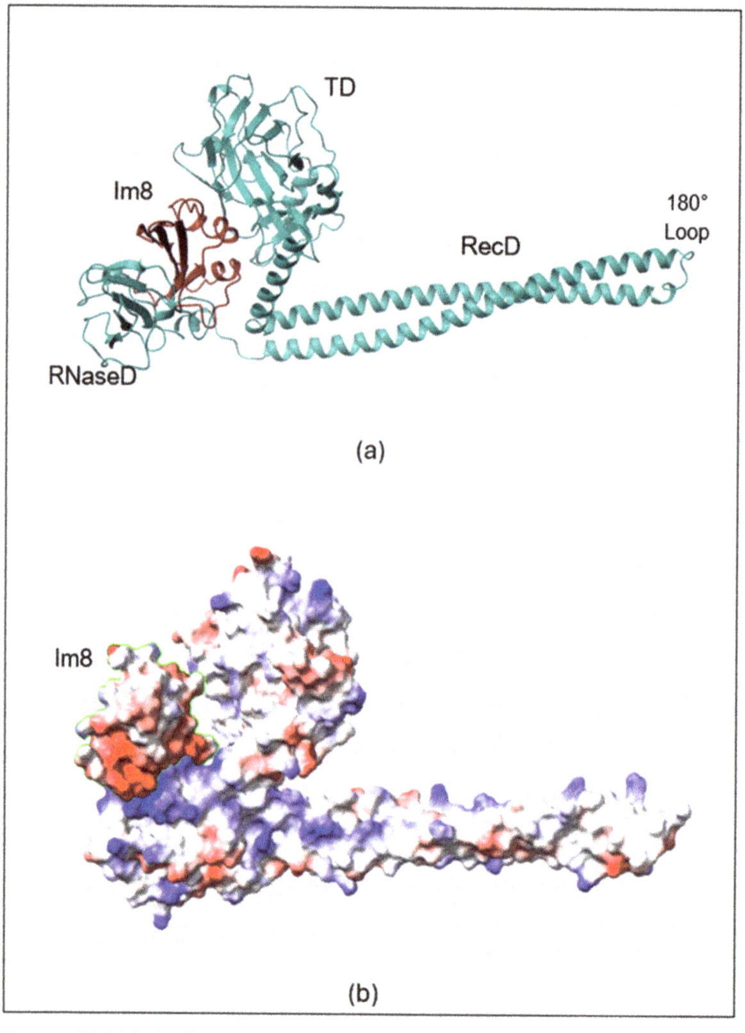

Figure 3. (**a**) AlphaFold2 predicted structure of Col_E8-Im8. Receptor domain (RecD), translocation domain (TD) and RNase domain (RNaseD). (**b**) Electrostatic mapping on the complex with Im8 outlined by the green line.

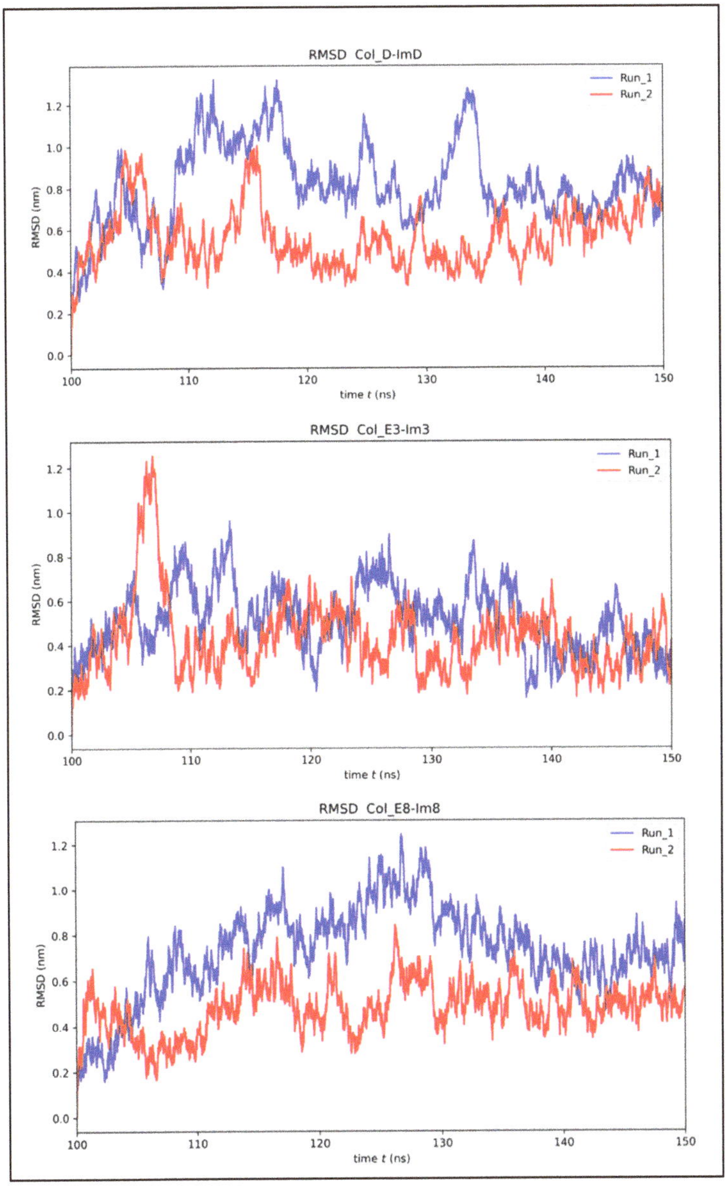

Figure 4. RMSD analysis of colicin/immunity complexes over 150 ns, showing stability in the 100–150 ns windows used for free energy analysis.

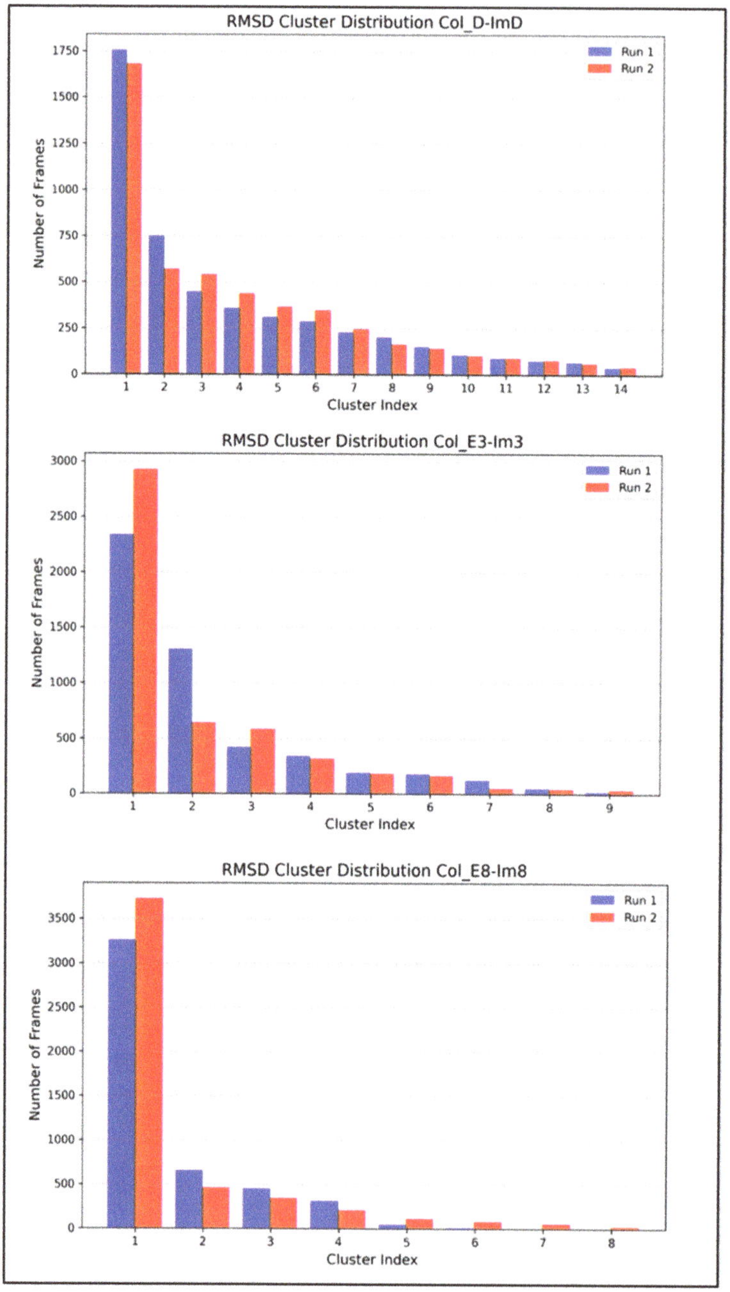

Figure 5. RMSD cluster analysis of colicin/immunity complexes in the 100–150 ns windows used for free energy analysis.

The hydrogen bond (H-bond) analysis and average number of buried water molecules provide a deeper insight into the intermolecular interactions within the colicin-immunity complexes, as depicted in Table 1 and Figure 6. In Col_D-ImD, the average number of hydrogen bonds increases from 13 before the simulation to 15.83 during the last 50 ns, with notable fluctuations between 15 and 25 and indicates stable interaction with some flexibility. The average number of 20.15 buried water molecules reveals that interfacial stabilization depends on water-mediated hydrogen bonds to compensate for its flexibility. In Col_E3-Im3, the average number of hydrogen bonds rises from 11 initially to 13.56, with variation between 10 and 22. This reflects a more compact interface as supported by the average 23.55 buried water molecules that balance hydration and direct interactions. Col_E8-Im8 shows a decrease in average hydrogen bonds from 10 to 8.5 during the same timeframe, ranging between 10 and 18 and exhibited the lowest hydration with just 11.81 average water molecules. This reflects a tightly packed and stable interface dominated by direct protein–protein interactions. Thus, a clear map of the relation between hydrogen bonds, hydration and interfacial compactness is shown, highlighting reduced hydration correlating with tighter packing as seen in Col_E8-Im8. On the other hand, moderate hydration contributes to stability and flexibility, as seen in Col_D-ImD and Col_E3-Im3. The interfacial hydrogen bonds are also visualized and compared (Figure 7) for the last frame of each complex. This structural representation illustrates hydrogen bond strength with labeled distance, providing insight into relative strength and stability. Stronger interactions are indicated by shorter distances (1.7–2.2 Å), while weaker bonds appear at greater distances (>2.5 Å). The difference in the interaction landscape across the complexes also highlights the different interfacial stabilization mechanisms.

The root-mean-square fluctuation (RMSF) analysis for the colicin/immunity complexes, provided in the Supplementary Materials, Figures S1–S3, demonstrates residue-level flexibility for both the colicin and immunity proteins across two independent simulation runs. In Col_D-ImD, the Col_D protein displays higher fluctuations at the N- and C-terminal regions, along with some loop regions, while the ImD protein shows lower fluctuations with peaks around loop and interaction regions. Col_E3-Im3 also demonstrates significant flexibility in the Col_E3 protein around residue 300, indicating the dynamic loop regions, while the Im3 protein exhibits moderate flexibility with peaks at specific residues. In Col_E8-Im8, the Col_E8 protein shows pronounced variations around residue 300, whereas the Im8 protein maintains relatively low flexibility, with a peak observed at the C-terminal region.

Additionally, the solvent-accessible surface area (SASA) analyses for the complexes are also provided in Figure S4, which give insights into the compactness and exposure of the protein complexes over the last 50 ns of simulation. In Col_D-ImD, SASA displays fluctuations between 395 and 420 nm^2, with consistent patterns across both runs, reflecting stable solvent exposure. With SASA values ranging from 285 to 315 nm^2, Col_E3-Im3 shows a steady exposure of surface area across the runs. Also, Col_E8-Im8 maintains relatively constant SASA values between 310 and 340 nm^2, highlighting consistent solvent interaction throughout the simulations.

The radius of gyration (Rg) in Figure S5 shows Col_D-ImD has Rg values fluctuating around 4.6 to 5.2 nm, indicating stable compactness with slight variations between two runs. Col_E3-Im3 complex shows Rg values varying from 3.6 to 3.8 nm, with consistent patterns reflecting a stable structural arrangement. Similarly, the Col_E8-Im8 maintains Rg values between 3.5 and 3.9 nm, reflecting the sustained compactness and structural integrity over the simulation timeframe.

These results, consistent across the two independent simulation runs, provide additional validation of the structural stability and dynamics of the colicin/immunity complexes and are presented in the Supplementary Materials for detailed reference.

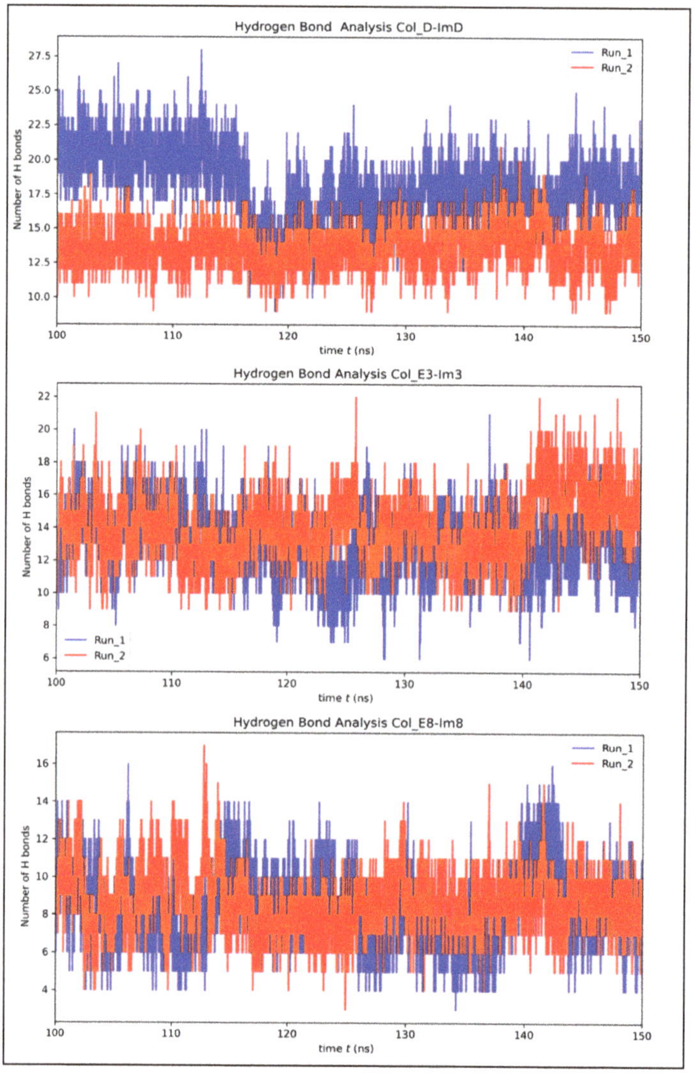

Figure 6. Hydrogen bond analysis of colicin/immunity complexes over 150 ns, showing stability in the 100–150 ns windows used for free energy analysis.

2.3. Binding Free Energy from gmx_MMPBSA

The binding free energy analysis of the colicin/immunity complexes obtained from gmx_MMPBSA presents the significant contributions of various energy components to their stability and interactions. Two independent runs used from two trajectories of random seeds show convergence and reproducibility. Col_D-ImD shows the total binding free energy of -95.9 ± 0.20 kcal/mol (Run 1) and -77.2 ± 0.17 kcal/mol (Run 2), as shown in Figure 8. This favorable binding is predominantly driven by strong van der Waals and electrostatic interactions, which are essential for stabilizing the complex. However, the polar solvation energy is the destabilizing factor with partial counteraction by nonpolar solvation contributions. Similarly, Col_E3-Im3 resulted in the most favorable total binding free energy of -135.8 ± 0.19 kcal/mol (Run 1) and -144.7 ± 0.18 kcal/mol (Run 2), emphasizing the significant role of robust van der Waals and electrostatic interactions for stabilization. Despite the destabilizing effects of polar solvation energy, these interactions dominate and ensure a highly stable binding. While Col_E8-Im8 complex shows comparative lower total binding free energies of −54.4 ± 0.16 kcal/mol (Run 1) and −57.9 ± 0.17 kcal/mol (Run 2), still, van der Waals and electrostatic interactions drive the binding, with destabilization counteraction from polar solvation energy. Per residue decomposition of the binding free energy of interfacial residues are shown in the Supplementary Materials, Tables S1–S6 highlighting the dominant contributions from van der Waals and electrostatic interactions.

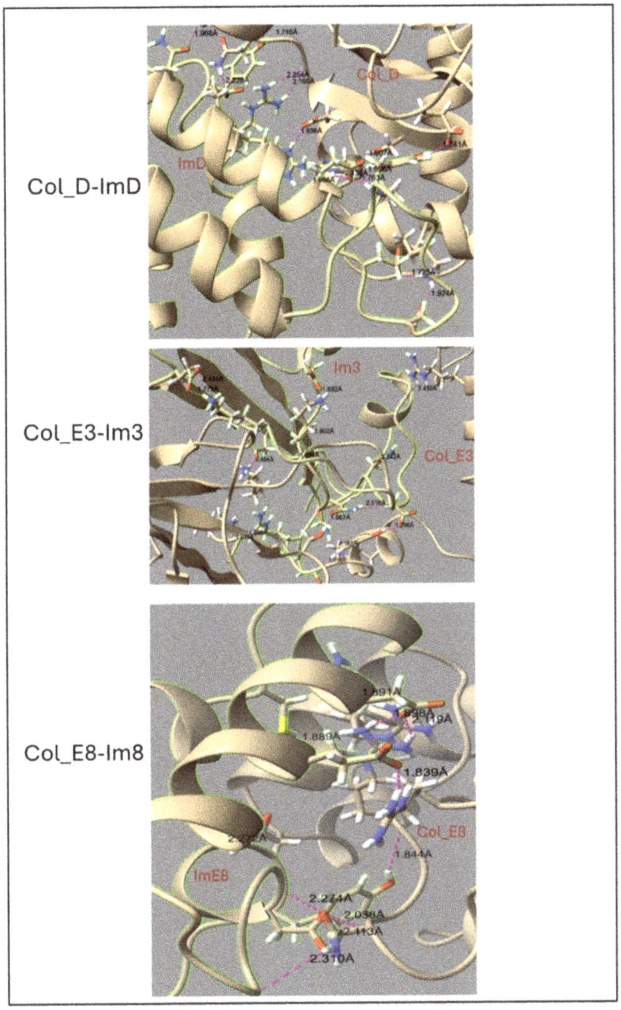

Figure 7. Hydrogen bonding interactions at the colicin/immunity interfaces. The hydrogen bonds are represented (dashed line) with their distance, and all immunity proteins are wrapped with a green selected line.

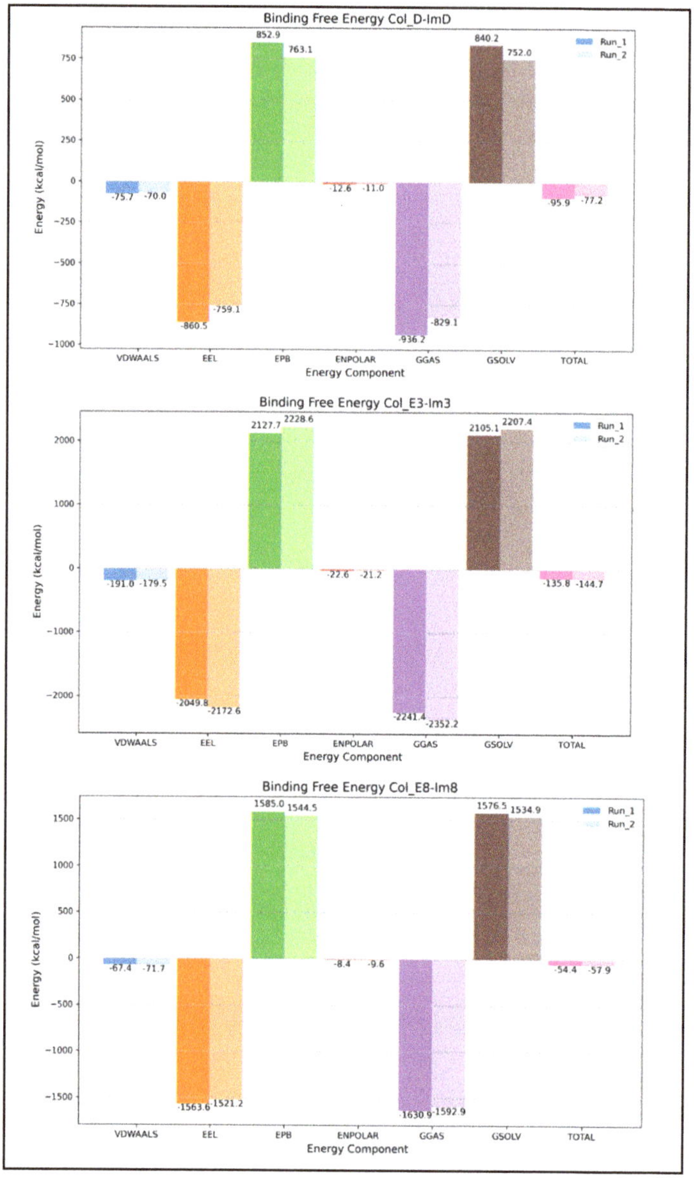

Figure 8. MM-PBSA results of colicin/immunity complexes.

The results shown provide fundamental insights into the binding mechanisms of colicin/immunity complexes, where van der Waals and electrostatic interactions emerge as the primary driving forces of the binding stability for each system, and polar solvation acts as a destabilizing factor but is mitigated by nonpolar solvation contributions. The most negative binding free energy of the Col_E3-Im3 underscores its robust intermolecular interactions, making it the most stable complex. These results help to highlight the roles of the molecular forces governing colicin-immunity binding, with potential implications for their biological roles and applications in food safety. The consistent trends observed across independent simulation runs further signify the reliability of the results and their broader applicability.

3. Discussion

We studied colicin-immunity protein interactions' structural integrity and dynamics through molecular dynamics simulations and free energy calculations approaches. The consistent binding behaviors highlighting the driving forces provide insights into molecular stability and activity. The RMSD-based cluster analysis confirms the sample convergence into a few dominant clusters within the last 100–150 ns, supporting the sufficiency of the 150 ns simulation for the relative binding energy comparison. The Col_E3-Im3 complex exhibited the most favorable binding free energy, driven by strong van der Waals interactions, electrostatic complementarity, and stable hydrogen bonding. These interactions contributed to a tightly packed protein/protein interface and low root-mean-square-deviation (RMSD) observed during the last 50 ns of the simulation [1,27]. Conversely, Col_D-ImD and Col_E8-Im8 complexes showed weaker binding energies, depicted by fewer stable hydrogen bonds and higher hydration dynamics [1,28]. This represents the importance of intermolecular forces and interface hydration in guiding binding stability. Dynamic structural behavior was observed in Colicin D and Colicin E8 complexes, especially during the early stages of simulation following positional restraint removal. RMSD and RMSF analyses highlighted significant deviations, mirroring relaxation and interface adjustment. These adjustments are fundamental for the biological roles of these systems. For Colicin D, flexibility at the N-termini facilitates efficient interaction with bacterial tRNA [1]. Colicin E8's dynamic nature aligns with its function as a DNase, which requires adaptability to engage bacterial DNA while avoiding host toxicity [1,24].

3.1. Overestimation of Computational Binding Energies Compared to the Experiment

Several factors may result in computational overestimation of the binding free energies of Col-Im complexes compared to that obtained from experimental data [22,29–31]. First, sampling limitations may contribute to the divergence of computational versus experimental binding energies [32]. The finite timescale of MD simulations and the number of independent simulations performed may result in gaps in the full exploration of relevant conformational space. Also, entropic contributions are, of course, incorporated into any experimental result. However, theoretical estimation of entropic contributions for large biomolecule complexes is difficult and computationally intensive, leading to simplifications (e.g., quasi-harmonic analysis [26]) or the exclusion of entropic contributions from simulations entirely. Additionally, standard force fields may not accurately reflect experimental conditions, leading to discrepancies between experimental binding energies and computational results [33]. Furthermore, differences between simulation parameters and experimental conditions (pH, ionic strength, etc.) could also contribute to discrepancies [34]. For example, water molecules buried at the interface of the protein complex may significantly affect binding energy as they may form a network of hydrogen bonds that shield and reduce electrostatic interactions between ligand (Im) and substrate (colicin). In addition, water has a fixed dipole (1.85 Debye) that can also act to shield electrostatic interactions between protein complex partners. These subtle dipole effects and hydrogen bond networks may be difficult to estimate with MD simulations and binding free energy calculations such that they accurately reflect experimental conditions. For instance, Zhou has raised the issue of the ionic strength of the solution having a significant effect on the association rates (k_a) of the colicin E3-Im3 complex [35,36].

3.2. RMSF Maxima

Of particular interest is the RMSF maxima at residue 300 (**M**300) of Col_E3 and Col_E8 shown in the top panels of Supplemental Figures S2 and S3, respectively. The maxima correspond to an unstructured loop region (AHDP**M**AG and AHDP**M**SG, respectively) resulting in a 180° hairpin turn of the colicin backbone chain that connects two 100 Å long coiled-coil antiparallel α-helices of the R-domain as shown in Figures 2a and 3a. It has been proposed that this unstructured, primarily hydrophobic loop region of the R-domain serves as the initial point of attachment of the Col_E3/Im3 complex to the hydrophobic pocket of the BtuB outer membrane receptor of a bacterial cell [2]. A model was proposed that the loop region also acts as a molecular *hinge*, allowing subsequent separation of the two globular regions (rRNase-domain and T-domain) that surround and bind Im3 [30]. Once rRNase and T-domains separates and become in physical contact with the outer and/or inner membranes of the cell, the immunity protein, being less tightly bound, detaches having served its primary purpose.

3.3. Contribution of Interfacial Buried Water Molecules

Analysis of buried water molecules at the protein/protein interfacial region disclosed hydration dynamics as an essential factor in stabilizing the protein–protein interface. The complexes of Col_D and Col_E3 with their immunity protein were dependent on interfacial water molecules to maintain flexibility and adaptability, with hydration contributing to transient hydrogen bonds that stabilize the interaction. In contrast, the Col_E8 complex exhibited fewer interfacial water molecules, revealing a compact, tightly packed interface but potentially condensed flexibility [21]. The close matching of the number of buried water molecules in Col_E3-Im3 with previously published work [37] Also signifies its relevance and importance. These hydration approaches align closely with the complexes' observed functional roles and differential stability. In addition, the consistent destabilizing effect of polar solvation energy in MM-PBSA calculations underlines a limitation of the Poisson-Boltzmann method, which overestimates the solvation contributions. Additionally, the cut-offs used for residue selection influence the per-residue decomposition, while the total energy accounts for the contributions from all residues. Improvements using tools like *DelPhi*, which provide grid-based solutions for electrostatic potentials, could help improve the solvation accuracy and, hence, binding energy predictions [12,21,38]. Expanding the studies with the available computational resources and integrating advanced sampling techniques with more accurate results help explore the conformational transitions more precisely [39–43]. Exploring these methods with some existing experimental validation will further elucidate the thermodynamic scenarios of these biologically relevant systems [21,24].

4. Materials and Methods

4.1. Bacterial Strains and Plasmids

The colicins and their cognate immunity genes were present in small plasmids carried by three different Shiga toxin-producing *E. coli* (STEC) strains that had been genomically sequenced previously [5,44]. The colicin E3 (PCG41550.1) and its cognate immunity protein (PCG41551.1) sequences were obtained from previous genomic sequencing of *E. coli* O113:H21 strain RM7788 [44]. The colicin E8 (AWJ52140.1) and its cognate immunity protein (AWJ52141.1) sequences were obtained from previous genomic sequencing of *E. coli* O43:H2 strain RM10042 [5]. The colicin D (AWJ36133.1) and its cognate immunity protein (AWJ36142.1) sequences were obtained from previous genomic sequencing of *E. coli* O103:H11 strain RM8385 [5]. Consistent with X-ray crystal structure data [2], eighty-three residues were removed from the N-terminus of both colicin E3 and E8. In addition, the N-terminal methionine was removed from the colicin E3 immunity protein. Sequences are provided in the Supplemental Materials.

4.2. Structure Prediction with AlphaFold2

The structure of colicins and their respective cognate immunity proteins was predicted using AlphaFold2 (version 2.3.2) [16,45], which utilizes state-of-the-art deep learning techniques to predict protein structures accurately. The predictions were conducted on high-performance GPU resources provided by the SciNet Ceres and Atlas clusters (https://scinet.usda.gov). The amino acid sequences of each protein complex were submitted in FASTA format and processed through the AlphaFold2 pipeline. The system routinely generates multiple sequence alignments (MSAs), which are used as input for structural modeling. Out of the predicted models, the structure with the highest pLDDT confidence score and performance metrics was selected for each colicin-immunity complex. This high-confidence structure was used for molecular dynamics simulations and additional computational analysis. The results from AlphaFold2 were essential in further investigations into these protein complexes' dynamics and interaction mechanisms. Additionally, the predicted structures were compared to experimental data to confirm their significance and precision for downstream applications [2,46].

4.3. Molecular Dynamics Simulations with GROMACS

Molecular dynamics (MD) simulations of the colicin-immunity protein complexes were conducted using GROMACS [28] (version 2024.1). The protein structures were obtained from AlphaFold2 predictions and prepared for simulations by adding hydrogens, solvating the system in a cubic box using the TIP3P water model, and neutralizing the system with appropriate counterions. All protonation states of polar residues were assigned based on the default CHARMM36 force field parameters at physiological pH (~7.0) without manual modification. This was done according to the standard protonation conditions and was consistent throughout the simulation. The CHARMM36 [47] force field was employed for the protein complexes.

Energy minimization was performed using the steepest descent algorithm until a convergence criterion of a maximum force of less than 1000 kJ/mol/nm was met. Following minimization, the system was equilibrated under constant volume (NVT) and constant pressure (NPT) conditions for 1 ns each. Temperature coupling at 300 °K was ensured using the V-rescale thermostat [48], while pressure coupling at 1 bar was applied using a Berendsen barostat [49]. Particle Mesh Ewald (PME) was employed to calculate long-range electrostatic interactions, and a 1.2 nm cutoff was applied for van der Waals and short-range electrostatic interactions. Subsequently, equilibration and production MD simulations were conducted for 150 ns using a 2-fs time step under periodic boundary conditions. Positional restraint was applied to Col_D-ImD and Col_E8-Im8 during the first 100 ns of the simulation to stabilize and prevent unrealistic structural deviation and facilitate system equilibration. System configurations were saved every 10 ps for further analysis. The stability of the protein complexes was confirmed by analyzing RMSD, RMSF, SASA, hydrogen bonds and radius of gyration. Interfacial residues were extracted from gmx_MMPBSA per-residue decomposition, and polar residues were counted. The equilibrated trajectory's final 50 ns (100–150 ns of the entire trajectory) was used for binding free energy calculations. The number of buried water molecules at the protein–protein interface was also calculated within the trajectory with a cutoff distance of 0.25 nm, ensuring only tightly bound water within interface residues was included [50].

4.4. Binding Free Energy Calculations Using gmx_MMPBSA

The binding free energy of the colicin/immunity protein complexes was estimated through MM-PBSA approach using gmx_MMPBSA [51]. The Poisson-Boltzmann approach was employed to calculate the polar solvation-free energy, while the non-polar solvation-free energy was determined using the solvent-accessible surface area (SASA) approach. The analysis was based on 5000 snapshots extracted from the final 50 ns of the MD trajectories with frames saved every 10 ps. The binding free energy of each colicin with its respective immunity protein (ΔG_{bind}) was computed using Equation (5).

$$\Delta G_{bind} = \Delta G_{gas} + \Delta G_{solv} \tag{5}$$

where,

$$\Delta G_{gas} = \Delta E_{elec} + \Delta E_{vdw}$$

$$\Delta G_{solv} = \Delta E_{polar} + \Delta E_{non-polar}$$

Here, the total energy (ΔG_{bind}) comprises the gas-phase molecular mechanics energy (ΔG_{gas}) and the solvation-free energy (ΔG_{solv}). The gas-phase energy consists of contribution from van der Waals (ΔE_{vdw}) and electrostatic (ΔE_{elec}). The solvation energy is separated into polar (ΔE_{polar}) and non-polar ($\Delta E_{non\text{-}polar}$) components. Entropy contributions were neglected in this study to diminish computational cost and complexity, focusing merely on the calculation of the effective free energy components (gas-phase interactions and solvation energy) [22,51]. This approach was chosen to streamline the analysis and focus on the key energetic interaction driving binding. Furthermore, residue decomposition analysis was performed to identify the contribution of individual residues to the binding free energy, offering insights into the key residues involved in the protein–protein interactions. The gmx_MMPBSA.sh [51] script was used to compute the energy components.

5. Conclusions

We investigated the molecular determinants of colicin-immunity protein interactions, emphasizing the roles of binding energetics, structural dynamics, and hydration in stabilizing these complexes. Col_E3-Im3 complex is the most stable system, with binding free energy driven by van der Waals forces and electrostatic interactions. Also, Col_D-ImD and Col_E8-Im8 showed the initial dynamic behavior due to positional restraint removal, stabilizing over time with hydration dynamics playing a vital role. With the buried water molecules contributing to interaction flexibility and stability, dynamics hydration can also be identified as an essential factor in the colicin-immunity interface. Col_E3-Im3 and Col_E8-Im8 showed higher hydration, and Col_E8-Im8 demonstrated reduced hydration as it resembled its densely packed interface. This can be aligned with the functional roles of the complexes about their adaptability and specificity in bacterial competition. It can significantly impact food safety and microbial control because colicins represent natural antibacterial agents that can target foodborne pathogens. This study of the structural dynamics and binding behavior of colicin-immunity can lead the way for their role in biocontrol strategies, advancing agricultural and food production systems. Additional studies involving experimental validation and alternative computational approaches could further strengthen our understanding of these biological systems.

Supplementary Materials: The following supporting information can be downloaded at: https://www.mdpi.com/article/10.3390/molecules30061277/s1, Sequence of Colicin and Immunity Proteins;

Figure S1. RMSF analysis of Col_D and ImD; Figure S2. RMSF analysis of Col_E3 and Im3; Figure S3. RMSF analysis of Col_E8 and Im8; Figure S4. SASA of colicin/immunity complexes over 150 ns, showing stability in the 100–150 ns windows used for free energy analysis; Figure S5. Radius of Gyration of colicin/immunity complexes over 150 ns, showing stability in the 100–150 ns windows used for free energy analysis; Table S1. Per-residue decomposition of the binding free energy for Col_D (chain B) in complex with ImD (chain A), obtained in Run1 using the MM-PBSA method with energy values given in kcal/mol; Table S2. Per-residue decomposition of the binding free energy for Col_D (chain B) in complex with ImD (chain A), obtained in Run2 using the MM-PBSA method with energy values given in kcal/mol; Table S3. Per-residue decomposition of the binding free energy for Col_E3 (chain B) in complex with Im3 (chain A), obtained in Run1 using the MM-PBSA method with energy values given in kcal/mol; Table S4. Per-residue decomposition of the binding free energy for Col_E3 (chain B) in complex with Im3 (chain A), obtained in Run2 using the MM-PBSA method with energy values given in kcal/mol; Table S5. Per-residue decomposition of the binding free energy for Col_E8 (chain B) in complex with Im8 (chain A), obtained in Run1 using the MM-PBSA method with energy values given in kcal/mol; Table S6. Per-residue decomposition of the binding free energy for Col_E8 (chain B) in complex with Im8 (chain A), obtained in Run2 using the MM-PBSA method with energy values given in kcal/mol.

Author Contributions: Conceptualization, M.K. and C.K.F.; Computational Work & Analysis, M.K.; Methodology, M.K.; Validation, M.K. and C.K.F.; Writing, M.K. and C.K.F.; Supervision, C.K.F. All authors have read and agreed to the published version of the manuscript.

Funding: This research was supported by USDA-ARS Project Numbers: 2030-42000-055-000-D and by an appointment to the Agricultural Research Service (ARS) Research Participation Program administered by the Oak Ridge Institute for Science and Education (ORISE) through an interagency agreement between the U.S. Department of Energy (DOE) and the U.S. Department of Agriculture (USDA). ORISE is managed by ORAU under DOE contract number DESC0014664. This research used resources provided by the SCINet project of the USDA Agricultural Research Service, ARS project number: 0500-00093-001-00-D.

Institutional Review Board Statement: Not applicable.

Informed Consent Statement: Not applicable.

Data Availability Statement: The data supporting this article have been included in the main text or as part of the Supplementary Materials. The molecular dynamics simulation and binding free energy calculations files can be available on request.

Conflicts of Interest: The authors declare no conflicts of interest.

Abbreviations

MM-GBSA, molecular mechanics/generalized Born surface area; MM-PBSA, molecular mechanics/Poisson Boltzmann surface area; CHARMM36, Chemistry at HARvard Macromolecular Mechanics; MD, molecular dynamics; TIP3P, Three Point Charge Model (TIP3P); NVT, number of particles (N)–volume (V)–temperature (T); NPT, number of particles–pressure–temperature; SASA, solvent-accessible surface area; RMSD, root mean square deviation; RMSF, root mean square fluctuation; PPI, protein–protein interactions; DNA, deoxyribonucleic acid; tRNA, transfer ribonucleic acid; RNA, ribonucleic acid.

References

1. Cascales, E.; Buchanan, S.K.; Duché, D.; Kleanthous, C.; Lloubes, R.; Postle, K.; Riley, M.; Slatin, S.; Cavard, D. Colicin biology. *Microbiol. Mol. Biol. Rev.* **2007**, *71*, 158–229. [CrossRef] [PubMed]
2. Soelaiman, S.; Jakes, K.; Wu, N.; Li, C.; Shoham, M. Crystal structure of colicin E3: Implications for cell entry and ribosome inactivation. *Mol. Cell* **2001**, *8*, 1053–1062. [CrossRef]

3. Michel, B. After 30 years of study, the bacterial SOS response still surprises us. *PLoS Biol.* **2005**, *3*, e255. [CrossRef] [PubMed]
4. Fagerquist, C.K.; Shi, Y.; Park, J. Unusual modifications of protein biomarkers expressed by plasmid, prophage, and bacterial host of pathogenic *Escherichia coli* identified using top-down proteomic analysis. *Rapid Commun. Mass Spectrom.* **2024**, *38*, e9667. [CrossRef]
5. Parker, C.T.; Cooper, K.K.; Huynh, S.; Smith, T.P.; Bono, J.L.; Cooley, M. Genome sequences of eight shiga toxin-producing *Escherichia coli* strains isolated from a produce-growing region in California. *Microbiol. Resour. Announc.* **2018**, *7*, e00807-18. [CrossRef]
6. Fagerquist, C.K.; Shi, Y.; Park, J. Plasmid-encoded colicin immunity proteins produced by pathogenic *Escherichia coli* identified by antibiotic induction, MALDI-TOF-TOF-MS/MS and top-down proteomic analysis. In Proceedings of the 72nd ASMS Conference on Mass Spectrometry and Allied Topics, Anaheim, CA, USA, 2–6 June 2024.
7. Fagerquist, C.K.; Shi, Y.; Koirala, M. Plasmid-encoded gene expression of pathogenic bacteria by antibiotic induction as detected by MALDI-TOF-TOF mass spectrometry and top-down proteomic analysis. *Int. J. Mass Spectrom.* **2025**, *551*, 117430. [CrossRef]
8. Safari-Alighiarloo, N.; Taghizadeh, M.; Rezaei-Tavirani, M.; Goliaei, B.; Peyvandi, A.A. Protein-protein interaction networks (PPI) and complex diseases. *Gastroenterol. Hepatol. Bed Bench* **2014**, *7*, 17–31.
9. Alexov, E. Protein-protein interactions. *Curr. Pharm. Biotechnol.* **2008**, *9*, 55–56. [CrossRef]
10. Arkin, M.R.; Tang, Y.; Wells, J.A. Small-molecule inhibitors of protein-protein interactions: Progressing toward the reality. *Chem. Biol.* **2014**, *21*, 1102–1114. [CrossRef]
11. Lu, H.; Zhou, Q.; He, J.; Jiang, Z.; Peng, C.; Tong, R.; Shi, J. Recent advances in the development of protein–protein interactions modulators: Mechanisms and clinical trials. *Signal Transduct. Target. Ther.* **2020**, *5*, 213. [CrossRef]
12. Koirala, M.; Alexov, E. Ab-initio binding of barnase–barstar with DelPhiForce steered Molecular Dynamics (DFMD) approach. *J. Theor. Comput. Chem.* **2020**, *19*, 2050016. [CrossRef]
13. Wang, S.; Zhao, Y.; Aguilar, A.; Bernard, D.; Yang, C.Y. Targeting the MDM2-p53 Protein-Protein Interaction for New Cancer Therapy: Progress and Challenges. *Cold Spring Harb. Perspect. Med.* **2017**, *7*, a026245. [CrossRef]
14. Koo, N.; Sharma, A.K.; Narayan, S. Therapeutics Targeting p53-MDM2 Interaction to Induce Cancer Cell Death. *Int. J. Mol. Sci.* **2022**, *23*, 5005. [CrossRef]
15. Jha, K.; Karmakar, S.; Saha, S. Graph-BERT and language model-based framework for protein–protein interaction identification. *Sci. Rep.* **2023**, *13*, 5663. [CrossRef] [PubMed]
16. Jumper, J.; Evans, R.; Pritzel, A.; Green, T.; Figurnov, M.; Ronneberger, O.; Tunyasuvunakool, K.; Bates, R.; Žídek, A.; Potapenko, A.; et al. Highly accurate protein structure prediction with AlphaFold. *Nature* **2021**, *596*, 583–589. [CrossRef]
17. Bijak, V.; Szczygiel, M.; Lenkiewicz, J.; Gucwa, M.; Cooper, D.R.; Murzyn, K.; Minor, W. The current role and evolution of X-ray crystallography in drug discovery and development. *Expert Opin. Drug Discov.* **2023**, *18*, 1221–1230. [CrossRef] [PubMed]
18. Bock, L.V.; Igaev, M.; Grubmüller, H. Single-particle Cryo-EM and molecular dynamics simulations: A perfect match. *Curr. Opin. Struct. Biol.* **2024**, *86*, 102825. [CrossRef]
19. Koehler Leman, J.; Künze, G. Recent advances in NMR protein structure prediction with ROSETTA. *Int. J. Mol. Sci.* **2023**, *24*, 7835. [CrossRef]
20. Saikusa, K. Advancing Native Mass Spectrometry Toward Cellular Biology. *J. Mass Spectrom.* **2024**, *59*, e5095. [CrossRef]
21. Kollman, P.A.; Massova, I.; Reyes, C.; Kuhn, B.; Huo, S.; Chong, L.; Lee, M.; Lee, T.; Duan, Y.; Wang, W.; et al. Calculating Structures and Free Energies of Complex Molecules: Combining Molecular Mechanics and Continuum Models. *Acc. Chem. Res.* **2000**, *33*, 889–897. [CrossRef]
22. Wang, E.; Sun, H.; Wang, J.; Wang, Z.; Liu, H.; Zhang, J.Z.H.; Hou, T. End-Point Binding Free Energy Calculation with MM/PBSA and MM/GBSA: Strategies and Applications in Drug Design. *Chem. Rev.* **2019**, *119*, 9478–9508. [CrossRef]
23. Swanson, J.M.; Henchman, R.H.; McCammon, J.A. Revisiting free energy calculations: A theoretical connection to MM/PBSA and direct calculation of the association free energy. *Biophys. J.* **2004**, *86*, 67–74. [CrossRef] [PubMed]
24. Schlitter, J. Estimation of absolute and relative entropies of macromolecules using the covariance matrix. *Chem. Phys. Lett.* **1993**, *215*, 617–621. [CrossRef]
25. Homeyer, N.; Gohlke, H. Free energy calculations by the molecular mechanics Poisson– Boltzmann surface area method. *Mol. Inform.* **2012**, *31*, 114–122. [CrossRef]
26. Karplus, M.; Kushick, J.N. Method for estimating the configurational entropy of macromolecules. *Macromolecules* **1981**, *14*, 325–332. [CrossRef]
27. Abraham, M.J.; Murtola, T.; Schulz, R.; Páll, S.; Smith, J.C.; Hess, B.; Lindahl, E. GROMACS: High performance molecular simulations through multi-level parallelism from laptops to supercomputers. *SoftwareX* **2015**, *1*, 19–25. [CrossRef]
28. Wang, Y.; Liu, J.; Li, J.; He, X. Fragment-based quantum mechanical calculation of protein–protein binding affinities. *J. Comput. Chem.* **2018**, *39*, 1617–1628. [CrossRef]

29. Wallis, R.; Moore, G.R.; James, R.; Kleanthous, C. Protein-protein interactions in colicin E9 DNase-immunity protein complexes. 1. Diffusion-controlled association and femtomolar binding for the cognate complex. *Biochemistry* **1995**, *34*, 13743–13750. [CrossRef]
30. Walker, D.; Moore, G.R.; James, R.; Kleanthous, C. Thermodynamic consequences of bipartite immunity protein binding to the ribosomal ribonuclease colicin E3. *Biochemistry* **2003**, *42*, 4161–4171. [CrossRef]
31. Keeble, A.H.; Kirkpatrick, N.; Shimizu, S.; Kleanthous, C. Calorimetric Dissection of Colicin Dnase—Immunity Protein Complex Specificity. *Biochemistry* **2006**, *45*, 3243–3254. [CrossRef]
32. Shirts, M.R.; Mobley, D.L. An introduction to best practices in free energy calculations. *Biomol. Simul. Methods Protoc.* **2013**, *924*, 271–311.
33. Lindorff-Larsen, K.; Maragakis, P.; Piana, S.; Eastwood, M.P.; Dror, R.O.; Shaw, D.E. Systematic validation of protein force fields against experimental data. *PLoS ONE* **2012**, *7*, e32131. [CrossRef] [PubMed]
34. Awoonor-Williams, E.; Rowley, C.N. Molecular simulation of nonfacilitated membrane permeation. *Biochim. Biophys. Acta (BBA) Biomembr.* **2016**, *1858*, 1672–1687. [CrossRef]
35. Zhou, H.X. Association and dissociation kinetics of colicin E3 and immunity protein 3: Convergence of theory and experiment. *Protein Sci.* **2003**, *12*, 2379–2382. [CrossRef] [PubMed]
36. Zhou, H.X. Disparate ionic-strength dependencies of on and off rates in protein–protein association. *Biopolym. Orig. Res. Biomol.* **2001**, *59*, 427–433. [CrossRef]
37. Kleanthous, C.; Walker, D. Immunity proteins: Enzyme inhibitors that avoid the active site. *Trends Biochem. Sci.* **2001**, *26*, 624–631. [CrossRef]
38. Li, C.; Jia, Z.; Chakravorty, A.; Pahari, S.; Peng, Y.; Basu, S.; Koirala, M.; Panday, S.K.; Petukh, M.; Li, L. DelPhi suite: New developments and review of functionalities. *J. Comput. Chem.* **2019**, *40*, 2502–2508. [CrossRef]
39. Deng, Y.; Roux, B. Computations of standard binding free energies with molecular dynamics simulations. *J. Phys. Chem. B* **2009**, *113*, 2234–2246. [CrossRef]
40. Phillips, J.C.; Braun, R.; Wang, W.; Gumbart, J.; Tajkhorshid, E.; Villa, E.; Chipot, C.; Skeel, R.D.; Kale, L.; Schulten, K. Scalable molecular dynamics with NAMD. *J. Comput. Chem.* **2005**, *26*, 1781–1802. [CrossRef]
41. Abrams, C.; Bussi, G. Enhanced sampling in molecular dynamics using metadynamics, replica-exchange, and temperature-acceleration. *Entropy* **2013**, *16*, 163–199. [CrossRef]
42. Kästner, J. Umbrella sampling. *Wiley Interdiscip. Rev. Comput. Mol. Sci.* **2011**, *1*, 932–942. [CrossRef]
43. Straatsma, T.; Berendsen, H. Free energy of ionic hydration: Analysis of a thermodynamic integration technique to evaluate free energy differences by molecular dynamics simulations. *J. Chem. Phys.* **1988**, *89*, 5876–5886. [CrossRef]
44. Quiñones, B.; Yambao, J.C.; Lee, B.G. Draft genome sequences of *Escherichia coli* O113: H21 strains recovered from a major produce production region in California. *Genome Announc.* **2017**, *5*, e01203-17. [CrossRef] [PubMed]
45. Yang, Z.; Zeng, X.; Zhao, Y.; Chen, R. AlphaFold2 and its applications in the fields of biology and medicine. *Signal Transduct. Target. Ther.* **2023**, *8*, 115. [CrossRef]
46. Chang, J.-W.; Sato, Y.; Ogawa, T.; Arakawa, T.; Fukai, S.; Fushinobu, S.; Masaki, H. Crystal structure of the central and the C-terminal RNase domains of colicin D implicated its translocation pathway through inner membrane of target cell. *J. Biochem.* **2018**, *164*, 329–339. [CrossRef] [PubMed]
47. Huang, J.; MacKerell, A.D., Jr. CHARMM36 all-atom additive protein force field: Validation based on comparison to NMR data. *J. Comput. Chem.* **2013**, *34*, 2135–2145. [CrossRef]
48. Bussi, G.; Donadio, D.; Parrinello, M. Canonical sampling through velocity rescaling. *J. Chem. Phys.* **2007**, *126*, 014101. [CrossRef]
49. Polymorphic transitions in single crystals: A new molecular dynamics method. *J. Appl. Phys.* **1981**, *52*, 7182–7190. [CrossRef]
50. Chong, S.-H.; Ham, S. Dynamics of Hydration Water Plays a Key Role in Determining the Binding Thermodynamics of Protein Complexes. *Sci. Rep.* **2017**, *7*, 8744. [CrossRef]
51. Valdés-Tresanco, M.S.; Valdés-Tresanco, M.E.; Valiente, P.A.; Moreno, E. gmx_MMPBSA: A new tool to perform end-state free energy calculations with GROMACS. *J. Chem. Theory Comput.* **2021**, *17*, 6281–6291. [CrossRef]

molecules

MDPI

Article

Delineating Host–Guest–Solvent Interactions in Solution from Gas-Phase Host–Guest Configurations: Thermodynamic Reversal and Structural Correlation of 24-Crown-8/H⁺/Diaminopropanol Non-Covalent Complexes in Aqueous Solution vs. in the Gas Phase

Young-Ho Oh [1], So Yeon Lee [2], Han Bin Oh [2,*] and Sungyul Lee [1,*]

[1] Department of Applied Chemistry, Kyung Hee University, 1732, Deogyeong-daero, Giheung-gu, Yongin-si 17104, Republic of Korea
[2] Department of Chemistry, Sogang University, 35 Baekbeom-ro, Mapo-gu, Seoul 04107, Republic of Korea
* Correspondence: hanbinoh@sogang.ac.kr (H.B.O.); sylee@khu.ac.kr (S.L.)

Abstract: We study the structures of 24-crown-8/H^+/diaminopropanol (CR/$DAPH^+$) and 24-crown-8/CsF/H^+/diaminopropanol (CR/CsF/$DAPH^+$) non-covalent host–guest complexes in both the gas phase and aqueous solution using the density functional theory (DFT) method. We examine the environment (complexation with CR vs. solvation) around the guest functional groups (ammoium, hydroxyl, and amino) in the CR/$DAPH^+$ and CR/CsF/$DAPH^+$ complexes. We find that the gas-phase configurations with the 'naked' hydroxyl/amino devoid of H-bonding with CR or CR/CsF are structurally correlated with the lowest Gibbs free energy conformers in aqueous solution in which the functional groups are solvated off the CR or CR/CsF host. We predict that the latter thermodynamically disadvantageous host–guest configurations would be identified in the gas phase by infrared multiphoton dissociation (IRMPD) spectroscopy, originating from the complexes in aqueous solution. This predicted 'thermodynamic reversal' and 'structural correlation' of the host–guest configurations in the gas phase vs. in solution are discussed in relation to the possibility of obtaining information on host–guest–solvent interactions in the solution phase from the gas-phase host–guest configurations.

Keywords: host–guest; crown ether; diaminopropanol; structural correlation

1. Introduction

As a technique for the selective production of non-covalent ionic host–guest complexes [1–9] in a gas phase, electrospray ionization/mass spectroscopy [10–13] (ESI/MS) has proven to be highly pivotal. The differentiation of a guest molecule by a chiral host [14–18] has been one of the most intensively studied topics in host–guest chemistry, aimed at shedding light on the processes of molecular recognition and self-assembly. Determining the structures of host–guest pairs in the gas phase is often challenging, but infrared multiple photon dissociation (IRMPD) spectroscopy [14,19–28] has been shown to be a powerful tool to elucidate the features of non-covalent interactions in gas-phase complexes. When combined with quantum chemical calculations, IRMPD spectra of biomolecules bound to a macrocycle host can reveal the detailed interactions in the host–guest pairs, as exemplified in our previous studies [29–32] of protonated AAH^+ (AA = alanine, isoleucine, lysine) bound to the permethylated *β*-cyclodextrin (perm-CD) host in the gas phase.

Here, we theoretically examine the 24-crown-8/H^+/diaminopropanol (CR/$DAPH^+$) and 24-crown-8/CsF/H^+/diaminopropanol (CR/CsF/$DAPH^+$) complexes (Scheme 1) both in the gas phase and in aqueous solution to examine the structural connectivity of the non-covalent host–guest complexes in the two phases. We probe the environment (complexation with CR vs. solvation) around the guest functional groups (ammonium, hydroxyl, and amino) in the CR/$DAPH^+$ complex, illustrating that the ammonium only binds at the CR host, whereas both -OH and -NH_2 are solvated by water molecules off the host. We also study the CR/CsF/$DAPH^+$ host–guest system to propose a new scheme of introducing F^- as a 'probe' for determining (by ^{19}F-NMR spectroscopy) the environment (specifically, complexation vs. solvation) around the functional groups in the guest in solution, which is very difficult to observe by ^{1}H-NMR spectroscopy due to the very rapid proton/proton and proton/deuterium exchange and the consequent extreme broadening of the proton peaks. Since the H-bonds between F^- and the guest hydroxyl/ammonium groups would be stronger than those with the CR host, using F^- is also anticipated to give a more salient case of thermodynamic reversal and a more conspicuous example of the structural correlation [31] of host–guest pairs in solution and in the gas phase. We find that the hydroxyl (and/or ammonium) in the $DAPH^+$ guest chooses to be solvated by water molecules in the solution phase, whereas in the lowest Gibbs free energy gas-phase conformers, it prefers complexation with CR/CsF. We compare these findings with those for the CR/$DAPH^+$ complex (which is devoid of CsF) to show that the host–guest interactions in the CR/CsF/$DAPH^+$ system are very similar to those in the CR/$DAPH^+$ complex, so that the influence of (perturbations due to) the presence of the CsF probe is minimal, in that the CR/CsF host behaves very much like the CR unit.

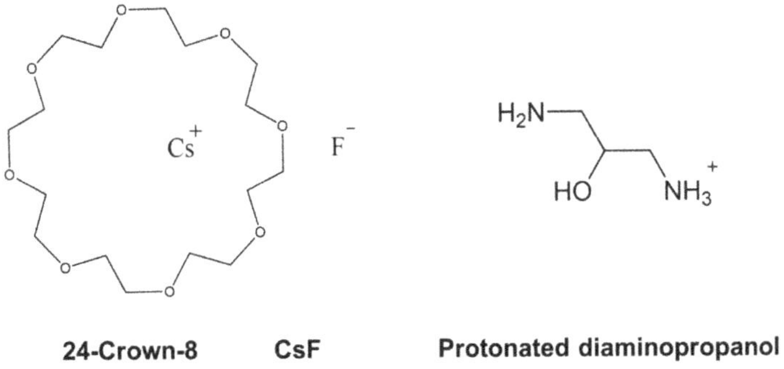

Scheme 1. Components of 24-Crown-8/protonated diaminopropanol and 24-Crown-8/CsF/protonated diaminopropanol covalent host–guest complexes.

We also predict that the thermodynamic stability of the (CR/$DAPH^+$) and (CR/CsF/$DAPH^+$) host–guest configurations in the gas phase is reversed in aqueous solution ('thermodynamic reversal') [32], and that the lowest Gibbs free energy conformer with solvated (without interactions with the CR or CR/CsF host) guest functional groups in solution is structurally related [32] to the thermodynamically disadvantageous gas-phase complex with the 'naked' -OH (and/or -NH_2). From these observations, we consolidate our previous proposal [32] that the gas-phase conformers of the CR/$DAPH^+$ and CR/CsF/$DAPH^+$ non-covalent complexes produced and to be identified by the ESI/MS and IRMPD procedure, respectively, may originate from the thermodynamically most favored host–guest complexes in solution, and that this close structural connection implicates the possibility of using the ESI/MS/IRMPD protocol for elucidating host–guest–solvent interactions in the solution phase directly from host–guest configurations in the gas phase.

2. Results

2.1. Structures and Thermodynamic Stability of CR/DAPH⁺ Non-Covalent Complex in Aqueous Solution

Figure 1 (For Cartesian coordinates, see Supplementary) depicts the calculated lower Gibbs free energy structures (CR/DAPH$^+$-S1, CR/DAPH$^+$-S2) of the CR/DAPHCl complex in aqueous solution. We find that nine water molecules completely envelop the DAPH$^+$ guest in the first shell; thus, we employ the (CR/DAPH$^+$ + 9 H_2O + water continuum) as the model system for the complex in aqueous solution. The two conformers exhibit distinct structural features: In CR/DAPH$^+$-S1, the CR host and the DAPH$^+$ guest form the complex through the [CR. . .ammonium]/[CR. . .hydroxyl] H-bonds, with the amino positioned off the CR unit. In CR/DAPH$^+$-S2, on the other hand, -NH_3^+ only interacts with the CR unit, while the hydroxyl and the amino groups are solvated away from the CR host. Since the H-bonds between the two guest functional groups (-OH and -NH_2) and the water molecules are stronger than those with the CR host, the lower (by 1.2 kcal/mol) Gibbs free energy of the CR/DAPH$^+$-S2 conformer is easily understood.

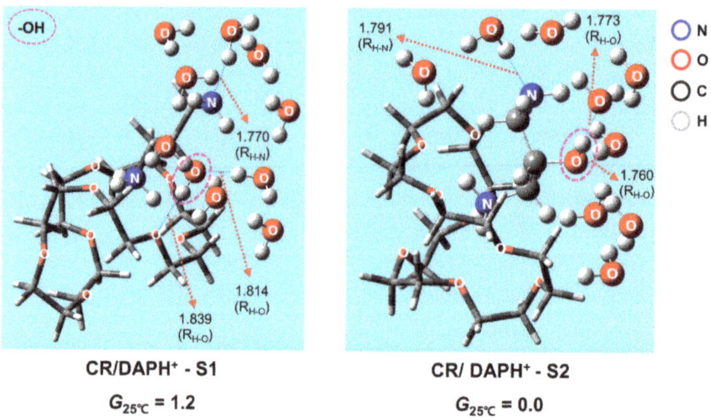

Figure 1. Calculated structures of 24-Crown-8/DAPH$^+$ in aqueous solution. Relative Gibbs free energy in kcal/mol and distance in Å. Blue background represents solvent continuum.

2.2. Structures and Thermodynamic Stability of CR/CsF/DAPHCl in Aqueous Solution

The use of the 24-crown-8/CsF/DAPH$^+$ host–guest system is motivated by two presumptions: First, it may allow the examination of the environment (solvation vs. complexation) around the DAPH$^+$ guest functional groups by ^{19}F-NMR spectroscopy (observing the characteristic positions (chemical shifts) of ^{19}F-NMR peaks), as demonstrated in our previous investigation [33]. Second, the [F^-. . .NH_3^+-], [F^-. . .HO-], and/or [-OH. . .F^-. . .NH_3^+-] H-bonds, which are stronger than the corresponding [CR. . .NH_3^+-], [CR. . .HO-], and/or [-OH. . .CR. . .NH_3^+-] H-bonds, respectively, are expected to 'magnify' the reversal of the thermodynamic stability of the solution-phase host–guest complex vs. that in the gas phase, as discussed below. Thus, by presenting theoretical analysis of the CR/CsF/DAPH$^+$ system, we propose a scheme (^{19}F-NMR spectroscopy for probing the [F^-. . .NH_3^+-], [F^-. . .HO-], [-OH. . .F^-. . .NH_3^+-], and [F^-. . .solvent] hydrogen bonds) for the unambiguous experimental vindication of the structural connectivity of host–guest configurations in solution to those in the gas phase.

Figure 2 (For Cartesian coordinates, see Supplementary) illustrates the calculated structures (CR/CsF/DAPH$^+$-S1, CR/CsF/DAPH$^+$-S2, and CR/CsF/DAPH$^+$-S3) of the 24-Crown-8/CsF/DAPHCl complex in solution, modeled as the supramolecule CR/CsF/DAPHCl/$(H_2O)_7$ in water continuum for the reason discussed below. In all structures, NH_3^+ is the main contact on the CR/CsF host. The conformers, however, differ in the interactions between the two other functional groups (-OH and -NH_2) and the CR/CsF unit. In CR/CsF/DAPH$^+$-S1, NH_3^+ forms a H-bond with F^-, whereas the two other guest functional groups (-OH and -NH_2) interact with the O atoms in the CR ring. In the CR/CsF/DAPH$^+$-S2 conformer, NH_3^+ forms H-bonds both with F^- and the CR unit, while -OH and -NH_2 are solvated off the CR/CsF host. In CR/CsF/DAPH$^+$-S3, NH_3^+ interacts with F^- and with the CR ring, the hydroxyl interacts with the solvent and with F^-, and the amino is fully solvated off the CR/CsF host. These strong H-bonds between the guest functional groups and the CR/solvent seem to be the origin of high thermodynamic advantage (Gibbs free energy lower than CR/CsF/DAPH$^+$-S1 and CR/CsF/DAPH$^+$-S2 by 10.2 and 5.8 kcal/mol, respectively). Compared with the CR/DAPH$^+$ conformers presented in Figure 1, it can be seen that the structures of the lowest Gibbs free energy configurations of the CR/CsF/DAPH$^+$ and CR/DAPH$^+$ systems are very similar in that the two guest functional groups (hydroxyl and amino) are solvated by water molecules. On the other hand, the differences in the Gibbs free energies of the most favored host–guest configurations relative to other conformers are calculated to be 'magnified'. These two observations may validate the use of the CsF 'probe' for determining the features of H-bonding between the CR host and the DAPH$^+$ guest in solution, which may be achievable by ^{19}F-NMR spectroscopy [33].

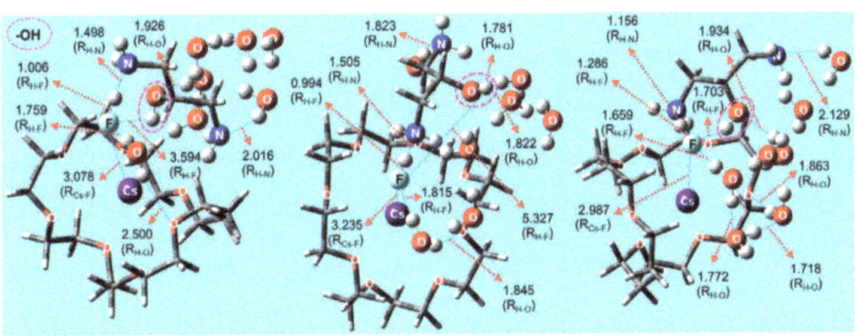

CR/CsF/DAPH$^+$ - S1	CR/CsF/DAPH$^+$ - S2	CR/CsF/DAPH$^+$ - S3
$G_{25^\circ C}$ = 10.2	$G_{25^\circ C}$ = 5.8	$G_{25^\circ C}$ = 0.0

Figure 2. Calculated structures of 24-Crown-8/CsF/DAPHCl in aqueous solution. Relative Gibbs free energy in kcal/mol and distance in Å. Blue background represents solvent continuum.

2.3. *Structures and Relative Gibbs Free Energy of CR/DAPH$^+$ in the Gas Phase*

Figure 3 (For Cartesian coordinates, see Supplementary) depicts the calculated lowest Gibbs free energy structures of the CR/DAPH$^+$ complex in the gas phase obtained by extensive searching over the landscape of the potential surface. CR/DAPH$^+$-G1 and CR/DAPH$^+$-G2 structurally correspond to the host–guest configuration CR/DAPH$^+$-S1 and CR/DAPH$^+$-S2 in solution, respectively, depicted in Figure 1. It is highly intriguing to observe that the gas-phase Gibbs free energy of CR/DAPH$^+$-G2, which is structurally connected to the lowest Gibbs free energy solution-phase complex CR/DAPH$^+$-S2, is now the highest (higher than that of CR/DAPH$^+$-G1 by 3.9 kcal/mol), illustrating a 'thermodynamic reversal' [32] of 5.1 kcal/mol. In the most stable solution-phase complex CR/DAPH$^+$-S2, the hydroxyl, which was fully surrounded by nine explicit H_2O molecules in the first shell around the complex, becomes 'naked' in the gas phase. This 'thermodynamic reversal' was the basis of the proposed possibility of elucidating the host–guest–solvent interactions from the host–guest configurations in the gas phase that is connected with the solution phase by the ESI/MS procedure [32].

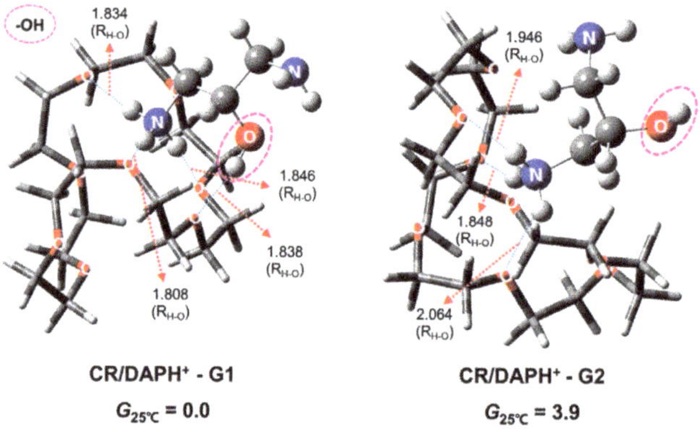

Figure 3. Calculated lowest Gibbs free energy structures of 24-Crown-8/DAPH$^+$ in the gas phase. Relative Gibbs free energy (at 25 °C) in kcal/mol and distance in Å.

These structural features are also manifested in the corresponding infrared spectra (Figure 4): The strong band at 3661 cm^{-1} indicates the absorption of the isolated -OH in CR/DAPH$^+$-G2, whereas the band for CR/DAPH$^+$-G1 at 3501 cm^{-1} corresponds to the red-shifted -OH, indicating the formation of H-bonds between -OH and the CR unit. The strong red-shifts (3080~3200 cm^{-1}) of the -NH stretch modes of the ammonium in the two complexes illustrate the strong interactions between -NH_3^+ and the CR host. The -NH stretch modes of the amino in the 3320~3400 cm^{-1} region signify the isolated -NH_2 in the two configurations. These IR spectra are to be compared with the experimental IRMPD spectra for structural identification of the gas-phase CR/CsF/DAPH$^+$ complex generated from the solution phase by the ESI/MS procedure.

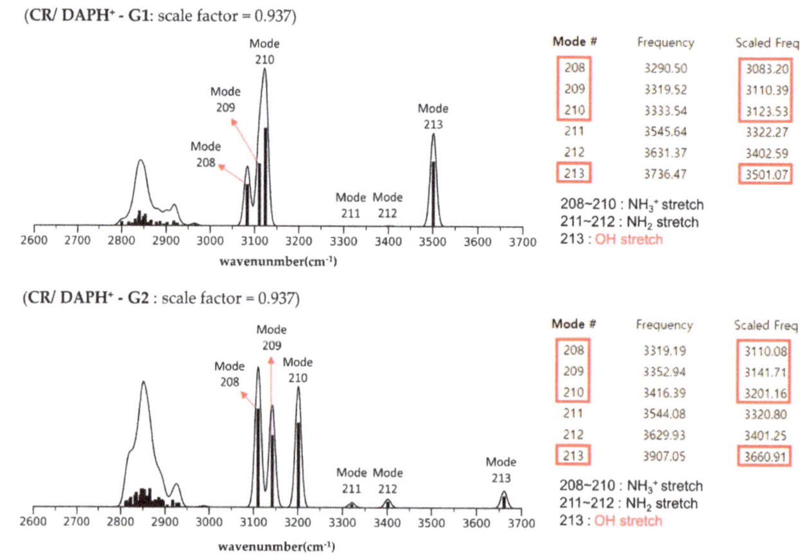

Figure 4. Calculated IR spectra of gas-phase 24-Crown-8/DAPH$^+$ in the gas phase. Frequency in cm^{-1}.

2.4. Structures and Relative Gibbs Free Energy of CR/CsF/DAPH$^+$ in the Gas Phase

Figure 5 (For Cartesian coordinates, see Supplementary) presents the calculated gas-phase structures of the CR/CsF/DAPH$^+$ complex (modeled as CR/DAPH$^+$ + 7 H_2O + water continuum) with distinct structural features. In the lowest Gibbs free energy (at 25 °C) configuration CR/CsF/DAPH$^+$-G1, the [F^-. . .-NH_3^+], [-OH. . .CR], and [NH_2. . .CR]] H-bonds ($R_{F\text{-}HN}$ = 1.006 Å, $R_{OH\text{-}O}$ = 1.977 Å, and $R_{NH\text{-}O}$ = 2.123 Å, respectively) provide the thermodynamic advantage, while the hydroxyl does not interact with F^-. On the other hand, in the thermodynamically less favorable CR/CsF/DAPH$^+$-G2, NH_3^+ forms H-bonds both with F^- and CR ($R_{F\text{-}HN}$ = 0.994 Å; $R_{NH\text{-}O}$ = 2.174 Å), while -OH and -NH_2 are positioned off the CR/CsF host without interactions. In the highest relative Gibbs free energy structure, the ammonium and hydroxyl form H-bonds with F^- ($R_{F\text{-}HN}$ = 1.286 Å; $R_{OH\text{-}O}$ = 1.703 Å), whereas the amino is completely isolated.

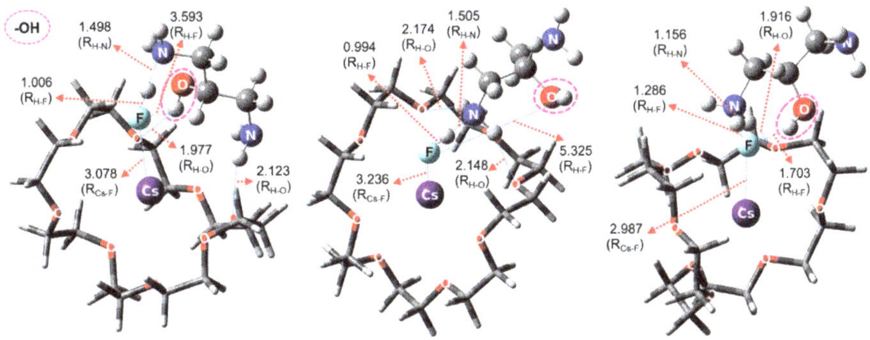

CR/CsF/DAPH$^+$ - G1 $G_{25°C}$ = 0.0

CR/CsF/DAPH$^+$ - G2 $G_{25°C}$ = 1.7

CR/CsF/DAPH$^+$ - G3 $G_{25°C}$ = 4.4

Figure 5. Calculated lower Gibbs free energy structures of the gas-phase 24-Crown-8/CsF/DAPH$^+$. Relative Gibbs free energy (at 25 °C) in kcal/mol and distance in Å.

For the present CR/CsF/DAPH$^+$ system, we also predict a 'thermodynamic reversal' in the relative CR/CsF/DAPH$^+$-G1 Gibbs free energies of the thermodynamically most favorable complex in the gas phase (CR/CsF/DAPH$^+$-G1) versus that in aqueous solution (CR/CsF/DAPH$^+$-S1) by as much as 10.2 kcal/mol. Thus, the presence of CsF indeed seems to result in a more salient 'thermodynamic reversal' than that observed for the CR/DAPH$^+$ system.

Figure 6 presents the calculated IR spectra of the CR/CsF/DAPH$^+$ complex in the gas phase, which can be interpreted by these structural features of the gas-phase configurations. The -OH stretch frequencies are the most conspicuous feature, as they may represent the environment of the guest functional group, specifically concerning the presence/absence of the H-bonding with F^- and/or with the CR ring. The red-shifted frequencies of 3546, and 3325 cm^{-1} for CR/CsF/DAPH$^+$-G1 and CR/CsF/DAPH$^+$-G3, respectively, are due to the [-OH. . .CR] and [-OH. . . F^-] H-bonds, whereas the 3679 cm^{-1} band for the CR/CsF/DAPH$^+$-G2 conformer represents the 'naked' OH.

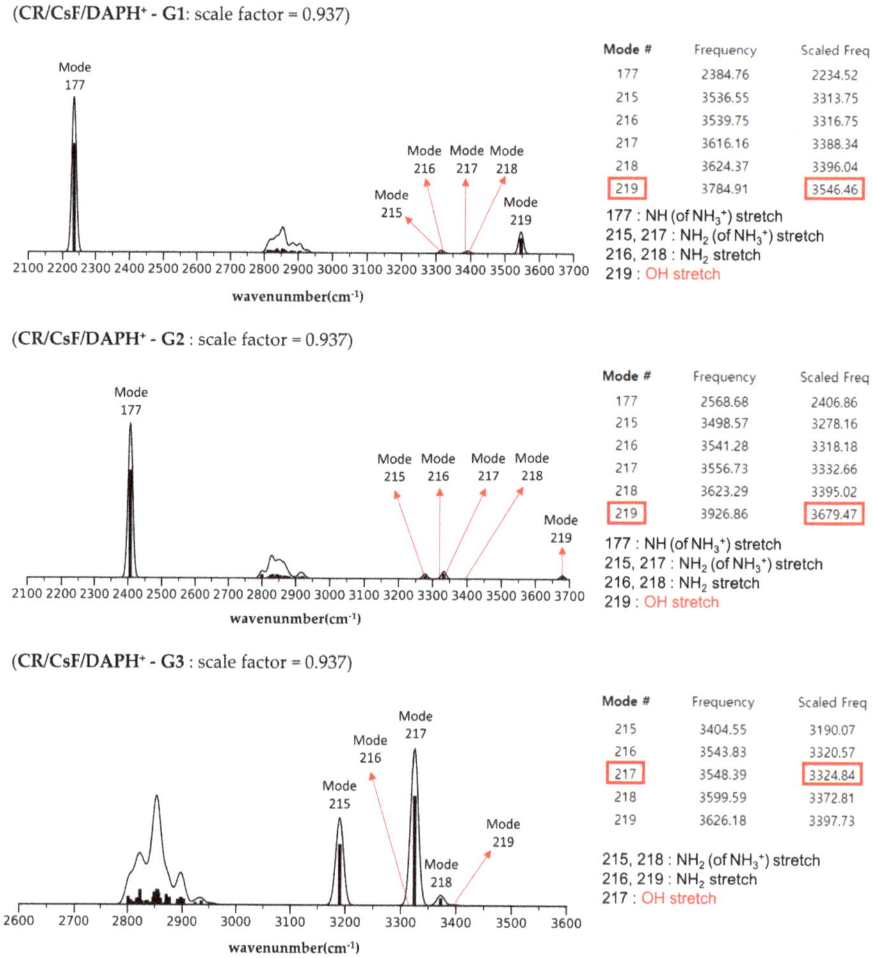

Figure 6. Calculated IR spectra of 24-Crown-8/CsF/DAPH+ in the gas phase. Frequency in cm^{-1}.

The low-frequency ammonium N-H stretch modes (#177 (2235 cm^{-1}) and #177 (2407 cm^{-1}) for CR/CsF/DAPH+-G1 and CR/CsF/DAPH+-G2, respectively) deserve scrutiny. In these two configurations, the ammonium H is almost transferred to F^-, so that the elongated N-H bonds (1.498 and 1.505 Å) give rise to very low IR frequencies. In CR/CsF/DAPH+-G3, on the other hand, the N-H bond is short (1.156 Å), and the stretch of this strong bond appears at a higher frequency (>3200 cm^{-1}). The -NH_2 stretch bands for the three CR/CsF/DAPH+ configurations are of low intensity and appear at similar positions (~3313 and ~3396 cm^{-1}), indicating that they are either isolated or in very weak interaction with the CR ring (but not with F^-). Comparison of these IR spectra with the experimental IRMPD spectra would lead to an unambiguous identification of the CR/CsF/DAPH+ complex in the gas phase.

Finally, Figure 7 depicts the structural connectivity of CR/DAPH+-S2 and CR/DAPH+-G2, and of CR/CsF/DAPH+-S3 and CR/CsF/DAPH+-G3. The host–guest configurations in solution and in the gas phase are shown to be very similar.

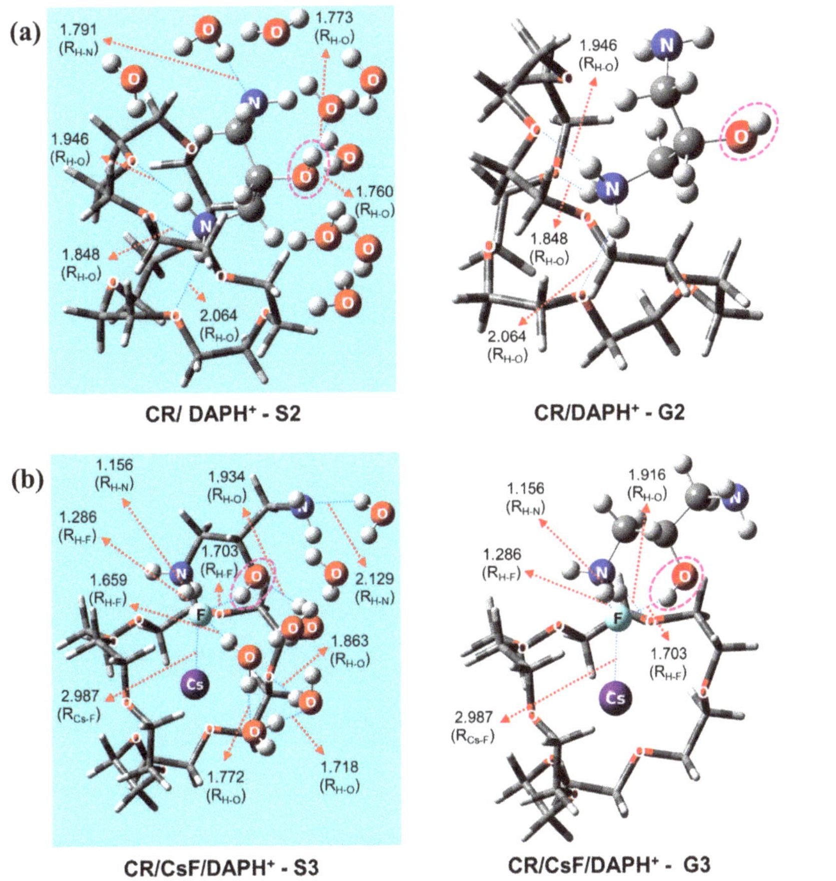

Figure 7. Comparison of the lowest Gibbs free energy for (**a**) 24-Crown-8/DAPH$^+$ and (**b**) 24-Crown 8/CsF/DAPH$^+$ configurations in solution with their structurally correlating gas-phase structures. Blue background represents the solvent continuum.

Based on the present findings, the most relevant question concerning the gas-phase CR/DAPH$^+$ and CR/CsF/DAPH$^+$ complexes would be the following: which conformers of the gas-phase CR/DAPH$^+$ and CR/CsF/DAPH$^+$ complexes are to be observed and identified by the IRMPD technique? In our previous studies, it was demonstrated that the relative thermodynamic stability (relative Gibbs free energy) of the gas-phase host–guest complexes may not determine the populations of the complexes if the gas-phase environment produced in the ESI/MS experiments is not in thermal equilibrium. In this situation, the most stable host–guest configurations in the solution phase may give rise to the structures of the host–guest in the gas phase.

If the lowest Gibbs free energy CR/DAPH$^+$ and CR/CsF/DAPH$^+$ complexes in solution are stripped off the solvent molecules by the ESI/MS, and if the produced gas-phase host–guest complexes are detected in the gas phase before thermal equilibrium is attained, then the thermodynamically least favorable (highest Gibbs free energy) CR/DAPH$^+$-G2 and CR/CsF/DAPH$^+$-G3 with the "naked" -OH and -NH_2 would be observed in the gas phase as produced by the ESI/MS protocol, originating from the thermodynamically most favorable solution-phase configurations CR/DAPH$^+$-S2 and CR/CsF/DAPH$^+$-S3.

We may now discuss the implications of the present findings with regard to the ESI/MS/IRMPD experiments for host–guest complexes. If the proposition from our previous works on the perm-CD/L-AAH$^+$ (AA = Ala, Ile, Lys) system [29–32,34–36] also applies to our present 24-Crown-8/DAPH$^+$ and 24-Crown-8/CsF/DAPH$^+$ host–guest complexes, then thermodynamically, the most stable solution-phase complex CR/DAPH$^+$-S2 [CR/CsF/DAPH$^+$-S3] would be instantly 'frozen' into CR/DAPH$^+$-G2 [CR/CsF/DAPH$^+$-G3] in the gas phase, produced under thermal non-equilibrium conditions by the ESI/MS techniques. Thus, the identification of the gas-phase host–guest complexes CR/DAPH$^+$-G2 and CR/CsF/DAPH$^+$-G3, which are thermodynamically least favorable, would further consolidate our proposition of the structural correlation between the host–guest complexes in solution and in the gas phase, and of the possibility of obtaining information on host–guest–solvent interactions in solution from host–guest structures in the gas phase via the ESI/MS/IRMPD protocol.

3. Computational Details

All calculations were performed using the wB97X-D functional [34], which may treat weak interactions (including hydrogen bonding) effectively, with the 6-311G** basis set implemented in the Gaussian16 suite of programs [35]. For the complexes in aqueous solution, the supramolecule/continuum approach was adopted, treating the solvent molecules directly interacting with the guest functional groups in the first shell around the complex as explicit molecules (CR/DAPH$^+$/$(H_2O)_n$, CR/CsF/DAPH$^+$$(H_2O)_n$), and other numerous H_2O molecules in the second shell and beyond as a water continuum modeled by the SMD method [36]. A scale factor of 0.927 was used to fit the calculated IR frequency of the hydroxyl and carboxyl -OH stretch modes to the experimental value (3660 cm^{-1}) [37]. All structures were obtained by verifying that all vibrational frequencies were real.

4. Conclusions

We have presented the calculated structures of the 24-Crown-8/DAPH$^+$ and 24-Crown-8/CsF/DAPH$^+$ complexes in both the gas phase and solution. Our main focus was on exploring the structural connectivity of the complexes in the two phases. We have proposed that the most stable (lowest Gibbs free energy) configurations of the complexes in the solution phase might produce higher Gibbs free energy host–guest complexes in the gas phase under thermal non-equilibrium gas-phase environments via the ESI/MS procedure, whose structures could subsequently be identified by the IRMPD technique. We also provided a theoretical basis for using F$^-$ as a 'probe' for determining the conformation of the guest functional groups. We believe that these findings offer further support for our approach of characterizing host–guest–solvent interactions in solution by using the gas-phase structures of host–guest complexes. We are working on unambiguous experimental validation for this proposition of structural correlation between the host–guest configurations in solution and in the gas phase. Further studies of these fascinating macrocycle/biomolecule host–guest systems would be an extremely intriguing avenue for exploration.

Supplementary Materials: The following supporting information can be downloaded at: https://www.mdpi.com/article/10.3390/molecules30081723/s1.

Author Contributions: Conceptualization, S.L. and H.B.O.; Methodology, S.L., S.Y.L. and Y.-H.O.; Software, S.Y.L. and Y.-H.O.; Validation, Y.-H.O.; Formal analysis, S.L., Y.-H.O. and H.B.O.; Investigation, S.Y.L. and Y.-H.O.; Resources, H.B.O.; Data curation, S.Y.L. and Y.-H.O.; Writing—original draft, Y.-H.O.; Writing—review & editing, S.L. and H.B.O.; Supervision, S.L.; Project administration,

H.B.O.; Funding acquisition, H.B.O. All authors have read and agreed to the published version of the manuscript.

Funding: This work was supported by grants from the National Research Foundation of Korea funded by the Ministry of Education (2018R1A6A1A03024940 and 2021R1A2C2007397) and the KISTI Supercomputing Center. H.B.O. is thankful to the research grant supported by Korea Environment Industry & Technology Institute (KEITI) through the "Advanced Technology Development Project for Predicting and Preventing Chemical Accident" program, funded by the Korea Ministry of Environment (MOE) (RS-2023-00219144).

Institutional Review Board Statement: Not applicable.

Informed Consent Statement: Not applicable.

Data Availability Statement: The original contributions presented in this study are included in the article. Further inquiries can be directed to the corresponding authors.

Conflicts of Interest: The authors declare no conflicts of interest.

References

1. Cram, D.J.; Cram, J.M. Host-Guest Chemistry: Complexes between organic compounds simulate the substrate selectivity of enzymes. *Science* **1974**, *183*, 803–809. [CrossRef] [PubMed]
2. Ma, X.; Zhao, Y. Biomedical applications of supramolecular systems based on host–guest interactions. *Chem. Rev.* **2015**, *115*, 7794–7839. [CrossRef] [PubMed]
3. Yang, H.; Yuan, B.; Zhang, X.; Scherman, O.A. Supramolecular chemistry at interfaces: Host–guest interactions for fabricating multifunctional biointerfaces. *Acc. Chem. Res.* **2014**, *47*, 2106–2115. [CrossRef] [PubMed]
4. Schalley, C.A. Supramolecular chemistry goes gas phase: The mass spectrometric examination of noncovalent interactions in host–guest chemistry and molecular recognition. *Int. J. Mass Spectrom.* **2000**, *194*, 11–39. [CrossRef]
5. Ortega-Caballero, F.; Rousseau, C.; Christensen, B.; Petersen, T.E.; Bols, M. Remarkable supramolecular catalysis of glycoside hydrolysis by a cyclodextrin cyanohydrin. *J. Am. Chem. Soc.* **2005**, *127*, 3238–3239. [CrossRef]
6. Jiao, D.; Geng, J.; Loh, X.J.; Das, D.; Lee, T.; Scherman, O.A. Supramolecular peptide amphiphile vesicles through host–guest complexation. *Angew. Chem. Int. Ed.* **2012**, *51*, 9633–9637. [CrossRef]
7. Zhang, J.; Zhou, Z.-H.; Li, L.; Luo, Y.-L.; Xu, F.; Chen, Y. Dual Stimuli-Responsive Supramolecular Self-Assemblies Based on the Host–Guest Interaction Between β-Cyclodextrin and Azobenzene for Cellular Drug Release. *Mol. Pharm.* **2020**, *17*, 1100–1113. [CrossRef]
8. Qu, D.-H.; Wang, Q.-C.; Zhang, Q.-W.; Ma, X.; Tian, H. Photoresponsive host–guest functional systems. *Chem. Rev.* **2015**, *115*, 7543–7588. [CrossRef]
9. Hu, Q.-D.; Tang, G.-P.; Chu, P.K. Cyclodextrin-based host–guest supramolecular nanoparticles for delivery: From design to applications. *Acc. Chem. Res.* **2014**, *47*, 2017–2025. [CrossRef]
10. Fenn, J.B.; Mann, M.; Meng, C.K.; Wong, S.F.; Whitehouse, C.M. Electrospray ionization for mass spectrometry of large biomolecules. *Science* **1989**, *246*, 64–71. [CrossRef]
11. Kebarle, P.; Verkerk, U.H. Electrospray: From ions in solution to ions in the gas phase, what we know now. *Mass Spectrom. Rev.* **2009**, *28*, 898–917. [CrossRef] [PubMed]
12. Konermann, L. Unraveling the Mechanism of Electrospray Ionization. *Anal. Chem.* **2013**, *85*, 2–9. [CrossRef] [PubMed]
13. Takáts, Z.; Wiseman, J.M.; Gologan, B.; Cooks, R.G. Mass Spectrometry Sampling Under Ambient Conditions with Desorption Electrospray Ionization. *Science* **2004**, *306*, 471–473. [CrossRef] [PubMed]
14. Sen, A.; Barbu-Debus, K.L.; Scuderi, D.; Zehnecker-Rentien, A. Mass Spectrometry Study and Infrared Spectroscopy of the Complex Between Camphor and the Two Enantiomers of Protonated Alanine: The Role of Higher-Energy Conformers in the Enantioselectivity of the Dissociation Rate Constants. *Chirality* **2013**, *25*, 436–443. [CrossRef]
15. Heo, K.S.; Ryoo, J.J. Synthesis and Application of New Calix[4]Arenes-containing (R)-phenylglycinol Chiral Stationary Phases for Enantioseparation. *Bull. Korean Chem. Soc.* **2020**, *41*, 753–758. [CrossRef]
16. Shahgaldian, P.; Pieles, U. Cyclodextrin derivatives as chiral supramolecular receptors for enantioselective sensing. *Sensors* **2006**, *6*, 593–615. [CrossRef]
17. Avilés-Moreno, J.R.; Quesada-Moreno, M.M.; López-González, J.J.; Martínez-Haya, B. Chiral Recognition of Amino Acid Enantiomers by a Crown Ether: Chiroptical IR-VCD Response and Computational Study. *J. Phys. Chem. B* **2013**, *117*, 9362–9370. [CrossRef]

18. Thomas, M.C.; Mitchell, T.W.; Harman, D.G.; Deeley, J.M.; Murphy, R.C.; Blanksby, S.J. Elucidation of double bond position in unsaturated lipids by ozone electrospray ionization mass spectrometry. *Anal. Chem.* **2007**, *79*, 5013–5022. [CrossRef]
19. Yeni, O.; Ollivier, S.; Moge, B.; Ropartz, D.; Rogniaux, H.; Legentil, L.; Ferrières, V.; Compagnon, I. Ring-Size Memory of Galactose-Containing MS/MS Fragments: Application to the Detection of Galactofuranose in Oligosaccharides and Their Sequencing. *J. Am. Chem. Soc.* **2023**, *145*, 15180–15187. [CrossRef]
20. Oomens, J.; Sartakov, B.G.; Meijer, G.; von Helden, G. Gas-phase infrared multiple photon dissociation spectroscopy of mass-selected molecular ions. *Int. J. Mass Spectrom.* **2006**, *254*, 1–19. [CrossRef]
21. Bush, M.F.; O'Brien, J.T.; Prell, J.S.; Saykally, R.J.; Williams, E.R. Infrared spectroscopy of cationized arginine in the gas phase: Direct evidence for the transition from nonzwitterionic to zwitterionic structure. *J. Am. Chem. Soc.* **2007**, *129*, 1612–1622. [CrossRef] [PubMed]
22. Azargun, M.; Fridgen, T.D. Guanine tetrads: An IRMPD spectroscopy, energy resolved SORI-CID, and computational study of M (9-ethylguanine)$_4$$^+$(M= Li, Na, K, Rb, Cs) in the gas phase. *Phys. Chem. Chem. Phys.* **2015**, *17*, 25778–25785. [CrossRef] [PubMed]
23. Kou, M.; Oh, Y.-H.; Lee, S.; Kong, X. Distinguishing Gas Phase Lactose and Lactulose Complexed with Sodiated L-Arginine by IRMPD Spectroscopy and DFT Calculations. *Phys. Chem. Chem. Phys.* **2023**, *25*, 25116–25121. [CrossRef] [PubMed]
24. Yao, H.; Steill, J.D.; Oomens, J.; Jockusch, R.A. Infrared multiple photon dissociation action spectroscopy and computational studies of mass-selected gas-phase fluorescein and 2′,7′-dichlorofluorescein ions. *J. Phys. Chem. A* **2011**, *115*, 9739–9747. [CrossRef]
25. Cheng, R.; Loire, E.; Fridgen, T.D. Hydrogen bonding in alkali metal cation-bound i-motif-like dimers of 1-methyl cytosine: An IRMPD spectroscopic and computational study. *Phys. Chem. Chem. Phys.* **2019**, *21*, 11103–11110.24. [CrossRef]
26. Seo, J.; Hoffmann, W.; Warnke, S.; Bowers, M.T.; Pagel, K.; von Helden, G. Retention of native protein structures in the absence of solvent: A coupled ion mobility and spectroscopic study. *Angew. Chem. Int. Ed.* **2016**, *55*, 14173–14176. [CrossRef]
27. Bush, M.F.; Forbes, M.W.; Jockusch, R.A.; Oomens, J.; Polfer, N.C.; Saykally, R.J.; Williams, E.R. Infrared spectroscopy of cationized lysine and ε-N-methyllysine in the gas phase: Effects of alkali-metal ion size and proton affinity on zwitterion stability. *J. Phys. Chem. A* **2007**, *111*, 7753–7760. [CrossRef]
28. Burt, M.; Wilson, K.; Marta, R.; Hasan, M.; Hopkins, W.S.; McMahon, T. Assessing the impact of anion–π effects on phenylalanine ion structures using IRMPD spectroscopy. *Phys. Chem. Chem. Phys.* **2014**, *16*, 24223–24234. [CrossRef]
29. Lee, S.-S.; Park, S.; Hong, Y.; Lee, J.U.; Kim, J.H.; Yoon, D.; Kong, X.; Lee, S.; Oh, H.B. Chiral differentiation of d- and l-alanine by permethylated β-cyclodextrin: IRMPD spectroscopy and DFT methods. *Phys. Chem. Chem. Phys.* **2017**, *19*, 14729–14737. [CrossRef]
30. Lee, S.-S.; Lee, J.U.; Oh, J.H.; Park, S.; Hong, Y.; Min, B.K.; Lee, H.H.L.; Kim, H.I.; Kong, X.; Lee, S.; et al. Chiral differentiation of D- and L-isoleucine using permethylated β-cyclodextrin: Infrared multiple photon dissociation spectroscopy, ion-mobility mass spectrometry, and DFT calculations. *Phys. Chem. Chem. Phys.* **2018**, *20*, 30428–30436. [CrossRef]
31. Oh, Y.-H.; Lee, S.Y.; Kong, X.; Oh, H.B.; Lee, S. Thermodynamic Reversal and Structural Correlation of 24-Crown-8/Protonated Tryptophan and 24-Crown-8/Protonated Serine Non-Covalent Complexes in the Gas Phase vs. in Solution: Quantum Chemical Analysis. *ACS Omega* **2024**, *9*, 23793–23801. [CrossRef] [PubMed]
32. Choi, H.; Oh, Y.-H.; Park, S.; Lee, S.-S.; Oh, H.B.; Lee, S. Unveiling Host-Guest-Solvent Interactions in Solution by Identifying Highly Unstable Host-Guest Configurations in Thermal Non-Equilibrium Gas Phase. *Sci. Rep.* **2022**, *27*, 5702. [CrossRef] [PubMed]
33. Jung, J.G.; Oh, Y.-H.; Park, T.H.; Lee, S.-S.; Kim, D.W.; Lee, S. Contact ion-pair SN2 reactions activated by Lewis Base Phase transfer catalysts. *Nat. Commun.* **2025**, *16*, 1236. [CrossRef] [PubMed]
34. Chai, D.; Head-Gordon, M. Long-range corrected hybrid density functionals with damped atom–atom dispersion corrections. *Phys. Chem. Chem. Phys.* **2008**, *10*, 6615–6620. [CrossRef]
35. Frisch, M.J.; Trucks, G.W.; Schlegel, H.B.; Scuseria, G.E.; Robb, M.A.; Cheeseman, J.R.; Scalmani, G.; Barone, V.; Petersson, G.A.; Nakatsuji, H.; et al. *Gaussian*; Gaussian, Inc.: Wallingford, CT, USA, 2016.
36. Marenich, A.V.; Cramer, C.J.; Truhlar, D.G. Universal solvation model based on solute electron density and on a continuum model of the solvent defined by the bulk dielectric constant and atomic surface tensions. *J. Phys. Chem. B* **2009**, *113*, 6378–6396. [CrossRef]
37. Stedwell, C.N.; Galindo, J.F.; Gulyuz, K.; Roitberg, A.E.; Polfer, N.C. Crown Complexation of Protonated Amino Acids: Influence on IRMPD Spectra. *J. Phys. Chem. A* **2013**, *117*, 1181–1188. [CrossRef]

molecules

MDPI

Article

Vibrational Spectroscopic and Quantum-Chemical Study of Indole–Ketone Hydrogen-Bonded Complexes

Branislav Jović [1], Nataša Negru [1], Dušan Dimić [2,*] and Branko Kordić [1]

[1] Faculty of Sciences, University of Novi Sad, Trg Dositeja Obradovića 3, 21000 Novi Sad, Serbia
[2] Faculty of Physical Chemistry, University of Belgrade, Studentski Trg 12-16, 11000 Belgrade, Serbia
* Correspondence: ddimic@ffh.bg.ac.rs

Abstract: This study investigates the structural and energetic properties of hydrogen-bonded complexes between indole and a range of aliphatic, cyclic, and aromatic ketones using a combined vibrational spectroscopic and quantum-chemical approach. FTIR measurements in CCl_4 revealed redshifts in the N-H stretching vibration of indole upon complexation, with formation constants (K_a) ranging from 0.3 to 6.6 M^{-1}. Cyclohexanone displayed the strongest binding, while benzophenone exhibited the weakest interaction. Quantum-chemical calculations, employing CREST and MMFF94 conformational sampling, along with M06-2X/6-311++G(d,p) optimizations, confirmed the formation of hydrogen bonds and additional weak interactions that govern the stability of the complex. QTAIM analysis revealed moderate closed-shell hydrogen bonds with electron densities at the bond critical points (ρ) ranging from 0.010 to 0.019 a.u. and potential energy densities (V) from −18.4 to −36.4 kJ mol^{-1}. Multivariate regression analysis established strong correlations (R^2 = 0.928 and 0.957) between experimental binding constants and theoretical descriptors, including binding energy, NBO charge on oxygen atom, ionization potential, and electrophilicity index, highlighting the interplay between geometric, electronic, and global reactivity factors. This comprehensive study underlines the predictive power of spectroscopic and quantum descriptors for assessing hydrogen bonding in biologically relevant systems.

Keywords: indole; hydrogen bond; ketones; FTIR; DFT; QTAIM

1. Introduction

The role of indole and indole-type molecules in biochemical processes remains a significant area of study. Examining the amino acid residues of proteins and the mechanisms of enzymatic reactions requires further clarification and expanded knowledge regarding indole. From a fundamental spectrochemical perspective, indole is particularly interesting due to its planar π-electronic system and N-H proton donor group, which allow for various interactions. Additionally, the carbonyl group represents an indispensable structural segment of many biochemical macromolecules, and the potential interactions with this group continue to be an important field of research in terms of fundamentals and applications.

Since the indole ring represents a side chain of the amino acid tryptophan, it is an integral part of proteins. Transmembrane proteins and peptides are known to have an exceptionally high content of tryptophan. Its ability to form hydrogen bonds significantly influences protein structures within cell membranes, where the indole ring interacts with other molecules, especially the abundant (lipid) carbonyl group. This amino acid participates in both hydrophobic and hydrophilic interactions, which explains its role in the transmembrane space and its capacity to form hydrogen bonds [1]. Furthermore, indole ring-containing compounds exhibit a wide range of biological activities, such as antidepressant, antifungal, antimicrobial, antiviral, anticancer, antihypertensive, anti-HIV, and more [2]. Therefore, it is a favorable structural motif in many medical therapeutics [3–9]. Aside from pharmaceuticals, indole and its derivatives have been used in solar cell production and similar applications [10–16]. The indole ring is a suitable electron donor because of its aromatic structure and the presence of nitrogen atoms [17]. Recent research reported novel use of its charge transfer properties in organic synthesis [18–20]. Studying the intermolecular interactions of indoles is also very important for understanding similar bioactive molecules, such as pyrrole [21–23].

The significance of the N-H···O hydrogen bond interaction in the crystal structure of indole derivatives has been documented [24]. While various crystal symmetries have been investigated, the hydrogen bond formed between the indole N-H group and a carbonyl group has emerged as a dominant intermolecular interaction that significantly influences the crystal packing. Further, N-H···O hydrogen bonding between polymer chains of the poly (iminoimino ketone ketone) has been reported [25]. The good thermal stability of poly (iminoimino ketone ketone) compared to similar polymers can be attributed to the formation of this hydrogen bond. Therefore, determining the physicochemical characteristics of N-H···O intermolecular interactions could provide better insights into the structure and potential modifications of polymer materials of this type.

There have been many experimental and computational studies concerning indole intermolecular interactions via the N-H group, but not with specific carbonyl compounds [26–29]. Further investigation of indole hydrogen bonding and charge transfer complexes is needed. This study represents a continuation of our ongoing systematic research of indoles and amides as important model systems for biological molecules [30–32]. The structures of indole and six ketones are presented in Figure 1. The ketones were chosen to investigate the effects of the alkyl chain, aliphatic ring, and aromatic systems, while also considering the influence of steric factors. Other polar groups in the proton acceptor molecules besides the carbonyl group were avoided to ensure that only one type of hydrogen bond can form between the proton donor and the acceptor, in accordance with the limitations of Becker's method [33].

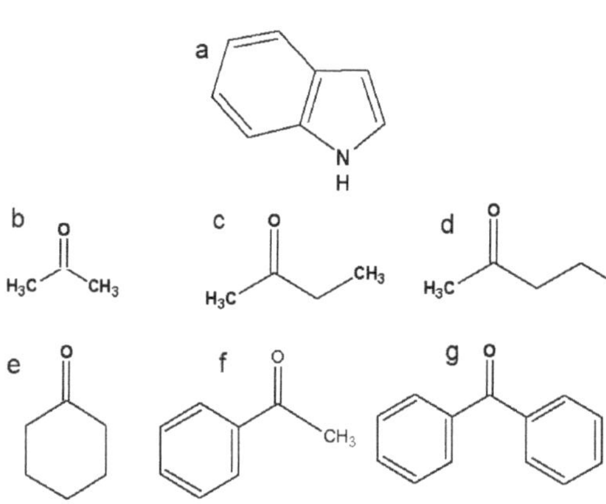

Figure 1. Structural formulas of investigated molecules: (**a**) indole, (**b**) acetone, (**c**) 2-butanone, (**d**) 2-pentanone, (**e**) cyclohexanone, (**f**) acetophenone, and (**g**) benzophenone.

2. Results and Discussion

2.1. FTIR Spectroscopic Analysis

Spectroscopic parameters for N-H···O hydrogen-bonded complexes and the equilibrium constants from infrared spectroscopy (IR) experiments are summarized in Table 1, where $\Delta\nu$(NH) and $\nu_{1/2}$ are wavenumber shift and halfwidth of the N-H complex band, respectively. Figure 2 relates to the IR spectra of indole in the presence of different ketones. The wavenumber shift represents the difference between the N-H group band position of the free indole and the band position of the same indole N-H group hydrogen-bonded to a ketone.

Table 1. Spectroscopic parameters for indole–ketone N-H···O hydrogen-bonded complexes obtained in IR measurements; $\Delta\nu$ (NH) and $\nu_{1/2}$ are wavenumber shift and halfwidth of the N-H complex band, respectively; ε and B^0 are linear and integral molar absorption coefficients, respectively.

Proton Donor	$\Delta\nu$ [cm^{-1}]	$\nu_{1/2}$ [cm^{-1}]	$\varepsilon \times 10^{-3}$ [$cm^2\ mol^{-1}$]	$B^0 \times 10^{-6}$ [$cm\ mol^{-1}$]	K_a [1] [M^{-1}]
Acetone	72	10	578.8	5.123	4.3
2-Butanone	68	12	434.1	11.32	4.1
2-Pentanone	69	11	394.4	93.42	4.4
Cyclohexanone	77	13	370.1	10.44	6.6
Acetophenone	86	13	3014	127.3	1.1
Benzophenone	23	10	6172	133.7	0.3

[1] Calculated by Equation (3).

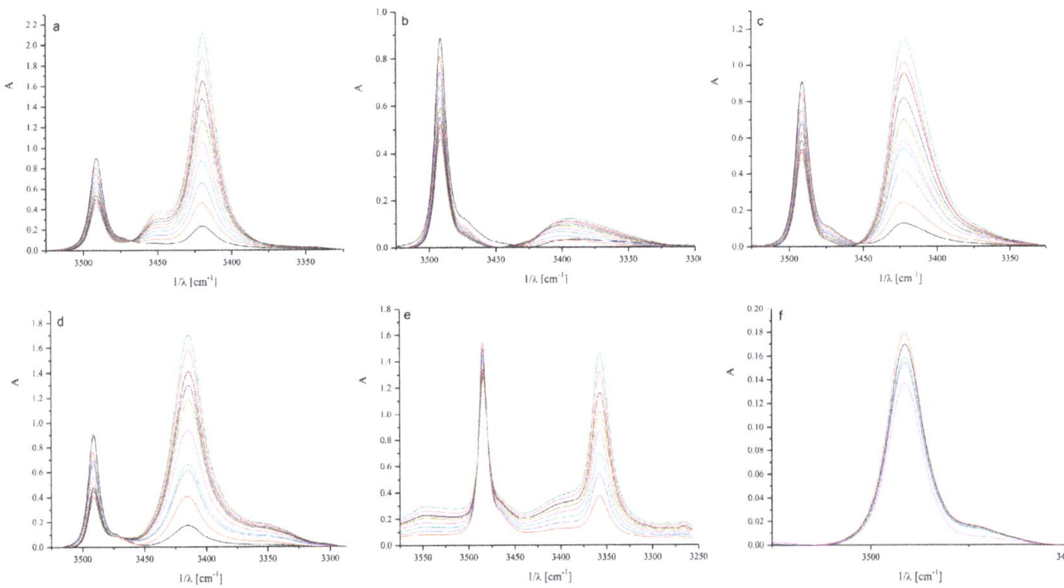

Figure 2. Infrared spectra of investigated systems: (**a**) acetone, (**b**) 2-butanone, (**c**) 2-pentanone, (**d**) cyclohexanone, (**e**) acetophenone, and (**f**) benzophenone. (The different colors of the spectra represent increasing concentrations of ketones in the presence of indole in the range of 0.01 to 0.3 M).

As shown in the data from Figure 2 and Table 1, lower alkyl ketones form similar hydrogen-bonded complexes with indole regarding structure and stability, whereas cyclic ketones and aromatics behave somewhat differently. The band of the free, non-bonded N-H indole group appears at 3491 cm^{-1} (Figure S1), while the band of the hydrogen-bonded N-H group appears at lower wavenumbers. The most noticeable difference occurs in complexes with the aromatic benzophenone. In the case of benzophenone, two bulky phenyl rings limit the accessibility of the carbonyl group for direct hydrogen bonding with the N-H group of indole and favor other types of interactions. It can be assumed that π–π interactions prevail in the case of acetophenone and benzophenone. Cyclohexanone exhibits a higher stability constant than the other tested ketones, which could be attributed to the sterically accessible carbonyl group of cyclohexanone (Table 1).

If the strengths of the obtained complexes of indole with ether and carbonyl oxygen are compared, the following can be observed: indole with different ethers, both cyclic and acyclic, forms complexes with hydrogen bond constants ranging from 1.5 to 3 M^{-1}, which are significantly lower than the constants in complexes with ketones. The greater stability of the indole complex with ketones is likely a consequence of the higher polarity of the carbonyl oxygen compared to the ether group. Hydrogen bonding of indole with aromatic π-systems has been reported. Muñoz et al. [34] obtained values for formation constants for hydrogen-bonded complexes of indole with different aromatic molecules in the range of 0.3–0.9 dm^3 mol^{-1}. As may be expected, these values are much lower than those obtained for systems with ketones in this work.

The interaction of the carbonyl group of ketones has been a research subject for a long time [35] and remains relevant. Specifically, in the spectroscopic examination of the complexes formed by acetone, acetophenone, and benzophenone with phenol derivatives, similar to this study, it was determined that, due to steric hindrances, the most stable complexes with phenols are formed with acetone, followed by acetophenone and benzophenone. More recent studies [36] on the interactions of the carbonyl group using spectroscopic methods include exploring the interactions of ketone carbonyl groups with alcohols. Interestingly, and similar to the results obtained in this study, it has been observed that the redshifts of the carbonyl band of ketones are greater for cyclic ketones than for acyclic ones. It was found that for a given ketone, the magnitude of the shift increases monotonically with the acidity of the alcohol. Spectral shifts increased linearly with decreasing pKa of the donor alcohols. Conversely, when the alcohol remained constant, the spectral shift depended on the C=O bond strength of the ketones. Additionally, very intriguing theoretical research offering a realistic assessment of the distribution of dipole moments in the liquid state [37] can be included in this discussion. Specifically, this paper concluded that the increase in dipole moments of acetone and 2-butanone in the liquid phase compared to the gas phase is not significant due to the lack of hydrogen bonds in the pure ketones themselves. An increase in polarization effects is also anticipated in solutions and mixtures of ketones with proton donors. In this context, it is also possible to suggest that the highest constant obtained in this study for cyclohexanone, as shown in Table 1, is attributed to its relatively rigid structure, stronger polarization, and narrower distribution of dipole moments.

Given the importance of the dipole moments of the carbonyl group and their dynamic character in different solvents, it can also be assumed that the steric and inductive effects are in a dynamic balance that depends on the solvent. Considering the dynamic nature and the multitude of impacts related to the carbonyl group, as well as the indole π-electron condensed system with a strong proton donor group, it seems that the interaction of ketone and indole represents an open challenge for both experimental and theoretical approaches.

2.2. Quantum-Chemical Examination of the Interactions

The theoretical support for the previously discussed experimental results was obtained through the optimization of indole–ketone pairs. The CREST (conformer-rotamer ensemble sampling tool) [38] was used to explore the conformational space and identify the initial geometries of the indole–ketone complexes. This program applies a semi-empirical tight-binding quantum chemistry method (GFN2-xTB) [39] to screen the conformational space. During this process, a large number of conformers are located. Upon close inspection of the structures, it is evident that the carbonyl group of the ketone and the amino group of the indole are in close proximity for the most stable conformers. As the complex with cyclohexanone had the highest binding constant, several characteristic examples of the conformers are presented in Figure 3.

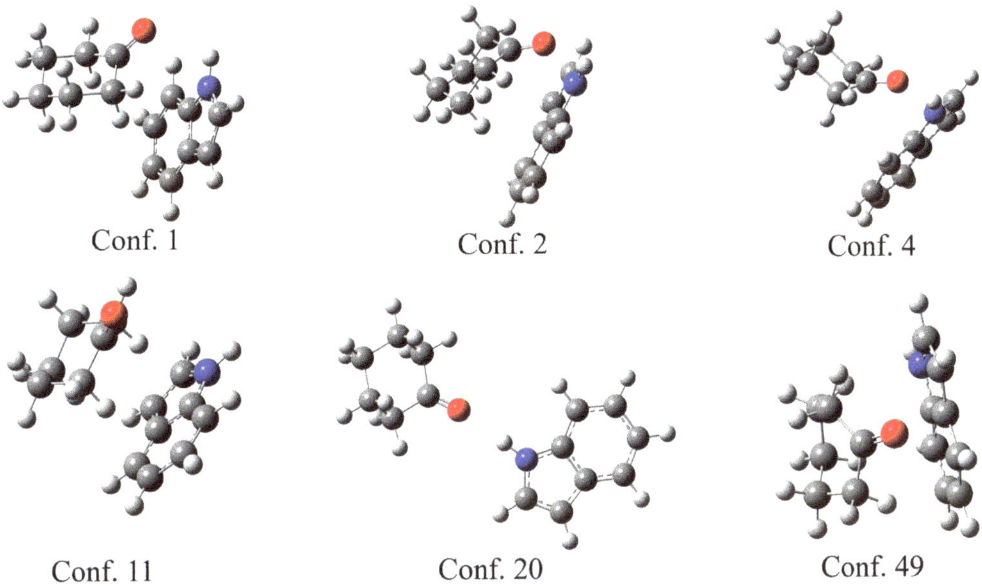

Figure 3. Selected conformers between cyclohexanol and indole, obtained in the CREST program, with the conformer number.

As shown in Figure 3, the six selected conformers exhibit distinct geometries governed by different interactions. Three conformers (1, 2, and 4) have the carbonyl group of the ketone directed towards the amino group of the indole, proving that the formation of a hydrogen bond is one of the most important stabilization parameters, as expected from the experimental results. The N-H···H distances are 2.20 (conf. 1), 2.43 (conf. 2), and 2.05 Å (conf. 4). Another distinctive feature of these structures is the position of cyclohexanone above the planar structure of indole, which allows for the formation of additional weak interactions between the two ring structures. Once the cyclohexanone carbonyl group is distanced from the amino group, only the weak interactions between rings stabilize the structure, as in the case in conformer 11. Conformer 20 is interesting because the compounds are positioned in a way that only a hydrogen bond is formed, with a bond distance of 2.05 Å. Conformer 49 is included for illustrative purposes, as the distance between the two compounds allows for the formation of only weak interactions between the hydrogen atoms of the ketone and the carbon atoms of the indole. A similar observation was made in complexes with all other ketones.

Although CREST calculations allowed identification of a wide variety of possible conformers, only the geometries were used as starting points for further calculations. No quantitative conclusions were drawn from semi-empirical energy differences. Ten conformers of each pair were then subjected to the MMFF94 (Merck Molecular Force Field)-based calculations [40] in Open Babel [41]. The MMFF94 method demonstrated strong performance and was recommended for hydrogen-bond-donating compounds in a study by Lewis-Atwell and coworkers [42]. The initial search by MMFF94 for the most stable conformers formed between the water molecule and aspirin, paracetamol, and caffeine was presented in the reference [43]. The relative energies of these conformers for each of the pairs are given in Tables S2–S7, together with the optimized geometries in the Supplementary Material. The geometries of these complexes are not identical, but rather very similar, as reflected in their energy values. For example, in the case of acetone, the relative energies range from 0 to 1.6 kJ mol^{-1}. With an increase in the complexity of ketones, the relative energies span a much wider range, up to 8.8 kJ mol^{-1} (acetophenone) and 21.2 kJ mol^{-1} (benzophenone). Based on the optimization of the cyclohexanone–indole pair, it can be concluded that the ring structure of cyclohexanone dictates the formation of stabilization interactions, as shown in the previous figure.

The lowest-energy conformers from MMFF94 calculations were selected and optimized at the M06-2X/6-311++G(d,p) level of theory without any geometric constraints. This was completed to ensure that the global minima on the potential energy surface are examined and correlated with the experimental results. The minima on the potential energy surface were proven by the absence of imaginary frequencies. Before exploring the results, it is essential to justify the use of the M06-2X functional for examining hydrogen bonds and other weak interactions formed between ketones and indole. In the paper by Walker and coworkers, it was shown that the M06 functional performed well in predicting general trends in conformer energies and global minima identification of halide ion-amino acid clusters when compared with MP2 [44]. Functional M06-2X was outlined for the optimization of hydrogen-bonded and halogen-bonded complexes by Zhang, Ma, and Wang in reference [45]. Paul examined the red- and blue-shifting hydrogen bonds of cyclic ketones with HF and CHF_3 and compared the structural and topological parameters obtained by B3LYP, O3LYP, and M06-2X functionals, and the choice of functional proved to be important in characterizing stabilizing effects [46]. Other systems in which intermolecular hydrogen bonds are important for stabilization and changes in spectra were optimized by the M06-2X functional in references [47–49].

This optimization step was taken as an additional proof that the minima are the most stable conformations on the potential energy surface. The optimized structures of the pairs are presented in Figure 4. The binding energy was calculated as the difference between the energy of the complex and the sum of the energies of the separate indole and ketone (Equation (4)), and is listed in Table 2, with the inclusion of the BSSE correction, as explained in Section 3.

Selected structural details and the Quantum theory of Atoms in Molecules (QTAIM) parameters of the important interactions are summarized in Tables 2 and S1. The total energy density was determined as the sum of kinetic (G(r)) and potential (V(r)) energy densities [50]. Previously, QTAIM analysis has proven helpful in examining inter- and intramolecular hydrogen bonds and predicting their strength [51,52]. The topological properties of hydrogen bonds were suggested by Koch and Popelier [53]. Based on this, the hydrogen bonds should exhibit relatively high electron density, between 0.002 and 0.034 a.u., and Laplacian in the range of 0.024–0.139 a.u.

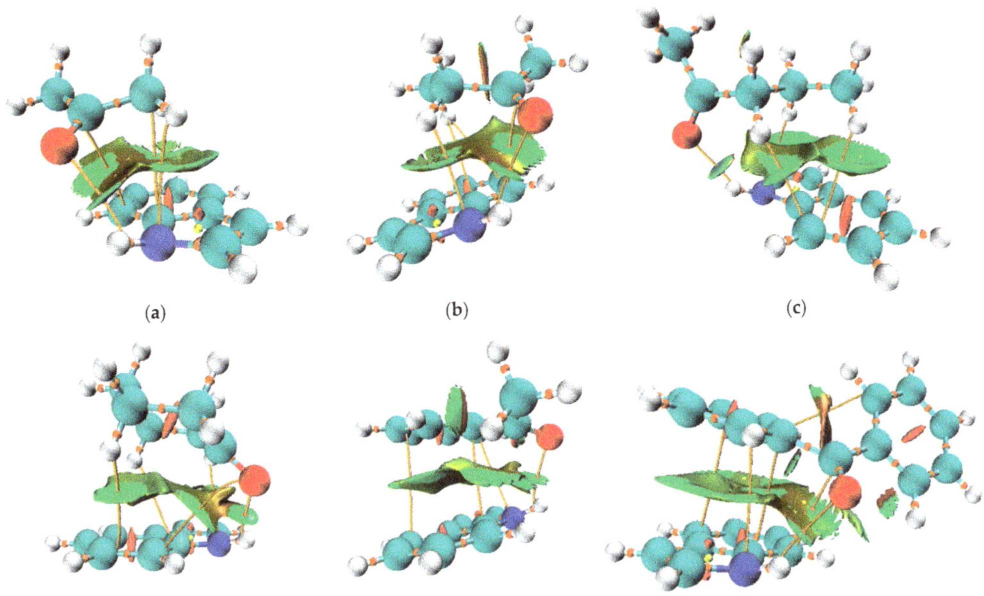

Figure 4. Non-Covalent Interaction (NCI) surface plot of the complexes between indole and (**a**) acetone, (**b**) 2-butanone, (**c**) 2-pentanone, (**d**) cyclohexanone, (**e**) acetophenone, and (**f**) benzophenone, computed based on the reduced density gradient (RDG) at the M06-2X/6-311++G(d,p) level. The isosurface is plotted at RDG = 0.5 a.u. and colored according to the sign(λ_2)ρ function: blue indicates strong attractive interactions (hydrogen bonding), green indicates van der Waals interactions, and red corresponds to steric repulsion. The surface reveals a π–π interaction region and a hydrogen bond consistent with the geometry optimization results.

Table 2. Selected structural and QTAIM parameters for indole–ketone N-H···O hydrogen-bonded complexes.

Proton Donor	E_b [kJ mol^{-1}]	d(H···O) [Å]	ρ(H···O) [a.u.]	V(H···O) [kJ mol^{-1}]	H(H···O) [kJ mol^{-1}]	Angle (N-H···O) [°]
Acetone	39.8	2.060	0.013	−22.6	3.8	137.5
2-Butanone	41.1	2.019	0.010	−18.4	2.7	141.9
2-Pentanone	47.6	2.100	0.019	−36.3	6.3	134.9
Cyclohexanone	47.3	2.244	0.016	−28.8	5.1	119.3
Acetophenone	43.6	2.384	0.011	−19.7	3.1	120.4
Benzophenone	44.3	2.250	0.014	−25.3	4.4	126.2

E_b—BSSE binding energy, d—bond length, ρ—electron density, V—potential energy density, H—total energy density.

In the theoretical examination of the binding process, the binding constant is expected to align with the structural and QTAIM parameters of bonds involved in the formation of N-H···O. As shown in Table S1, the bond lengths between hydrogen and nitrogen atoms of indole in hydrogen-bonded pairs are in a very narrow range of 1.010 to 1.016 Å (Table S1). The change in the N-H bond length cannot be regarded as a definitive criterion, as the formation of hydrogen bonds leads to significant structural deformation of both the donor and the acceptor, as discussed by Wang and coworkers [54]. This is also reflected in electron density values, with slight differences. The elongation of the C=O bond also depends on the strength of the hydrogen bond and the electron-donating/withdrawing effects of substituents on both sides. The C=O bonds are longer in pairs with higher binding constants, as in pairs with acetone, 2-butanone, 2-pentanone, and cyclohexanone (between 1.229 and 1.234 Å). In complexes with acetophenone and benzophenone, the C=O bond length is shorter (1.214 and 1.218 Å), indicating the presence of other interactions except a hydrogen bond. The electron densities of C=O bonds are inversely proportional to the bond lengths, as expected. The NBO charges on oxygen atoms are similar for all ketones and are between −0.607 and −0.597 *e*, reflecting the bond's elongation as dictated by the presence of different substituents. It should be noted that the lowest charges on oxygen atoms were found for the ketones with the extended delocalization. A simple correlation between the structural parameters of indole and ketones and the experimental values is not possible, suggesting that the formation of hydrogen bonds is not the sole parameter influencing stability.

On the other hand, the differences in H···O bond lengths are much more pronounced, between 2.019 (2-butanone) and 2.384 Å (acetophenone), as shown in Table 2. This result indicates the difference in hydrogen bond strength and orientation. The electron density at the H···O BCPs ranges from 0.011 to 0.019 a.u., and the potential energy density spans from −18.4 to −36.3 kJ mol^{-1}, both consistent with moderate closed-shell hydrogen bonds. The lowest potential energy density values were obtained for the ketones with the highest binding constant (2-pentanone and cyclohexanone). The total energy densities H(H···O) (3.1–6.3 kJ mol^{-1}) align with the classification proposed by Popelier [53], confirming weak-to-moderate interaction strength. Acetophenone exhibits the weakest H-bond, correlating with the longest H···O distance and lowest ρ(H···O). The angle between the hydrogen atom donor and acceptor is also included in Table 2. The highest value of the angle was determined for the 2-butanone–indole pair (140.9°). Other hydrogen bonds have much lower angles, which classify them as weak hydrogen bonds that are distorted due to steric or electronic factors.

The vibrational frequency shifts (Δν) show the highest value for acetophenone (86 cm^{-1}), potentially reflecting a significant perturbation of the N-H bond through vibrational coupling or local environment effects, rather than purely H-bonding strength. Benzophenone, despite its lower Δν (23 cm^{-1}), has a relatively strong binding energy, suggesting the influence of π–π interactions with indole.

The binding energies (E_b), calculated with BSSE correction, follow the trend 2-pentanone (47.6 kJ mol^{-1}) > cyclohexanone (47.3 kJ mol^{-1}) > benzophenone (44.3 kJ mol^{-1}) > acetophenone (43.6 kJ mol^{-1}) > 2-butanone (41.1 kJ mol^{-1}) > acetone (39.8 kJ mol^{-1}). These results partially correlate with the experimental binding constants (K_a), which are highest for cyclohexanone (6.6 M^{-1}) and lowest for benzophenone (0.3 M^{-1}). The binding energies obtained by DFT methods have to be carefully evaluated [55] and the discrepancies, such as between E_b and K_a for benzophenone, imply the necessity of examining additional interactions, notably π–π stacking, that influence overall complex stability.

In addition to the previously discussed hydrogen bond, each pair is characterized by further interactions between ketone substituents and the indole, as illustrated in Figure 4. The same figure presents the Non-Covalent Interaction surface plot in which the green area indicates weak interactions. The strength of these interactions was assessed using the approach suggested by Espinosa, in which the interatomic interaction energy is determined as half the value of V(r) [56]. The complex formed between acetone and indole is characterized by the presence of an additional N···C bond with an electron density of 0.008 a.u. and an interaction energy of −7.0 kJ mol^{-1}. Additionally, there is an interaction between carbon atoms with an energy of −4.0 kJ mol^{-1}. When the energies of these interactions are combined, it can be concluded that they significantly influence the stability of the formed complex, in addition to the previously discussed hydrogen bond. In the case of complexes with 2-butanone and 2-pentanone, the interactions between the carbon/hydrogen atom of the ketone and the nitrogen atom of indole are stronger (−7.0 and −7.5 kJ mol^{-1}). There are interactions between the hydrogen atom of the ketone and the carbon atom of the indole. The number of these interactions and their strength explain the higher binding energies of 2-butanone and 2-pentanone, at 41.1 and 47.6 kJ mol^{-1}, respectively, when compared to acetone (39.8 kJ mol^{-1}). The same types of interactions are found in the cyclohexanone–indole pair, with energies of −7.1 (H···N), −4.8 (O···C), and −5.5/−2.9 kJ mol^{-1} (H···C), leading to the binding energy of 47.3 kJ mol^{-1}. These interactions also explain the highest binding constant of 6.6 M^{-1}.

When acetophenone and benzophenone are considered, an additional type of interaction can be observed, involving the carbon atoms of aromatic moieties. These interactions can be denoted as C···C or π–π. In the case of acetophenone, there are two interactions of this type with binding energies of −2.6 and −6.5 kJ mol^{-1}. As expected, a higher number of C···C interactions was found for the benzophenone–indole pair, with energies of −2.8, −5.9, and −5.2 kJ mol^{-1}. The interactions between the nitrogen of indole and the carbon atoms were also observed. This is an additional reason why the experimental shift in N-H band position cannot be used for simple comparison, as several interactions influence the change in the bond strength.

The relationship between binding constant and theoretical parameters was examined by multiple two-parameter linear regression models. In the quantitative assessment of the factors influencing the binding constant K_a, particular attention was given to DFT, NBO charge, and QTAIM parameters, as presented in Tables 2 and S1. The multivariate regression analysis confirms that a combination of binding energy E_b and oxygen atom partial charge ($q(O)$) provides the strongest statistical correlation with the experimentally determined equilibrium constants.

$$K_a = -369.63 - 0.205 \cdot E_b - 634.90 \cdot q(O) \quad (1)$$

This model yielded a coefficient of determination (R^2) value of 0.928, indicating that over 92.8% of the observed variation in K_a can be attributed to changes in these two parameters. The average absolute difference between the experimental and calculated value is 0.56 M^{-1}. From a chemical perspective, this result aligns with established principles governing the formation of hydrogen bonds. The binding energy E_b reflects the overall stabilization of the complex and integrates contributions from all Non-Covalent Interactions. A more negative E_b correlates with stronger binding and thus larger equilibrium constants. Meanwhile, the partial charge on the oxygen atom $q(O)$ is a direct indicator of its electronegativity and capacity to act as a hydrogen bond acceptor—a more negative $q(O)$ enhances the interaction strength with the N-H donor group.

Importantly, these two descriptors remain statistically independent (correlation < 0.5), and together they provide a physically grounded interpretation of the binding process. While E_b captures the energetic consequence of the interaction, $q(O)$ reflects the electronic predisposition for hydrogen bond formation, making the model both predictive and interpretable.

Nevertheless, limitations must be acknowledged. The dataset comprises only six ketone–indole complexes, which restricts statistical robustness and may exaggerate the apparent precision of the regression. The linearity assumption may also oversimplify the complex interplay of forces involved in hydrogen bonding, especially in solution.

Moreover, the current model does not account for solvation effects, entropic contributions, or secondary interactions such as dispersion forces or π-stacking, which could be especially relevant for aromatic ketones. Despite these limitations, the current analysis underscores the value of combining global energetic parameters with localized electronic descriptors in modeling molecular recognition phenomena.

The global reactivity parameters are a set of metrics used to assess the reactivity of compounds based on the energies of the Highest Occupied Molecular Orbital (HOMO) and Lowest Unoccupied Molecular Orbital (LUMO) [57–59]. These parameters (Table S3) include ionization potential (IP), vertical electron affinity (EA), chemical potential (μ), electronegativity (χ), chemical hardness (η), and electrophilicity index (ω), calculated as detailed in the Supplementary Material. The HOMO- LUMO gap can be regarded as a measure of the overall reactivity of compounds. The lowest HOMO- LUMO gap values were recorded for benzophenone (7.47 eV) and acetophenone (7.82 eV) due to extensive delocalization within the structure. Acetone, 2-butanone, and 2-pentanone exhibit values for this parameter within a narrow range (between 8.74 and 8.88 eV), as expected from their structural features [60]. Global hardness describes the resistance of the electron cloud to deformation under small perturbations. All the examined compounds possess values for this parameter within 1 eV, ranging from 3.74 (benzophenone) to 4.44 eV (acetone). This parameter is also significantly influenced by delocalization within the structure. Electronegativity, defined by Mulliken, is the average of the ionization potential and electron affinity [61]. Regarding this parameter, the lowest values were observed for cyclohexanone, acetone, 2-butanone, and 2-pentanone. The reactivity of these molecules can be attributed to the absence of steric effects that would hinder reactions with surrounding molecules. The electrophilicity index is another descriptor of reactivity [62]. A higher value of this parameter indicates stronger electrophiles (weaker nucleophiles). Acetophenone and benzophenone are stronger electrophiles, while the other compounds are stronger nucleophiles. Since the interaction with indole involves engagement with positively charged amino groups, the lower reactivity of acetophenone and benzophenone can be understood through this effect. The discussion of the global reactivity parameters demonstrates that the reactivity and overall geometry of the included ketones influence interactions with indole.

To further explore the molecular determinants of hydrogen bond-driven binding affinity, a multiple linear regression analysis was performed using quantum-chemical descriptors derived from frontier molecular orbital theory. Among all tested two-variable combinations, the model incorporating ionization potential (IP) and electrophilicity index (ω) emerged as the most statistically robust, yielding a correlation coefficient of $R^2 = 0.957$ and an average absolute error of 0.37 M^{-1}. The resulting equation

$$K_a = 78.224 - 6.856 \cdot \mathrm{IP} - 5.77 \cdot \omega \quad (2)$$

indicates that both a lower ionization potential and lower electrophilicity favor stronger complex formation. This is chemically intuitive, as a lower IP reflects a more readily polarizable electron cloud, enhancing donor–acceptor interactions, while a lower ω suggests reduced electronic stabilization upon accepting charge density, allowing more effective participation in intermolecular bonding.

The predicted K_a values closely match the experimental measurements for all compounds in the study, including subtle distinctions between structurally similar ketones. For example, cyclohexanone, which exhibits the highest experimental K_a (6.6 M^{-1}), was also correctly identified by the model as the strongest binder, with a predicted value of 6.16 M^{-1}.

On the other end of the spectrum, benzophenone, characterized by high IP and ω values due to the extended aromatic system, exhibited the lowest predicted and experimental affinities (0.27 M^{-1} and 0.3 M^{-1}, respectively). These findings reinforce the value of global reactivity descriptors—particularly those derived from HOMO–LUMO characteristics—in capturing the intrinsic electronic features that govern non-covalent binding.

Overall, this model provides a reliable and physically interpretable framework for estimating hydrogen bond strength in structurally diverse ketone systems. Its predictive accuracy surpasses that of previous models based solely on QTAIM or geometric parameters, suggesting that donor–acceptor charge flow, as quantified by frontier orbital theory, plays a central role in modulating intermolecular interaction energies in solution.

3. Materials and Methods

3.1. FTIR Spectroscopic Measurements

Indole (Merck, Darmstadt, Germany, >99%), carbon tetrachloride (Merck, Darmstadt, Germany, >99.9%), acetone, 2-butanone, 2-pentanone, cyclohexanone (Merck, Darmstadt, Germany, >98%), acetophenone, and benzophenone (Merck, Darmstadt, Germany, >98%) were used without further purification. Structural formulas of selected ketones are shown in Figure 1.

Indole certainly dimerizes in a CCl_4 solution, which has been proven spectroscopically and theoretically [63,64]. However, dimerization occurs at higher indole concentrations in CCl_4. In the present study, we used very low indole concentrations at 10^{-3}–10^{-4} mol dm^{-3} where dimerization can be neglected. The indole concentration in the carbon tetrachloride solutions was sufficient to observe measurable bands in the IR spectrum, thereby preventing self-association. The concentration of the ketone proton acceptors was varied between 0.01 and 0.3 M. The IR spectra were obtained using a Thermo Scientific Nicolet iS20 instrument. A Deuterated Triglycine Sulfate (DTGS) detector was employed in the IR measurements. The samples were placed in a 1 cm quartz cell, and the spectra were recorded at 298 K. Using Becker's procedure [33], the hydrogen bonding formation constant can be determined from the difference in monomer band absorbency of pure amide in carbontetrachloride solution and carbon tetrachloride solution in the presence of a proton acceptor.

$$\mathrm{K} = \frac{A^0 - A}{AC^0_{PA}} \tag{3}$$

where A^0 is the monomer absorbance for carbon tetrachloride solution containing indole, and A is the absorbance of the same band for the solution with both indole and the ketone proton acceptor concentration (C^0_{PA}). The reported frequencies and half-widths were reproducible to within 0.2 and 1 cm^{-1}, respectively. The integrated molar absorption coefficients were determined with an accuracy of ±10%. The equilibrium constants were determined with an average relative standard deviation of 6%.

3.2. Theoretical Calculations

The experimental findings were supported by theoretical calculations performed on optimized indole–ketone complexes. Low-energy conformers were generated using the CREST (Conformer–Rotamer Ensemble Sampling Tool) [38], which efficiently explores conformational space through the GFN2-xTB semi-empirical tight-binding method [39]. Ten conformers of each pair were then subjected to the MMFF94 (Merck Molecular Force Field)-based calculations [40] in Open Babel [41]. The structures of indole and six ketones were optimized using the Gaussian 09 Program package [65] at the M06-2X/6-311++G(d,p) level of theory [66,67]. The atomic coordinates of indole–ketone pairs and separate compounds are presented in the Supplementary Information. The number of imaginary frequencies was zero in the individual structures and the interacting pairs. Several starting geometries were selected for each pair, and only the most stable ones are examined in the text. As the shifts in FTIR spectra were monitored in carbon tetrachloride solution, the optimizations of structures were carried out in a simulated solution using the Conductor-like Polarizable Continuum (CPCM) [68] model implemented in Gaussian 09. The discussion of vibrations mainly focuses on the purely N-H stretching vibration, as visualized in GaussView 5 [69]. The stabilization interactions formed between indole and ketones were examined using the Natural Bond Orbital (NBO) charges [70] and Quantum Theory of Atoms in Molecules [71,72], similar to previous work [73]. In the last-mentioned reference, interactions between N-monosubstituted benzamides and three proton acceptors (benzene, acetonitrile, and tetrahydrofuran) were examined by the combination of FTIR spectroscopy, DFT calculation, QTAIM, and NBO analysis. The AIMAll program package [74] was utilized for the QTAIM calculations, starting from the wfx file obtained in Gaussian 09. The Multiwfn program [75,76] was used to prepare a Non-Covalent Interaction (NCI) surface plot that was later visualized in the VMD program [77]. The binding energy was determined as the difference between separate units and their complex, as suggested by Parlak and coworkers [78]:

$$E_b = E_{complex} - (E_{indole} + E_{ketone}) \quad (4)$$

In the previous equation, $E_{complex}$, E_{indole}, and E_{ketone} are the electronic energies of the optimized complex formed between indole and ketone, and separate indole and ketone, respectively. The basis set superposition errors (BSSEs) were included in the counterpoise correction method.

3.3. Statistical Analysis

To explore the relationship between experimental N-H vibrational redshifts and theoretical descriptors, multiple two-parameter linear regression models were constructed. Each model correlated $\Delta\nu$(NH) values with pairs of quantum-chemical parameters, such as binding energy, hydrogen bond distance, electron density, and energy density at the bond critical point. The regression analysis was performed in Microsoft Excel, and model performance was evaluated based on the coefficient of determination (R^2) and the average deviation from experimental data.

4. Conclusions

This study presents a comprehensive vibrational spectroscopic and quantum-chemical analysis of hydrogen-bonded complexes formed between indole and six ketones of varying structural features. Experimentally determined binding constants (K_a) from FTIR spectroscopy ranged from 0.3 M^{-1} for benzophenone to 6.6 M^{-1} for cyclohexanone. While alkyl ketones showed consistent interaction patterns, aromatic ketones such as benzophenone exhibited reduced hydrogen bond strength due to steric hindrance and potential π–π interactions, despite their inherent electron-withdrawing character.

CREST conformational sampling followed by MMFF94 and M06-2X/6-311++G(d,p) optimization of the most stable structure revealed that hydrogen bonding is a dominant stabilizing interaction, with N-H···O distances ranging from 2.019 Å (2-butanone) to 2.384 Å (acetophenone). Binding energies (Eb), corrected for BSSE, varied between 39.8 kJ mol^{-1} (acetone) and 47.6 kJ mol^{-1} (2-pentanone). QTAIM parameters confirmed closed-shell hydrogen bonding, with electron density at the bond critical points (ρ(H···O)) between 0.010–0.019 a.u. and potential energy densities from −18.4 to −36.3 kJ mol^{-1}.

Multivariate regression analysis established a strong correlation (R^2 = 0.928) between K_a and the BSSE-corrected binding energy and the NBO charge on the oxygen atom. A second model involving global reactivity parameters—ionization potential (IP) and electrophilicity index (ω)—yielded an even higher correlation (R^2 = 0.957), with an average absolute error of only 0.37 M^{-1}. This model accurately predicted cyclohexanone as the strongest binder and benzophenone as the weakest, in agreement with experiment.

Altogether, the results demonstrate that both local hydrogen bond characteristics and global molecular reactivity govern the strength and geometry of Non-Covalent Interactions, offering a predictive framework for rationalizing complex formation in biologically and chemically relevant systems.

Supplementary Materials: The following supporting information can be downloaded at https://www.mdpi.com/article/10.3390/molecules30132685/s1, Figure S1: Four indole spectra at the concentration range from 0.0001; 0.001; 0.002 and 0.005 M; Table S1: Selected structural and QTAIM parameters for indole–ketone N-H···O hydrogen-bonded complexes; Table S2: The relative conformer energies for acetone–indole complex; Table S3: The relative conformer energies for 2-butanone–indole complex; Table S4: The relative conformer energies for 2-pentanone–indole complex; Table S5: The relative conformer energies for cyclohexanone–indole complex; Table S6: The relative conformer energies for acetophenone–indole complex; Table S7: The relative conformer energies for benzophenone–indole complex; Table S8: The global reactivity indexes for the ketones used in the study (in eV); Output .xyz file geometries of indole, six ketones, and their pairs (MMFF94 and M06-2X/6-311++G(d,p)).

Author Contributions: Conceptualization, B.J., N.N., D.D. and B.K.; methodology, B.J., D.D. and B.K.; software, D.D.; validation, B.J. and N.N.; formal analysis, B.J., N.N. and D.D.; investigation, N.N. and B.K.; resources, B.J. and D.D.; data curation, N.N.; writing—original draft preparation, N.N. and B.K.; writing—review and editing, B.J. and D.D.; visualization, B.J.; supervision, B.J.; project administration, D.D. and B.K.; funding acquisition, D.D. All authors have read and agreed to the published version of the manuscript.

Funding: This research was funded by the Ministry of Science, Technological Development, and Innovation of the Republic of Serbia, grant numbers 451-03-137/2025-03/200146, 451-03-136/2025-03/200146, 451-03-137/2025-03/200125, and 451-03-136/2025-03/200125.

Institutional Review Board Statement: Not applicable.

Informed Consent Statement: Not applicable.

Data Availability Statement: The data are contained in this article. Further inquiries can be directed to the corresponding author.

Conflicts of Interest: The authors declare no conflicts of interest. The funders had no role in the design of the study; in the collection, analyses, or interpretation of data; in the writing of the manuscript; or in the decision to publish the results.

References

1. Chaudhuri, A.; Haldar, S.; Sun, H.; Koeppe, R.E.; Chattopadhyay, A. Importance of Indole NH Hydrogen Bonding in the Organization and Dynamics of Gramicidin Channels. *Biochim. Biophys. Acta (BBA) Biomembr.* **2014**, *1838*, 419–428. [CrossRef]
2. Sharma, V.; Kumar, P.; Pathak, D. Biological Importance of the Indole Nucleus in Recent Years: A Comprehensive Review. *J. Heterocycl. Chem.* **2010**, *47*, 491–502. [CrossRef]
3. Hamid, H.A.; Ramli, A.N.M.; Yusoff, M.M. Indole Alkaloids from Plants as Potential Leads for Antidepressant Drugs: A Mini Review. *Front. Pharmacol.* **2017**, *8*, 96. [CrossRef] [PubMed]

4. Mehra, A.; Sharma, V.; Verma, A.; Venugopal, S.; Mittal, A.; Singh, G.; Kaur, B. Indole Derived Anticancer Agents. *ChemistrySelect* **2022**, *7*, e202202361. [CrossRef]
5. Gumus, I.; Solmaz, U.; Gonca, S.; Arslan, H. Molecular Self-Assembly in Indole-Based Benzamide Derivative: Crystal Structure, Hirshfeld Surfaces and Antimicrobial Activity. *Eur. J. Chem.* **2017**, *8*, 349–357. [CrossRef]
6. Ibad, M.F.; Zinad, D.S.; Hussain, M.; Ali, A.; Villinger, A.; Langer, P. One-Pot Synthesis of Arylated 1-Methyl-1H-Indoles by Suzuki–Miyaura Cross-Coupling Reactions of 2,3-Dibromo-1-Methyl-1H-Indole and 2,3,6-Tribromo-1-Methyl-1H-Indole. *Tetrahedron* **2013**, *69*, 7492–7504. [CrossRef]
7. Bie, J.; Liu, S.; Zhou, J.; Xu, B.; Shen, Z. Design, Synthesis and Biological Evaluation of 7-Nitro-1H-Indole-2-Carboxylic Acid Derivatives as Allosteric Inhibitors of Fructose-1,6-Bisphosphatase. *Bioorganic Med. Chem.* **2014**, *22*, 1850–1862. [CrossRef]
8. Yamuna, E.; Kumar, R.A.; Zeller, M.; Prasad, K.J.R. Synthesis, Antimicrobial, Antimycobacterial and Structure–Activity Relationship of Substituted Pyrazolo-, Isoxazolo-, Pyrimido- and Mercaptopyrimidocyclohepta[b]Indoles. *Eur. J. Med. Chem.* **2012**, *47*, 228–238. [CrossRef]
9. Bie, J.; Liu, S.; Li, Z.; Mu, Y.; Xu, B.; Shen, Z. Discovery of Novel Indole Derivatives as Allosteric Inhibitors of Fructose-1,6-Bisphosphatase. *Eur. J. Med. Chem.* **2015**, *90*, 394–405. [CrossRef]
10. Fujisawa, J. Interfacial Charge-Transfer Transitions between TiO_2 and Indole. *Chem. Phys. Lett.* **2020**, *739*, 136974. [CrossRef]
11. Nitha, P.R.; Soman, S.; John, J. Indole Fused Heterocycles as Sensitizers in Dye-Sensitized Solar Cells: An Overview. *Mater. Adv.* **2021**, *2*, 6136–6168. [CrossRef]
12. Chen, Z.; Dong, R.; Wang, X.; Huang, L.; Qiu, L.; Zhang, M.; Mi, N.; Xu, M.; He, H.; Gu, C. Efficient Decomposition of Perfluoroalkyl Substances by Low Concentration Indole: New Insights into the Molecular Mechanisms. *Environ. Sci. Technol.* **2024**, *58*, 3530–3539. [CrossRef]
13. Chen, Z.; Teng, Y.; Mi, N.; Jin, X.; Yang, D.; Wang, C.; Wu, B.; Ren, H.; Zeng, G.; Gu, C. Highly Efficient Hydrated Electron Utilization and Reductive Destruction of Perfluoroalkyl Substances Induced by Intermolecular Interaction. *Environ. Sci. Technol.* **2021**, *55*, 3996–4006. [CrossRef]
14. Qian, X.; Yan, R.; Hang, Y.; Lv, Y.; Zheng, L.; Xu, C.; Hou, L. Indeno[1,2-*b*]Indole-Based Organic Dyes with Different Acceptor Groups for Dye-Sensitized Solar Cells. *Dye. Pigment.* **2017**, *139*, 274–282. [CrossRef]
15. Zhao, C.; Guo, Y.; Guan, L.; Ge, H.; Yin, S.; Wang, W. Theoretical Investigation on Charge Transport Parameters of Two Novel Heterotetracenes as Ambipolar Organic Semiconductors. *Synth. Met.* **2014**, *188*, 146–155. [CrossRef]
16. Sakthivel, P.; Song, H.S.; Chakravarthi, N.; Lee, J.W.; Gal, Y.-S.; Hwang, S.; Jin, S.-H. Synthesis and Characterization of New Indeno[1,2-*b*]Indole-Co-Benzothiadiazole-Based π-Conjugated Ladder Type Polymers for Bulk Heterojunction Polymer Solar Cells. *Polymer* **2013**, *54*, 4883–4893. [CrossRef]
17. Carbas, B.B.; Noori, H.A.; Kavak, E.; Kaya, Y.; Kıvrak, A. Optical, Electrochemical and DFT Studies of Donor-Acceptor Typed Indole Derivatives. *J. Mol. Struct.* **2023**, *1271*, 134129. [CrossRef]
18. Zhu, M.; Zhou, K.; Zhang, X.; You, S.-L. Visible-Light-Promoted Cascade Alkene Trifluoromethylation and Dearomatization of Indole Derivatives via Intermolecular Charge Transfer. *Org. Lett.* **2018**, *20*, 4379–4383. [CrossRef]
19. Ho, H.E.; Pagano, A.; Rossi-Ashton, J.A.; Donald, J.R.; Epton, R.G.; Churchill, J.C.; James, M.J.; O'Brien, P.; Taylor, R.J.K.; Unsworth, W.P. Visible-Light-Induced Intramolecular Charge Transfer in the Radical Spirocyclisation of Indole-Tethered Ynones. *Chem. Sci.* **2020**, *11*, 1353–1360. [CrossRef]
20. Wang, D.; Garra, P.; Fouassier, J.P.; Graff, B.; Yagci, Y.; Lalevée, J. Indole-Based Charge Transfer Complexes as Versatile Dual Thermal and Photochemical Polymerization Initiators for 3D Printing and Composites. *Polym. Chem.* **2019**, *10*, 4991–5000. [CrossRef]
21. Kumar, S.; Biswas, P.; Kaul, I.; Das, A. Competition between Hydrogen Bonding and Dispersion Interactions in the Indole···Pyridine Dimer and $(Indole)_2$···Pyridine Trimer Studied in a Supersonic Jet. *J. Phys. Chem. A* **2011**, *115*, 7461–7472. [CrossRef]
22. Balevicius, V.; Bariseviciute, R.; Aidas, K.; Svoboda, I.; Ehrenberg, H.; Fuess, H. Proton Transfer in Hydrogen-Bonded Pyridine/Acid Systems: The Role of Higher Aggregation. *Phys. Chem. Chem. Phys.* **2007**, *9*, 3181. [CrossRef]
23. Melikova, S.M.; Rutkowski, K.S.; Gurinov, A.A.; Denisov, G.S.; Rospenk, M.; Shenderovich, I.G. FTIR Study of the Hydrogen Bond Symmetry in Protonated Homodimers of Pyridine and Collidine in Solution. *J. Mol. Struct.* **2012**, *1018*, 39–44. [CrossRef]
24. Kerr, J.R.; Trembleau, L.; Storey, J.M.D.; Wardell, J.L.; Harrison, W.T.A. Crystal Structures of Four Indole Derivatives with a Phenyl Substituent at the 2-Position and a Carbonyl Group at the 3-Position: The C(6) N–H...O Chain Remains the Same, but the Weak Reinforcing Interactions Are Different. *Acta Crystallogr. Sect. E Crystallogr. Commun.* **2016**, *72*, 363–369. [CrossRef] [PubMed]
25. Chang, G.; Xu, Y.; Zheng, Y.; Zhang, L. Intermolecular Hydrogen Bonding of Poly(Imino Imino Ketone Ketone). *J. Macromol. Sci. Part B* **2014**, *53*, 749–755. [CrossRef]
26. Kumar, S.; Pande, V.; Das, A. π-Hydrogen Bonding Wins over Conventional Hydrogen Bonding Interaction: A Jet-Cooled Study of Indole···Furan Heterodimer. *J. Phys. Chem. A* **2012**, *116*, 1368–1374. [CrossRef]

27. Wasio, N.A.; Quardokus, R.C.; Brown, R.D.; Forrest, R.P.; Lent, C.S.; Corcelli, S.A.; Christie, J.A.; Henderson, K.W.; Kandel, S.A. Cyclic Hydrogen Bonding in Indole Carboxylic Acid Clusters. *J. Phys. Chem. C* **2015**, *119*, 21011–21017. [CrossRef]
28. Almeida, A.R.R.P.; Monte, M.J.S. Vapour Pressures of 1-Methyl Derivatives of Benzimidazole, Pyrazole and Indole. The Energy of the Intermolecular Hydrogen Bond NH···N. *J. Chem. Thermodyn.* **2014**, *77*, 46–53. [CrossRef]
29. Liu, Y.; Yang, Y.; Jiang, K.; Shi, D.; Sun, J. Excited-State N–H···S Hydrogen Bond between Indole and Dimethyl Sulfide: Time-Dependent Density Functional Theory Study. *Phys. Chem. Chem. Phys.* **2011**, *13*, 15299. [CrossRef]
30. Kordić, B.; Kovačević, M.; Sloboda, T.; Vidović, A.; Jović, B. FT-IR and NIR Spectroscopic Investigation of Hydrogen Bonding in Indole-Ether Systems. *J. Mol. Struct.* **2017**, *1144*, 159–165. [CrossRef]
31. Nikolić, A.; Jović, B.; Vraneš, M.; Dožić, S.; Gadžurić, S. Volumetric Properties of Binary Mixtures of N -Ethylformamide with Tetrahydropyran, 2-Pentanone, and Propylacetate from (293.15 to 313.15) K. *J. Chem. Eng. Data* **2013**, *58*, 1070–1077. [CrossRef]
32. Nikolić, A.; Jović, B.; Krstić, V.; Tričković, J. N–H...O Hydrogen Bonding. FT-IR, NIR and 1H NMR Study of N-Methylpropionamide—Dialkyl Ether Systems. *J. Mol. Struct.* **2008**, *889*, 328–331. [CrossRef]
33. Becker, E.D. Infrared Studies of Hydrogen Bonding in Alcohol-Base Systems. *Spectrochim. Acta* **1961**, *17*, 436–447. [CrossRef]
34. Muñoz, M.A.; Ferrero, R.; Carmona, C.; Balón, M. Hydrogen Bonding Interactions between Indole and Benzenoid-π-Bases. *Spectrochim. Acta Part A Mol. Biomol. Spectrosc.* **2004**, *60*, 193–200. [CrossRef] [PubMed]
35. Iogansen, A.V. Direct Proportionality of the Hydrogen Bonding Energy and the Intensification of the Stretching ν(XH) Vibration in Infrared Spectra. *Spectrochim. Acta Part A Mol. Biomol. Spectrosc.* **1999**, *55*, 1585–1612. [CrossRef]
36. Borisenko, V.E.; Zavjalova, Y.A.; Tretjakova, T.G.; Kozlova, Z.S.; Koll, A. Thermodynamic Properties of the Hydrogen Bonded Complexes between N-Substituted Anilines and Proton Acceptors. *J. Mol. Liq.* **2004**, *109*, 125–135. [CrossRef]
37. Borisenko, V.E.; Kuzmin, D.S.; Morev, A.V.; Koll, A. Thermodynamics of the Formation of Complexes between Aniline Derivatives and Proton Acceptors in Solution. *J. Mol. Liq.* **2000**, *88*, 259–276. [CrossRef]
38. Pracht, P.; Bohle, F.; Grimme, S. Automated Exploration of the Low-Energy Chemical Space with Fast Quantum Chemical Methods. *Phys. Chem. Chem. Phys.* **2020**, *22*, 7169–7192. [CrossRef]
39. Grimme, S.; Bannwarth, C.; Shushkov, P. A Robust and Accurate Tight-Binding Quantum Chemical Method for Structures, Vibrational Frequencies, and Noncovalent Interactions of Large Molecular Systems Parametrized for All Spd-Block Elements (Z = 1–86). *J. Chem. Theory Comput.* **2017**, *13*, 1989–2009. [CrossRef]
40. Halgren, T.A. Merck Molecular Force Field. I. Basis, Form, Scope, Parameterization, and Performance of MMFF94. *J. Comput. Chem.* **1996**, *17*, 490–519. [CrossRef]
41. O'Boyle, N.M.; Banck, M.; James, C.A.; Morley, C.; Vandermeersch, T.; Hutchison, G.R. Open Babel: An Open Chemical Toolbox. *J. Cheminformatics* **2011**, *3*, 33. [CrossRef]
42. Lewis-Atwell, T.; Townsend, P.A.; Grayson, M.N. Comparing the Performances of Force Fields in Conformational Searching of Hydrogen-Bond-Donating Catalysts. *J. Org. Chem.* **2022**, *87*, 5703–5712. [CrossRef] [PubMed]
43. Zeinalipour-Yazdi, C.D. A DFT Study of the Interaction of Aspirin, Paracetamol and Caffeine with One Water Molecule. *J Mol Model* **2022**, *28*, 285. [CrossRef] [PubMed]
44. Walker, M.; Harvey, A.J.A.; Sen, A.; Dessent, C.E.H. Performance of M06, M06-2X, and M06-HF Density Functionals for Conformationally Flexible Anionic Clusters: M06 Functionals Perform Better than B3LYP for a Model System with Dispersion and Ionic Hydrogen-Bonding Interactions. *J. Phys. Chem. A* **2013**, *117*, 12590–12600. [CrossRef] [PubMed]
45. Zhang, Y.; Ma, N.; Wang, W. Assessment of the Performance of the M05-Class and M06-Class Functionals for the Structure and Geometry of the Hydrogen-Bonded and Halogen-Bonded Complexes. *J. Theor. Comput. Chem.* **2012**, *11*, 1165–1173. [CrossRef]
46. Paul, B.K. Red- and Blue-shifting Hydrogen Bonds of Cyclic Ketones of Varying Ring Sizes with HF and CHF_3: Comparison of the Results of B3LYP, O3LYP, and M06-2X Functionals. *ChemPhysChem* **2025**, 2500057. [CrossRef]
47. Ragi, C.; Muraleedharan, K. Hydrogen Bonding Interactions between Hibiscetin and Ethanol/Water: DFT Studies on Structure and Topologies. *Chem. Phys. Impact* **2024**, *8*, 100416. [CrossRef]
48. Natarajan Sathiyamoorthy, V.; Suvitha, A.; Abdul Rahim, S.; Sahara, R. Intermolecular Hydrogen Bond Interactions in Water Clusters of Zwitterionic Glycine: DFT, MESP, AIM, RDG, and Molecular Dynamics Analysis. *J. Mol. Liq.* **2024**, *396*, 123932. [CrossRef]
49. Adu-Effah, N.; Saikia, N. Intermolecular Hydrogen Bonding Delineates the Stability of Non-Canonical Adenine Base Pairs: A First-Principles Study. *Phys. Chem. Chem. Phys.* **2024**, *26*, 29150–29166. [CrossRef]
50. Kasalović, M.P.; Jelača, S.; Milanović, Ž.; Maksimović-Ivanić, D.; Mijatović, S.; Lađarević, J.; Božić, B.; Marković, Z.; Dunđerović, D.; Rüffer, T.; et al. Novel Triphenyltin(iv) Compounds with Carboxylato *N*-Functionalized 2-Quinolones as Promising Potential Anticancer Drug Candidates: *In Vitro* and *in Vivo* Evaluation. *Dalton Trans.* **2024**, *53*, 8298–8314. [CrossRef]
51. Fuster, F.; Grabowski, S.J. Intramolecular Hydrogen Bonds: The QTAIM and ELF Characteristics. *J. Phys. Chem. A* **2011**, *115*, 10078–10086. [CrossRef] [PubMed]

52. Dimić, D. The Importance of Specific Solvent–Solute Interactions for Studying UV–Vis Spectra of Light-Responsive Molecular Switches. *Comptes Rendus Chim.* **2018**, *21*, 1001–1010. [CrossRef]
53. Koch, U.; Popelier, P.L.A. Characterization of C-H-O Hydrogen Bonds on the Basis of the Charge Density. *J. Phys. Chem.* **1995**, *99*, 9747–9754. [CrossRef]
54. Wang, H.; Huang, Z.; Shen, T.; Guo, L. Hydrogen-Bonding Interactions in Adrenaline-Water Complexes: DFT and QTAIM Studies of Structures, Properties, and Topologies. *J. Mol. Model.* **2012**, *18*, 3113–3123. [CrossRef]
55. Bursch, M.; Mewes, J.; Hansen, A.; Grimme, S. Best-Practice DFT Protocols for Basic Molecular Computational Chemistry. *Angew. Chem.* **2022**, *134*, e202205735. [CrossRef]
56. Espinosa, E.; Alkorta, I.; Elguero, J.; Molins, E. From Weak to Strong Interactions: A Comprehensive Analysis of the Topological and Energetic Properties of the Electron Density Distribution Involving X–H··· F–Y Systems. *J. Chem. Phys.* **2002**, *117*, 5529–5542. [CrossRef]
57. Avdović, E.H.; Dimić, D.S.; Fronc, M.; Kožišek, J.; Klein, E.; Milanović, Ž.B.; Kesić, A.; Marković, Z.S. Structural and Theoretical Analysis, Molecular Docking/Dynamics Investigation of 3-(1-m-Chloridoethylidene)-Chromane-2,4-dione: The Role of Chlorine Atom. *J. Mol. Struct.* **2021**, *1231*, 129962. [CrossRef]
58. Karelson, M.; Lobanov, V.S.; Katritzky, A.R. Quantum-Chemical Descriptors in QSAR/QSPR Studies. *Chem. Rev.* **1996**, *96*, 1027–1044. [CrossRef]
59. Rajan, V.K.; Muraleedharan, K. A Computational Investigation on the Structure, Global Parameters and Antioxidant Capacity of a Polyphenol, Gallic Acid. *Food Chem.* **2017**, *220*, 93–99. [CrossRef]
60. Kar, R.; Chandrakumar, K.R.S.; Pal, S. The Influence of Electric Field on the Global and Local Reactivity Descriptors: Reactivity and Stability of Weakly Bonded Complexes. *J. Phys. Chem. A* **2007**, *111*, 375–383. [CrossRef]
61. Mulliken, R.S. A New Electroaffinity Scale; Together with Data on Valence States and on Valence Ionization Potentials and Electron Affinities. *J. Chem. Phys.* **1934**, *2*, 782–793. [CrossRef]
62. Parr, R.G.; Szentpály, L.V.; Liu, S. Electrophilicity Index. *J. Am. Chem. Soc.* **1999**, *121*, 1922–1924. [CrossRef]
63. Popović, G.V.; Sladić, D.M.; Stefanović, V.M.; Pfendt, L.B. Study on Protolytic Equilibria of Lorazepam and Oxazepam by UV and NMR Spectroscopy. *J. Pharm. Biomed. Anal.* **2003**, *31*, 693–699. [CrossRef] [PubMed]
64. Pejov, L. A Gradient-Corrected Density Functional Study of Indole Self-Association through N–H··· π Hydrogen Bonding. *Chem. Phys. Lett.* **2001**, *339*, 269–278. [CrossRef]
65. Frisch, M.J.; Trucks, G.W.; Schlegel, H.B.; Scuseria, G.E.; Robb, M.A.; Cheeseman, J.R.; Scalmani, G.; Barone, V.; Mennucci, B.; Petersson, G.A.; et al. *Gaussian 09*; Gaussian, Inc.: Wallingford, CT, USA, 2009.
66. Dunning, T.H. Gaussian Basis Sets for Use in Correlated Molecular Calculations. I. The Atoms Boron through Neon and Hydrogen. *J. Chem. Phys.* **1989**, *90*, 1007. [CrossRef]
67. Zhao, Y.; Truhlar, D.G. The M06 Suite of Density Functionals for Main Group Thermochemistry, Thermochemical Kinetics, Noncovalent Interactions, Excited States, and Transition Elements: Two New Functionals and Systematic Testing of Four M06-Class Functionals and 12 Other Function. *Theor. Chem. Acc.* **2007**, *120*, 215–241. [CrossRef]
68. Cossi, M.; Rega, N.; Scalmani, G.; Barone, V. Energies, Structures, and Electronic Properties of Molecules in Solution with the C-PCM Solvation Model. *J. Comput. Chem.* **2003**, *24*, 669–681. [CrossRef]
69. Dennington, R.; Todd, K.; Millam, J. *GausView*, Version 5; Semichem Inc.: Shawnee, KS, USA, 2009.
70. Foster, J.P.; Weinhold, F. Natural Hybrid Orbitals. *J. Am. Chem. Soc.* **1980**, *102*, 7211–7218. [CrossRef]
71. Bader, R.F.W. Atoms in Molecules. *Acc. Chem. Res.* **1985**, *18*, 9–15. [CrossRef]
72. Popelier, P.L.A.; Bader, R.F.W. The Existence of an Intramolecular CHO Hydrogen Bond in Creatine and Carbamoyl Sarcosine. *Chem. Phys. Lett.* **1992**, *189*, 542–548. [CrossRef]
73. Kordić, B.; Dimić, D.; Despotović, V.; Jović, B. Spectroscopic and Theoretical Investigation of Solvent Effect on N–H···O, N–H···N and N–H···π Interactions in Complexes of N-Monosubstituted Benzamides. *J. Mol. Liq.* **2024**, *399*, 124472. [CrossRef]
74. Keith, T.A. *AIMAll*, Version 19.02.13; TK Gristmill Software: Overland Park, KS, USA, 2016.
75. Lu, T.; Chen, F. Multiwfn: A Multifunctional Wavefunction Analyzer. *J. Comput. Chem.* **2012**, *33*, 580–592. [CrossRef] [PubMed]
76. Lu, T. A Comprehensive Electron Wavefunction Analysis Toolbox for Chemists, Multiwfn. *J. Chem. Phys.* **2024**, *161*, 082503. [CrossRef] [PubMed]
77. Humphrey, W.; Dalke, A.; Schulten, K. VMD: Visual Molecular Dynamics. *J. Mol. Graph.* **1996**, *14*, 33–38. [CrossRef]
78. Daboe, M.; Parlak, C.; Direm, A.; Alver, Ö.; Ramasami, P. Interaction between Escitalopram and Ibuprofen or Paracetamol: DFT and Molecular Docking on the Drug–Drug Interactions. *J. Biomol. Struct. Dyn.* **2024**, *42*, 672–686. [CrossRef]

molecules

MDPI

Article

Efficient Photo-Driven Electron Transfer from Amino Group-Decorated Adamantane to Water

Xiangfei Wang [1,2,*], Jonathan Remmert [3], Beate Paulus [1] and Annika Bande [2,4]

1 Institute of Chemistry and Biochemistry, Freie Universität Berlin, Arnimallee 22, 14195 Berlin, Germany; b.paulus@fu-berlin.de
2 Helmholtz-Zentrum Berlin für Materialien und Energie, Hahn-Meitner-Platz 1, 14109 Berlin, Germany; annika.bande@aci.uni-hannover.de
3 Department of Physics, Freie Universität Berlin, Arnimallee 14, 14195 Berlin, Germany; jonathan@remmert.berlin
4 Institute of Inorganic Chemistry, Leibniz University Hannover, Callinstr. 9, 30167 Hannover, Germany
* Correspondence: xfwangqc@zedat.fu-berlin.de

Abstract: Nanodiamonds in water can generate solvated electrons under ultraviolet (UV) excitation, but UV light constitutes only a small portion of solar energy. To harvest solar energy in the visible range, we investigate band gap reduction via surface amino functionalization and examine its impact on photo-excited charge transfer to water. Adamantane, the smallest nanodiamond, is used as a model due to its electron emission properties. Liquid water is first represented using water dimers and then complete solvation shell structures surrounding the adamantane. By systematically analyzing different functionalized adamantane structures, we find that nitrogen serves as the primary electron donor to nearby water molecules. Furthermore, the negative electron affinity of adamantane, which determines its emission capability, is preserved with half of the amino group functionalization on the surface. Our findings motivate further experimental verification using nanodiamonds with amino-functionalized surfaces.

Keywords: adamantane; amino group; electron transfer; excited states; linear response time-dependent density functional theory (LR-TDDFT)

1. Introduction

With the development of modern society, there is a growing demand for energy due to human activities. Currently, the primary source of energy relies on fossil fuels. Their combustion releases a large amount of CO_2, which is one of the leading causes of the greenhouse effect and, consequently, climate change [1]. To use energy in a sustainable way, research on photochemistry has been focusing on designing materials that can harvest energy from sunlight. One of the most promising candidates is the nanodiamond (ND), which has attracted considerable attention [2] due to its unique electronic properties [2], eco-friendly composition [3,4], and increasing availability through modern synthesis techniques [5–8]. Therefore, there has been widespread research into the application of NDs for generating green H_2 gas [9] from solar energy and for the reduction of CO_2 [10] and N_2 [11].

To understand NDs as effective catalysts for photochemistry, it is important to investigate their surface structure. NDs have an exceptionally high surface-to-volume ratio [12], providing abundant sites for chemical reactions. In addition, the surface of NDs allows for various functionalizations. The most common type of surface termination involves hydrogen atoms, which enable high chemical stability and resistance to degradation [13,14]. Furthermore, the hydrogen atoms on the surface carry a positive charge, while the nearby carbon atoms are negatively charged. The resulting charge separation creates dipole moments, leading to negative electron affinity (NEA) [15,16]. NEA is a property shared by all forms of NDs down to the smallest adamantane ($C_{10}H_{16}$). Specifically, when NDs absorb photon energy, electrons are excited to the conduction band and subsequently migrate to the surface. Due to NEA, the emission process can be highly efficient [15,17,18]. Therefore, the applications of NDs range from fluorescent imaging [19] to field emission displays [20]. In particular, recent experimental and theoretical studies [21,22] show that illuminated NDs in a water environment are an effective source of solvated electrons, which can activate a variety of high-energy reduction reactions, such as the reduction of N_2 to NH_3 [11] and CO_2 to CO [23]. Understanding the electron emission process into water is critical for the design of illuminated NDs as efficient catalysts for harvesting solar energy.

To investigate the mechanism of photo-excited electron transfer from NDs to liquid water, Wang et al. [24] used adamantane to represent NDs and small water clusters to model the electronic properties of liquid water. Using linear-response time-dependent density functional theory (LR-TDDFT), their calculations show that the photo-excited electron transfers from adamantane to various electronically bound states of the surrounding water clusters. Although the surface plays a determining role in the electron emission properties of NDs, the effects of functional groups remain less explored. Buchner et al. [21] showed that carboxyl and carbonyl terminations can suppress the emission of solvated electrons from NDs into liquid water. Kirschbaum et al. [25] suggested that amino functional groups can reduce the band gap of hydrogen-terminated NDs (approximately 5.4 eV) [16] to the visible range of the solar spectrum (1.6–3.3 eV) [26]. However, how amino functionalization determines the electron emission from NDs into liquid water remains unclear.

Following a similar approach, this study uses water clusters as models for the electronically bound states in liquid water. In particular, the simplest water dimer with C_s symmetry [27] is sufficient to host an electronically bound state. Furthermore, adamantane is used as a model for NDs due to its negative electron affinity (NEA) and similar electron emission properties. To understand the role of nitrogen atoms, various amino group-decorated adamantane structures are studied with respect to their frontier orbital energies, which reflect their ionization and NEA properties. Additionally, different relative positions of the water dimer are systematically explored around the amino-decorated adamantane. To consider a more complex solvation environment involving hydrogen bond networks, we also use molecular dynamics simulations to sample the solvation shell structures. By systematically exploring water models of increasing accuracy, we reveal the mechanism of photo-induced charge transfer into liquid water. Electronic structure calculations further clarify how amino-functionalized surfaces determine the photo-excited charge transfer process.

2. Results

2.1. *Electronic Properties of Amino-Functionalized Adamantane*

There are multiple possibilities for adding amino groups to adamantane; we selected the 10 systems presented in Figure 1 to model different chemical situations. To introduce the amino functional group, instead of increasing the number one by one, we added four amino groups at once and arranged them at equivalent positions on the surface. The functionalization sites were in a tetrahedral configuration compatible with the T_d symmetry group of adamantane. Since the four positions were equivalent, the symmetric structure served as a good representation of an evenly functionalized adamantane surface. Additional structures were systematically constructed based on this symmetrically modified structure. The total number of amino groups increased from 4 to 13. For the same number of amino groups, we selected representative isomers with distinct structures for investigation. A complete enumeration of all possible isomers was beyond the scope of this study.

The electrons in the highest occupied molecular orbital (HOMO) are the most active ones in chemical processes [28]. In particular, the ionization potential theorem of Kohn–Sham density functional theory (KS-DFT) establishes that the negative value of the HOMO energy is an approximation of the ionization energy [29–31]. Therefore, during the photoexcitation process, the electron is first removed from the HOMO. Thus, the position and shape of the HOMO provide a qualitative analysis of different molecular segments as electron donors. Figure 1 shows the structure of adamantane together with the systematically explored amino-functionalized structures. Beneath each structure, the HOMOs are also depicted accordingly. For adamantane, the HOMO localized on the carbon atoms. For amino group-functionalized structures, the HOMO localized primarily on the nitrogen atoms. One particular structure is Ad_7NH2_2, which has all the nitrogen atoms on the same side. Accordingly, the HOMO was also primarily located on the same side of the Ad_7NH2_2 molecule.

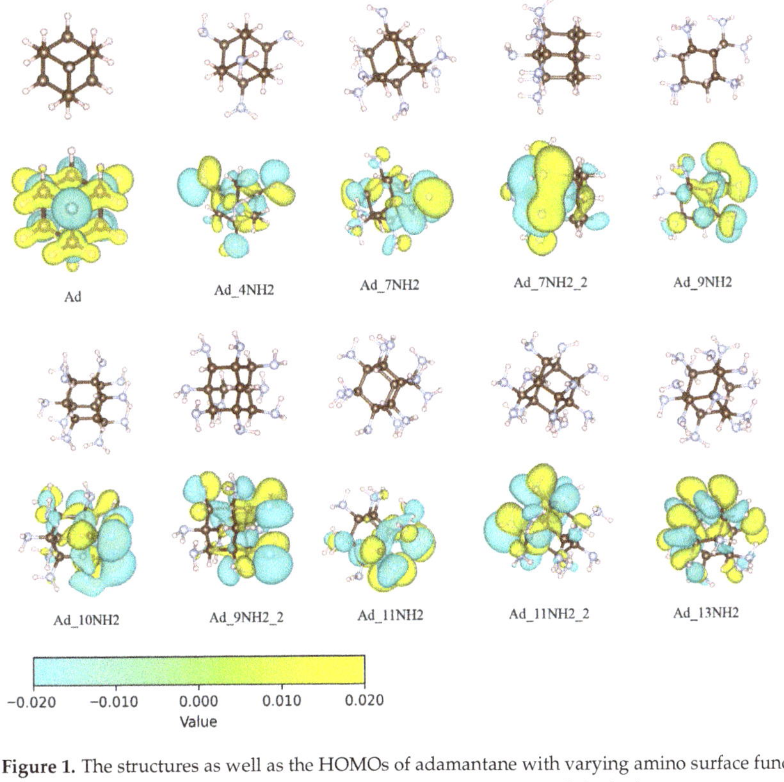

Figure 1. The structures as well as the HOMOs of adamantane with varying amino surface functionalization are shown. The upper row displays only the structures, while the lower row presents the HOMO of the corresponding structure. The naming convention for these structures includes "Ad" for adamantane followed by the number of amino groups. The final segment in each name distinguishes different isomers. The isovalue for all orbital plots was set to $2 \times 10^{-2}\ a_0^{-3/2}$. The bottom of the figure shows the color bar corresponding to the isovalue contour plot. The yellow and cyan colors represent the positive and negative values of the wave function, respectively.

To illustrate the role of nitrogen atoms as electron donors, we analyzed the HOMO energies. Table 1 shows that from adamantane to Ad_4NH_2, the amino group increased the HOMO energy from −9.131 eV to −8.162 eV. Furthermore, the HOMO energies continued to increase with a higher number of amino groups, as shown by other structures. This trend, shown in Figure 2a, confirms that amino groups effectively reduced the ionization energy, thereby facilitating the photoexcitation process. To avoid unintentional excitation of the water, we compared the HOMO energies of the water. For the water monomer and small water clusters, the CAM-B3LYP/aug-cc-pVDZ level of theory predicted the HOMO energy to be approximately −10.5 eV [24]. Furthermore, the experimentally measured ionization energy for liquid water was about 9.9 eV [32,33]. Because the HOMO energies of water in different states of matter were lower than those of amino-functionalized adamantanes, the water was not excited when amino-functionalized adamantane was available for excitation.

Table 1. HOMO and LUMO energies, along with atomic charges, summarized for adamantane molecules with various surface functionalizations. The exact values of the HOMO and LUMO energies are dependent on the exchange correlation functional, which was CAM-B3LYP in our study. The molecular structures are illustrated in Figure 1. The Hirshfeld population analysis was performed on all atoms. The atomic charge values are averages for each atom type. The charges are in atomic units.

Molecule	HOMO (eV)	LUMO (eV)	C (*e*)	H (*e*)	N (*e*)
Ad	−9.131	0.249	−0.122	0.082	0
Ad_4NH2	−8.162	0.087	−0.026	0.088	−0.380
Ad_7NH2	−7.982	0.067	0.038	0.094	−0.355
Ad_7NH2_2	−7.410	0.070	0.022	0.091	−0.355
Ad_9NH2	−7.473	−0.004	0.060	0.094	−0.344
Ad_9NH2_2	−7.624	0.078	0.077	0.094	−0.339
Ad_10NH2	−7.355	−0.095	0.080	0.095	−0.337
Ad_11NH2	−7.265	−0.103	0.098	0.096	−0.331
Ad_11NH2_2	−7.524	−0.032	0.113	0.096	−0.329
Ad_13NH2	−7.786	−0.019	0.149	0.097	−0.322

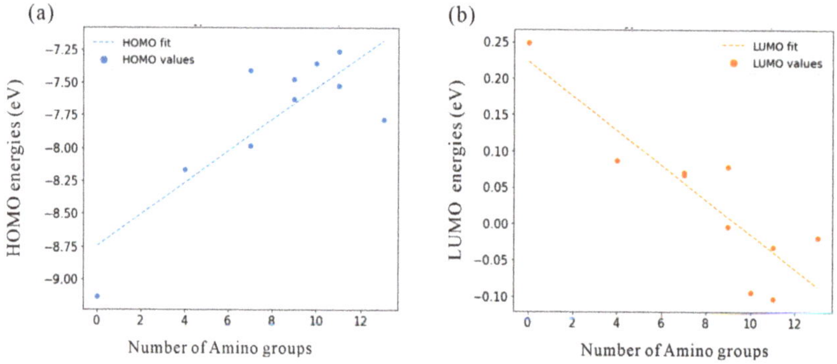

Figure 2. Dependence of HOMO and LUMO energies on the number of surface amino groups in functionalized adamantane. (**a**) The HOMO energy (in eV) as a function of the number of amino groups. The dashed line shows a linear fit. (**b**) The LUMO energy versus an increasing number of amino groups. Legend and units are the same as in (**a**).

For an estimate of the electron affinities of the amino-functionalized adamantane structures under investigation, we used the LUMO energies. Table 1 shows that the LUMO energies decreased with an increasing number of amino groups. In particular, the LUMO energy of adamantane was 0.249 eV, and that of Ad_4NH2 was 0.087 eV. The energies of both structures were above zero and therefore corresponded to an NEA. However, with a higher number of amino functional groups, some structures exhibited negative LUMO energies, including Ad_9NH2, Ad_10NH2, Ad_11NH2, Ad_11NH2_2, and Ad_13NH2. For a direct visualization of this trend, Figure 2b shows that the LUMO energies decreased systematically with an increasing number of amino groups. Consequently, the original NEA property of adamantane was reduced with a higher degree of amino functionalization. Nevertheless, most of the amino-functionalized structures had positive LUMO energies. Therefore, the photo-excited electron will be attracted to the water molecules rather than the differently decorated adamantane structures, which have a lower electron affinity.

To explain the loss of NEA properties upon amino functionalization, we performed Hirshfeld population analysis on all structures to determine the atomic charges. For each atom type, we averaged the charges over the same atoms. As shown in Table 1, the pure adamantane had carbon atoms with a charge of −0.122 and surface hydrogens with a charge of +0.082. The surface dipoles pointed uniformly from hydrogen to carbon, resulting in NEA. Introducing amino groups (e.g., Ad_4NH_2) reduced the negative charge on carbon and instead places a larger negative charge on nitrogen (−0.380). This is because nitrogen has a higher electronegativity than carbon. In particular, hydrogen had the lowest electronegativity and remained positively charged in all structures. With more than four amino groups, some carbon atoms became positively charged, disrupting the original surface dipole. Therefore, amino-functionalized structures feature dipoles that point from hydrogen to nitrogen in various directions and result in the loss of the NEA property.

2.2. Charge Transfer to Water Dimers

To study the photo-excited charge transfer to the liquid water, the water dimer was used as an effective representation of the electronic bound state. The most stable water dimer had the C_S symmetry with the hydrogen atoms in a cis configuration (C_SC) [27]. For the amino-functionalized adamantane, two specific structures were considered: one with amino groups on a single side (Ad_7NH2_2) and another with amino groups uniformly distributed across the entire surface (Ad_10NH2). Regarding the relative geometry effects, this study investigates various placements and orientations of the C_SC water dimer. Even though the number of possible relative geometries can be numerous, only select systematic structures were considered. Figure 3 shows the representative structures in which the C_SC water dimers were arranged around adamantane. For each orientation, the optimal center-of-mass (COM) distance between the water dimers and adamantane was set to minimize the total system energy. As an example, Figure 4 shows the energy as a function of the center-of-mass (COM) distance for the `Ad_7NH2_2_CsC_1` structure. The minimum energy occurred at a COM distance of 5.420 Å, which defines the optimal separation of the water dimer relative to the functionalized adamantane. The structure details are in Appendix A.

LR-TDDFT calculations were performed on all joint structures to obtain various excited states. Although all excited states up to 20 were computed, this study focused on those in which more than 90% of the hole density was localized on amino-functionalized adamantane. Among these selected states, only the lowest-energy excited state was analyzed in detail to investigate the spatial distribution of the photo-excited electron. Nevertheless, due to the presence of an electronic bound state associated with the water dimer, all of the preselected states exhibited partial charge transfer to the water dimer. Therefore, in

LR-TDDFT calculations, the lowest excited states with a hole only on the adamantane were referred to as the lowest-charge transfer states.

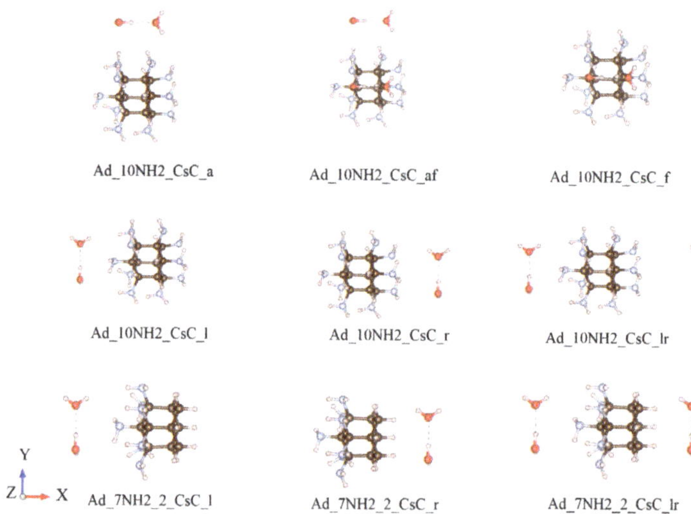

Figure 3. Schematics showing the structures used to calculate electron transfer from amino-functionalized adamantane to water molecules. C_SC water dimers are depicted in various positions and orientations relative to adamantane: above (a), in front (f), to the left (l), and to the right (r). Additionally, two types of amino-functionalized adamantane were considered, namely Ad_10NH2, with nitrogen atoms uniformly distributed, and Ad_7NH2_2, with nitrogen atoms concentrated on one side.

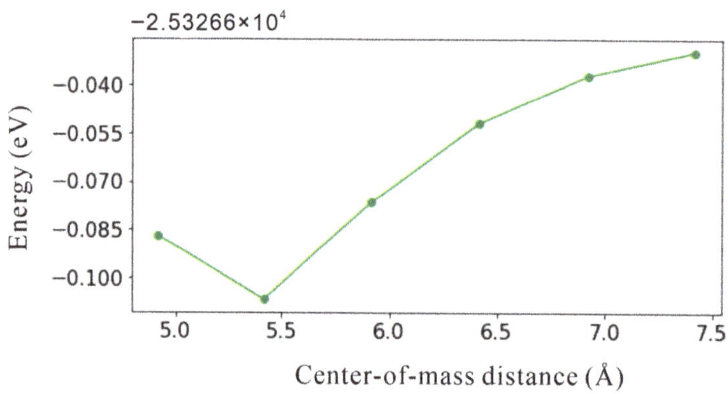

Figure 4. The total ground state energy as a function of the center-of-mass (COM) distance between the water dimer and the `Ad_7NH2_2` cluster in the `Ad_7NH2_2_CsC_l` structure is shown. The energy is given in eV units, and the COM distance is in Å units.

To demonstrate that photoexcitation and electron transfer to water molecules are feasible, Table 2 provides a quantitative analysis of the lowest-charge transfer states. For each of the structures, there was consistent partial charge transfer ($q_w > 0$) to the C_SC water dimers. Therefore, the photo-excited electron was, in any case, transferred to the water dimer, regardless of position and different surface functionalizations. However, all the states exhibited incomplete charge transfer ($q_w < 1.0$) due to the low electron affinity of the C_SC water dimer. For complete charge transfer with q_w close to 1.0, the water cluster must have a high electron affinity and form deep electronic bound states.

After confirming the charge transfer to the water dimer, the role of nitrogen atoms as electron donors needed to be explored. Table 2 shows that the electron NTO density minus the hole NTO density for all nitrogen atoms (q_{ad}^N) was negative. Therefore, the nitrogen atoms primarily contributed to the hole of the exciton. In addition, the absolute value of q_{ad}^N was larger than that of q_w. The large magnitude of q_{ad}^N indicates that charge transfer primarily occurred from the nitrogen atoms to the water molecules. Consequently, a small portion of the charge was transferred from the nitrogen atoms to the carbon atoms. The contribution of nitrogen atoms in donating electrons to water can be verified by the shape of the HOMO, which was primarily localized around the nitrogen atoms. In addition, the charge transfer states were all optically accessible with significant transition dipole moments.

The charge transfer excited states had an excitation energy of roughly 5 eV, which is about 2 eV lower than the absorption onset of pure adamantane at approximately 7 eV [24]. Similarly, large nanodiamonds exhibit excitation energies near 5 eV [34]. Thus, amino functionalization of nanodiamond surfaces could potentially lower the charge transfer excitation energy to about 3 eV, which lies in the visible range of the spectrum.

Finally, the last column of Table 2 shows that most charge transfer states corresponded to the first excited state. The association of charge transfer states with the lowest excited state is important, as higher excited states will rapidly relax to the lowest excited state. In the first example, the highly functionalized `Ad_10NH2_CsC_l` exhibited charge transfer at the second excited state due to the reduced electron emission properties caused by the excess amino groups. For another example, `Ad_7NH2_2_CsC_r` showed its charge transfer at its third excited state. Here, the amino groups were primarily on the left side of the structure, while the water dimer was on the right. Therefore, the geometric arrangement is unfavorable for charge transfer. The influence of geometry will be explored further in the following section.

Table 2. Properties of the lowest charge transfer states corresponding to the structures in Figure 3. For each structure, q_w represents the magnitude of the electron transferred to the water molecules, while q_{ad}^N denotes the electron-minus-hole density for all nitrogen atoms. Excitation energies of the lowest charge transfer states (E_{ct}) are given in eV, and transition dipole moments (d_{ct}) are in debyes (D). The order of the corresponding excited states is shown in the final column.

Molecule Structures	q_w	q_{ad}^N	E_{ct} (eV)	d_{ct} (D)	State
Ad_10NH2_CsC_a	0.426	−0.509	5.024	0.314	1
Ad_10NH2_CsC_af	0.420	−0.508	5.043	0.345	1
Ad_10NH2_CsC_f	0.368	−0.511	5.144	0.416	1
Ad_10NH2_CsC_l	0.312	−0.483	5.443	0.363	2
Ad_10NH2_CsC_r	0.333	−0.504	5.057	0.323	1
Ad_10NH2_CsC_rl	0.168	−0.484	5.164	0.330	1
Ad_7NH2_2_CsC_l	0.418	−0.494	5.056	0.508	1
Ad_7NH2_2_CsC_r	0.546	−0.549	5.705	0.193	3
Ad_7NH2_2_CsC_rl	0.418	−0.494	5.071	0.516	1

2.3. *Charge Transfer to Water Dimers in Proximity*

In the previous section, we found that nitrogen atoms can donate electrons during photoexcitation. Therefore, it is important to understand whether the water molecules adjacent to the amino functional groups are more likely to attract the electron than the water dimer that is farther away. To investigate the effect of different relative positions, the one-sided amino-functionalized adamantane molecule Ad_7NH2_2 is used as an example. To compare the influence of water molecules at different distances relative to the nitrogen atoms, two identical water C_SC clusters were placed symmetrically on the left and right sides. First, the water clusters were arranged at their respective equilibrium distances (d_0). Then, the two water clusters were systematically shifted away from their equilibrium positions. Figure 5 shows the structures, molecular orbitals, and lowest charge transfer excited-state properties for the left and right water dimers at different distances.

At the equilibrium distance, the water dimer interacted strongly with the surface of adamantane. This influence is reflected in the molecular orbitals and their energies. Specifically, the left panel of Figure 5a shows the HOMO, LUMO, and LUMO+1 of the system with two water dimers at a distance d_0. The HOMO was localized on the nitrogen atoms of the decorated adamantane, while the LUMO and LUMO+1 were primarily located on the left and right water dimers, respectively. Although both water dimers had identical structures, the left water dimer interacted more strongly with the nitrogen atoms, leading to a lower orbital energy. In contrast, the right water dimer was farther from the nitrogen atoms and consequently had a higher orbital energy. The charge transfer exciton featured orbital transitions primarily involving the HOMO-to-LUMO transition. For example, the charge transfer state to the left water dimer involved an orbital transition from the HOMO to the LUMO. The charge transfer state to the right water dimer involved a transition from the HOMO to the LUMO+1. In addition, the magnitude of charge transfer to the left water dimer was 0.420, and the charge transfer to the right water dimer was 0.413. The excitation energy of the left charge transfer state was also lower than that of the excited state corresponding to the right water dimer. Therefore, the photo-excited electron transfer was more favorable to the left water dimer, which was closer to the electron-donating nitrogen atoms.

The distances between the water dimers and the nitrogen atoms play a crucial role in the charge transfer process. Therefore, it is important to verify the charge transfer behavior at different distances. For example, when the water dimers were shifted 1 Å further away from their original positions, the influence of the nitrogen atoms decreased. Figure 5 shows the lowest charge transfer state with the distance increased to $d_0 + 1$ Å. As shown in Figure 5b, when the water molecules moved away from the amino-functionalized adamantane, the HOMO and LUMO of the two charge transfer states had similar shapes to the molecular orbitals in Figure 5a. Nevertheless, due to the larger distance, the transferred charge (q_w) from the amino-functionalized adamantane to the water dimer decreased, and the energy required for excitation increased. Consequently, charge transfer became more difficult at longer distances. The charge transfer excited state to the right water dimer was particularly sensitive to longer distances. The magnitude of the charge transferred to the right water dimer was only a fractional value of 0.160. Therefore, the photo-excited electron had a distribution predominantly around the amino-functionalized adamantane, which means that charge transfer to the right water dimer could be challenging.

For structures with greater distances, we summarize the charge transfer state properties in Table 3. The table includes the magnitude of the transferred charge (q_w) to the water dimers, the excited state energies (E), as well as the order of the excited states. The distance systematically increased from d_0 to d_0 + 4.0 Å. The charge transfer state exhibited a consistent trend favoring the left water dimer. Specifically, the excited state with charge transfer to the left water dimer had a lower excitation energy than the charge transfer state to the right water dimer. Furthermore, the magnitude of transferred charge to the left water dimer was also systematically larger than that to the right water dimer. Therefore, the photo-excited charge transfer was more favorable to the water cluster that was in close proximity to the nitrogen atoms. Furthermore, the charge-transfer state involving the left water dimer consistently corresponded to the first excited state, resulting in a monotonically decreasing charge transfer q_w with increasing distance. In contrast, the charge transfer states associated with the right water dimer had higher excitation energies, typically the fourth or fifth state. Therefore, the photo-excited charge transfer favored the left water dimer due to the lowest excitation energy and close proximity.

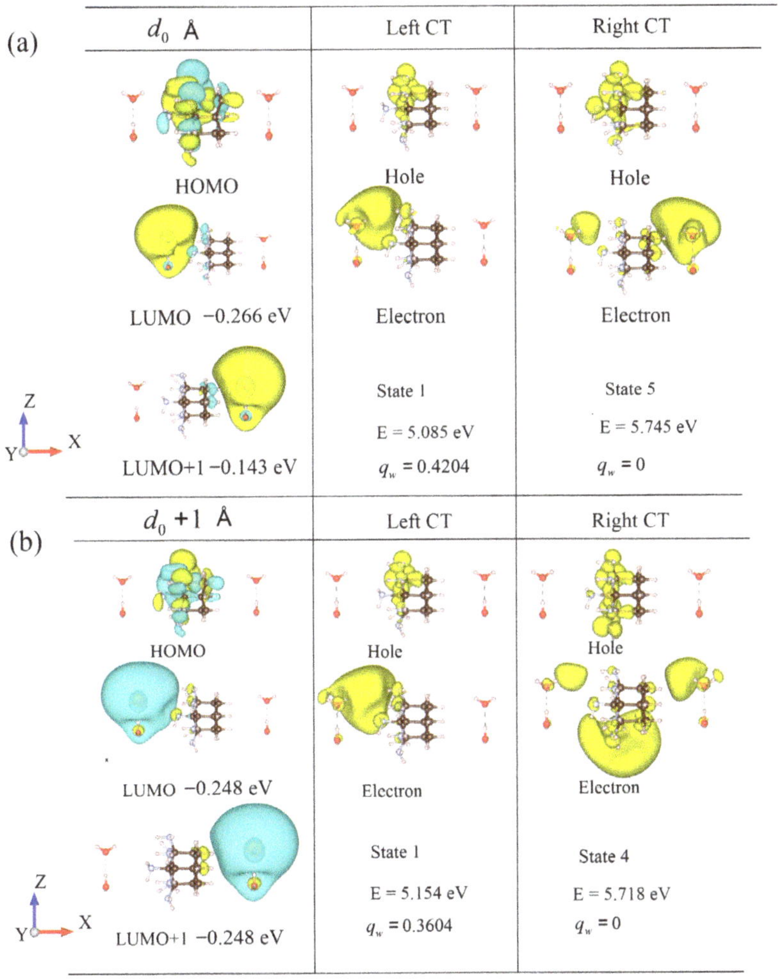

Figure 5. Photo-excited charge transfer from amino-functionalized adamantane to water dimers at various distances. The structure corresponds to the Ad_7NH2_2_CsC_lr configuration, as shown in Figure 3 in this study. (**a**) The left and right water dimers at their respective equilibrium distance d_0. (**b**) The structures after shifting both the left and right water dimers 1 Å further away from their respective equilibrium distances. The left column displays the molecular orbitals, including the HOMO, LUMO, and LUMO+1. The middle and right columns show the NTO densities for the left and right charge transfer states. Qualitative properties, including the order of the excited state, lowest charge transfer excitation energies (E_{ct}), and charge transferred to the water molecules (q_w), are also provided alongside the corresponding NTO plots. The isovalue for all orbital plots was set to $2 \times 10^{-2}\ a_0^{-3/2}$. The isovalue for the NTO plots was set to $4 \times 10^{-4}\ a_0^{-3}$.

Table 3. Transferred charge (q_w), lowest charge transfer excitation energies (E_{ct}), and the order of the charge transfer states, with the distance between the water and nitrogen-doped adamantane increasing from d_0 Å to $d_0 + 4.0$ Å. The structure corresponds to the Ad_7NH2_2_CsC_lr configuration shown in Figure 3. The left half of the table presents the properties of the charge transfer (CT) excited states for the left water dimer, while the right half shows the corresponding properties for the right water dimer.

	CT Left			CT Right		
d_0+	q_w	E_{ct} (eV)	State	q_w	E_{ct} (eV)	State
0.0 Å	0.420	5.085	1	0.413	5.745	5
1.0 Å	0.360	5.154	1	0.160	5.718	4
2.0 Å	0.365	5.259	1	0.254	5.818	5
3.0 Å	0.252	5.331	1	0.118	5.891	5
4.0 Å	0.173	6.121	7	0.156	6.170	8

2.4. *Charge Transfer to Water Molecules Within the Solvation Shell*

As shown previously with simple water dimers, charge transfer favored water molecules near the amino groups. However, the solvation environment around adamantane is far more complex than that in a dimer model. First, liquid water forms an extended hydrogen bond network with bond breaking and reforming, which can influence the charge transfer [35]. Although a dimer already contains a broken hydrogen bond, it does not fully capture the influence of the extended network. Second, water molecules can form hydrogen bonds directly with the amino groups, generating unique hydration structures [36]. Finally, DFT ground-state optimizations yield 0 K geometries and neglect thermal fluctuations, which may also affect charge transfer.

To sample more realistic hydration structures, we performed MD simulations on the one-sided functionalized Ad_7NH2_2 system. Three snapshots were selected at random for excited-state calculations. In all cases, the lowest excited state showed charge transfer to the solvation shell close to the amino groups. The excitation energies were 5.258, 5.552, and 5.792 eV, and the charge transferred to water q_W was 0.943, 0.959, and 0.933, respectively.

To illustrate the characteristics of the charge transfer state, Figure 6a shows one of the selected MD snapshots. The zoomed view in Figure 6b highlights a water molecule that formed two hydrogen bonds with the two nearest amino groups. Therefore, the amino group contributed to complex solvation structures with water molecules. For the hole NTO densities, Figure 6c reveals that the density was localized on the nitrogen atoms. In contrast, the electron NTO density in Figure 6d was confined within the water shell and close to the surface amino functional group. Other snapshots revealed similar hole and electron distributions. Hence, the MD-based simulation of the solvation shell confirmed the conclusion derived from the simple water dimer model. The charge transfer occurred primarily from the nitrogen atoms to the nearest water molecules.

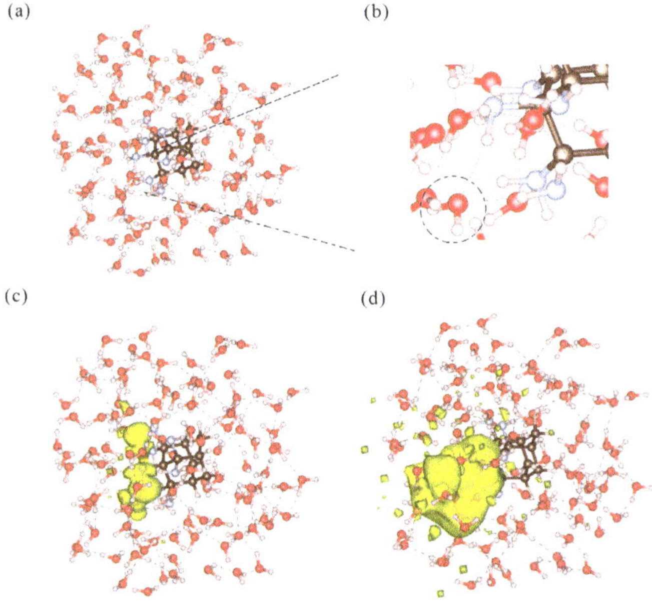

Figure 6. Solvation shell structures and charge transfer excited-state properties of one exemplary MD snapshot. (**a**) The water solvation structure around the amino group-functionalized adamantane (Ad_7NH2_2). (**b**) Zoomed-in view of the hydrogen bonds formed between the amino groups and water molecules. (**c**) Hole NTO density distribution of the excited states; the isovalue was $1 \times 10^{-3}\, a_0^{-3}$. (**d**) Electron NTO density distribution; the isovalue was $2 \times 10^{-4}\, a_0^{-3}$. The isovalues were selected based on the maximum density distribution. The excited-state energy was 5.258 eV. The charge transfer to the water molecules (q_w) was 0.943.

3. Theory and Methods

3.1. Electronic Structure Calculations

The electronic structure and properties of amino-decorated adamantane and water molecules were studied using density functional theory (DFT). The exchange correlation functional employed was the Coulomb-attenuated CAM-B3LYP [37], which is well suited for describing long-range and intermolecular charge-transfer processes. The basis set used was aug-cc-pVDZ [38,39], which includes diffuse functions to capture the diffuse surface states of adamantane and long-range charge transfer. In addition, Wang et al. [24] employed the CAM-B3LYP functional with the aug-cc-pVDZ basis set in a similar study. Their benchmarks show that this combination provides a more balanced description of the excited state compared with both range-separated and non-range-separated functionals. In particular, adding dispersion [40] as a perturbative correction to the total energy has a negligible impact on the computed excited states. Therefore, we used CAM-B3LYP without dispersion corrections. Furthermore, the aug-cc-pVDZ basis set shows convergence in excited-state energies, with the aug-cc-pVTZ basis set showing differences of about 1% for the same excited state.

Excited-state calculations were carried out via linear response time-dependent DFT (LR-TDDFT) using the same CAM-B3LYP/aug-cc-pVDZ set-up. For structures involving a shell of water molecules around amino-functionalized adamantane, the conductor-like polarizable continuum model (CPCM) [41] was used to account for the effect of water outside the solvation shell. All calculations were carried out using the ORCA software package (version 5.0.0) [42], while post-processing of the wavefunctions and population analysis were performed using the Multiwfn package (version 3.8) [43].

To investigate the charge transfer properties in different excited states, it is necessary to quantify the spatial separation between electrons and holes. In an excited state, an electron is promoted to a higher energy level, while the hole represents the positive charge left behind in the orbital originally occupied by the electron. The relative position of the hole and electron can therefore be used to classify different types of excitons. For example, in a charge transfer exciton, the hole is primarily localized on the adamantane molecule, whereas the electron is mainly localized on the water molecules. However, when both the hole and the electron are located either on adamantane or on the water clusters, no intermolecular charge transfer occurs.

Following one of our previous works [24], the natural transition orbitals (NTOs) [44] are used to characterize the spatial distributions of the electron and hole. The Hirshfeld population analysis [45] is then used to quantify the integrated electron and hole NTO densities on the atoms of the respective molecules.

For a charge transfer exciton, the amount of negative charge on the water cluster q_{w} resulting from the separation of the hole and electron is defined as follows:

$$q_{\mathrm{w}} = q_{\mathrm{w}}^{\mathrm{e}} - q_{\mathrm{w}}^{\mathrm{h}}, \tag{1}$$

where $q_{\mathrm{w}}^{\mathrm{e}}$ represents the integrated electron NTO density for the water cluster (w) and $q_{\mathrm{w}}^{\mathrm{h}}$ is the corresponding integrated hole NTO density for the same water cluster.

Similarly, the amount of positive charge on the adamantane q_{ad} is defined by

$$q_{\mathrm{ad}} = q_{\mathrm{ad}}^{\mathrm{e}} - q_{\mathrm{ad}}^{\mathrm{h}}, \tag{2}$$

where $q_{\mathrm{ad}}^{\mathrm{e}}$ denotes the integrated electron NTO density for adamantane (ad) and $q_{\mathrm{ad}}^{\mathrm{h}}$ denotes the corresponding integrated hole NTO density.

3.2. Molecular Dynamic Simulations

To characterize the solvation structure of water around amino-functionalized adamantane, we performed classical molecular dynamics (MD) simulations with the GROMACS (v2025.2) [46] package. The parameters for the solute were taken from the general Amber force field (GAFF), and atomic partial charges were generated with the AM1-BCC scheme in Antechamber (v24.8) [47]. The water was modeled with TIP3P [48].

A cubic simulation box with periodic boundary conditions was constructed such that the solute was at least 1 nm from each box face. The system was equilibrated in the NPT ensemble at $P = 1$ bar and $T = 300$ K until the box volume converged (17.0 nm^3). The temperature and pressure were controlled using a Langevin thermostat and a Parrinello–Rahman barostat [49]. Production trajectories were then run in the NVT ensemble for 1 ns with a 4 fs timestep. Snapshots were extracted every 100 fs. For each snapshot, water molecules within a 7 Å sphere centered on the solute's center of mass were extracted to define the hydration shell.

4. Conclusions

This study systematically explored the strategy of using amino group termination to reduce the energy gap of adamantane derivatives and particularly its impact on charge transfer to water molecules. The results show that amino functional groups can lower the LUMO energy levels, raise the HOMO energy levels, and reduce the energy gap, enabling better alignment with the solar spectrum and enhancing light absorption efficiency. Additionally, nitrogen atoms play a dominant role in charge transfer excited states, with the hole localized on nitrogen atoms rather than on carbon atoms. Therefore, nitrogen atoms act as the primary donors of solvated electrons rather than carbon atoms. However, excessive amino group functionalization can damage adamantane's original NEA properties and limit the surface's ability to emit electrons. Hence, maintaining a moderate level of amino functionalization is crucial to preserving NEA while optimizing the energy gap. Furthermore, water clusters near amino functional groups are prioritized to accept the photo-excited electron, with the closest nitrogen atoms acting as donors. Similarity, in a complex solvation environment, the charge transfer prioritizes the water molecules that are close to the nitrogen atoms. Thus, amino group functionalization offers a new approach to facilitating photo-excited charge transfer.

Author Contributions: Conceptualization, A.B.; formal analysis, X.W. and J.R.; investigation, X.W. and J.R.; resources, B.P.; data curation, J.R. and X.W.; writing—original draft preparation, X.W.; writing—review and editing, B.P. and A.B. All authors have read and agreed to the published version of the manuscript.

Funding: Research funded by China Scholarship Council (202006230131).

Institutional Review Board Statement: Not applicable.

Informed Consent Statement: Not applicable.

Data Availability Statement: Data are contained within the article. Further inquiries can be directed to the corresponding author(s).

Acknowledgments: The computing resources came from the Freie Universität Berlin (HPC cluster Curta) [50]. The authors gratefully acknowledge the use of ChatGPT (OpenAI GPT-4) for grammar corrections.

Conflicts of Interest: The authors declare no conflicts of interest.

Appendix A

Table A1. Molecular Cartesian coordinates for `Ad_7NH2_2_CsC_1`, with Å as the coordinate units.

Atom	X	Y	Z
C	−0.26012	0.00000	1.50157
C	1.28888	−0.01809	1.46912
C	1.81839	1.23647	0.75085
C	1.79424	−1.27426	0.73188
C	−0.80559	0.00000	0.00000
C	1.29996	1.26362	−0.69938
C	1.28301	−1.26520	−0.72145
C	−0.27047	−1.31433	−0.74716
C	−0.25417	1.33188	−0.76265
C	1.79612	0.00490	−1.43526
N	−0.72494	1.21580	2.19689
N	−0.72275	−1.21019	2.23315
H	1.65393	−0.03150	2.50872
H	2.91690	1.20887	0.74509
H	1.52961	2.14134	1.29244
H	1.47769	−2.18425	1.25083
H	2.89227	−1.27328	0.72853
N	−2.27358	0.04046	0.04692
H	1.66101	2.15977	−1.21379
H	1.65613	−2.16345	−1.24021

Table A1. *Cont.*

Atom	X	Y	Z
N	−0.72165	−1.34732	−2.15932
N	−0.87230	−2.51389	−0.13860
N	−0.69825	2.60405	−0.18562
N	−0.67312	1.40890	−2.16211
H	2.89480	−0.00638	−1.44780
H	1.46132	0.03630	−2.47627
H	−2.61452	−0.92277	0.05531
H	−2.60265	0.44797	−0.82970
H	−1.38307	2.11902	−2.29801
H	−0.92560	0.50192	−2.54862
H	−0.46708	2.61517	0.80656
H	−1.71953	2.62010	−0.20607
H	−1.60758	−1.85003	−2.17759
H	−0.06316	−1.89269	−2.71387
H	−0.66134	−2.55454	0.86029
H	−0.49104	−3.33998	−0.60039
H	−0.51891	1.12070	3.19163
H	−1.74092	1.24678	2.09664
H	−1.73552	−1.14811	2.34296
H	−0.33125	−1.18213	3.17557
O	−5.42045	−0.07161	−1.54517
H	−5.42045	−0.06665	−0.59887
H	−5.42045	0.84285	−1.78865
O	−5.42045	−0.07161	1.37114
H	−4.67166	0.18477	1.88988
H	−6.16925	0.18477	1.88988

References

1. Friedlingstein, P.; O'Sullivan, M.; Jones, M.W.; Andrew, R.M.; Hauck, J.; Olsen, A.; Peters, G.P.; Peters, W.; Pongratz, J.; Sitch, S.; et al. Global Carbon Budget 2020. *Earth Syst. Sci. Data Discuss.* **2020**, *2020*, 1–3. [CrossRef]
2. Su, L.X.; Cao, Y.; Hao, H.S.; Zhao, Q.; Zhi, J. Emerging Applications of Nanodiamonds in Photocatalysis. *Funct. Diam.* **2022**, *1*, 93–109. [CrossRef]
3. Nadeem, N.; Zahid, M.; Abbas, Q.; Jilani, A. Carbon-Based Materials: Powering the Future of Energy and Environmental Progress. *Front. Chem.* **2024**, *12*, 1481960. [CrossRef]
4. Hazen, R.M.; Hemley, R.J.; Mangum, A.J. Carbon in Earth's Interior: Storage, Cycling, and Life. *Eos Trans. Am. Geophys. Union* **2012**, *93*, 17–18. [CrossRef]
5. Danilenko, V.V. On the History of the Discovery of Nanodiamond Synthesis. *Phys. Solid State* **2004**, *46*, 595–599. [CrossRef]
6. Yang, G.W.; Wang, J.B.; Liu, Q.X. Preparation of Nano-Crystalline Diamonds Using Pulsed Laser Induced Reactive Quenching. *J. Phys. Condens. Matter* **1998**, *10*, 7923. [CrossRef]
7. Frenklach, M.; Howard, W.; Huang, D.; Yuanwen, J.; Spear, K.E.; Koba, R. Induced Nucleation of Diamond Powder. *Appl. Phys. Lett.* **1991**, *59*, 546–548. [CrossRef]
8. Boudou, J.P.; Curmi, P.A.; Jelezko, F.; Wrachtrup, J.; Aubert, P.; Sennour, M.; Balasubramanian, G.; Reuter, R.; Thorel, A.; Gaffet, E. High-Yield Fabrication of Fluorescent Nanodiamonds. *Nanotechnology* **2009**, *20*, 235602. [CrossRef] [PubMed]
9. Bagheri, S.; Julkapli, N.M. Nanodiamond-Based Photocatalysis for Solar Hydrogen Production. *Int. J. Hydrogen Energy* **2020**, *45*, 31538–31554. [CrossRef]
10. Gao, X.; Han, X.; Zhao, Z.; Huang, N.; Jiao, K.; Song, P.; Zhu, J.; Wang, Y. Lewis Functional Nanodiamonds for Efficient Metal-Free Photocatalytic CO_2 Reduction. *J. Mater. Chem. A* **2024**, *12*, 32745–32759. [CrossRef]
11. Zhu, D.; Zhang, L.; Ruther, R.E.; Hamers, R.J. Photo-Illuminated Diamond as a Solid-State Source of Solvated Electrons in Water for Nitrogen Reduction. *Nat. Mater.* **2013**, *12*, 836–841. [CrossRef]
12. Paci, J.T.; Man, H.B.; Saha, B.; Ho, D.; Schatz, G.C. Understanding the Surfaces of Nanodiamonds. *J. Phys. Chem. C* **2013**, *117*, 17256–17267. [CrossRef]

13. Galli, G. *Structure, Stability and Electronic Properties of Nanodiamonds*; Springer: Berlin/Heidelberg, Germany, 2010.
14. Su, S.; Li, J.; Kundrát, V.; Abbot, A.M.; Ye, H. Hydrogen-Terminated Detonation Nanodiamond: Impedance Spectroscopy and Thermal Stability Studies. *J. Appl. Phys.* **2013**, *113*, 023707. [CrossRef]
15. Drummond, N.; Williamson, A.; Needs, R.; Galli, G. Electron emission from diamondoids: A diffusion quantum Monte Carlo study. *Phys. Rev. Lett.* **2005**, *95*, 096801. [CrossRef]
16. Mochalin, V.; Shenderova, O.; Ho, D.; Gogotsi, Y. The Properties and Applications of Nanodiamonds. In *Nano-Enabled Medical Applications*; Jenny Stanford Publishing: Singapore, 2020; pp. 313–350.
17. Yamaguchi, H.; Masuzawa, T.; Nozue, S.; Kudo, Y.; Saito, I.; Koe, J.; Kudo, M.; Yamada, T.; Takakuwa, Y.; Okano, K. Electron emission from conduction band of diamond with negative electron affinity. *Phys. Rev. B—Condens. Matter Mater. Phys.* **2009**, *80*, 165321. [CrossRef]
18. Tafel, A.; Meier, S.; Ristein, J.; Hommelhoff, P. Femtosecond laser-induced electron emission from nanodiamond-coated tungsten needle tips. *Phys. Rev. Lett.* **2019**, *123*, 146802. [CrossRef]
19. Alkahtani, M.H.; Alghannam, F.; Jiang, L.; Almethen, A.; Rampersaud, A.A.; Brick, R.; Gomes, C.L.; Scully, M.O.; Hemmer, P.R. Fluorescent nanodiamonds: Past, present, and future. *Nanophotonics* **2018**, *7*, 1423–1453. [CrossRef]
20. Zhang, X.; Wei, S.; Ji, Q. Preparation of Nanodiamond Film by Sol–Gel Coating Method for Field Emission Display. *Adv. Nanoparticles* **2017**, *6*, 1–9. [CrossRef]
21. Buchner, F.; Kirschbaum, T.; Venerosy, A.; Girard, H.; Arnault, J.C.; Kiendl, B.; Krueger, A.; Larsson, K.; Bande, A.; Petit, T.; et al. Early Dynamics of the Emission of Solvated Electrons from Nanodiamonds in Water. *Nanoscale* **2022**, *14*, 17188–17195. [CrossRef]
22. Maier, F.; Ristein, J.; Ley, L. Electron Affinity of Plasma-Hydrogenated and Chemically Oxidized Diamond (100) Surfaces. *Phys. Rev. B* **2001**, *64*, 165411. [CrossRef]
23. Zhang, L.; Zhu, D.; Nathanson, G.M.; Hamers, R.J. Selective Photoelectrochemical Reduction of Aqueous CO_2 to CO by Solvated Electrons. *Angew. Chem.* **2014**, *126*, 9904–9908. [CrossRef]
24. Wang, X.; Krause, P.; Kirschbaum, T.; Palczynski, K.; Dzubiella, J.; Bande, A. Photo-excited charge transfer from adamantane to electronic bound states in water. *Phys. Chem. Chem. Phys.* **2024**, *26*, 8158–8176. [CrossRef]
25. Kirschbaum, T.; von Seggern, B.; Dzubiella, J.; Bande, A.; Noé, F. Machine Learning Frontier Orbital Energies of Nanodiamonds. *J. Chem. Theory Comput.* **2022**, *19*, 4461–4473. [CrossRef]
26. Gueymard, C.A. The Sun's Total and Spectral Irradiance for Solar Energy Applications and Solar Radiation Models. *Sol. Energy* **2004**, *76*, 423–453. [CrossRef]
27. Kim, J.; Suh, S.B.; Kim, K.S. Water Dimer to Pentamer with an Excess Electron: Ab Initio Study. *J. Chem. Phys.* **1999**, *111*, 10077–10087. [CrossRef]
28. Yu, J.; Su, N.Q.; Yang, W. Describing chemical reactivity with frontier molecular orbitalets. *JACS Au* **2022**, *2*, 1383–1394. [CrossRef]
29. Nazarov, V.U. Breakdown of the Ionization Potential Theorem of Density Functional Theory in Mesoscopic Systems. *J. Chem. Phys.* **2021**, *155*, 194105. [CrossRef]
30. Perdew, J.P.; Levy, M. Physical Content of the Exact Kohn-Sham Orbital Energies: Band Gaps and Derivative Discontinuities. *Phys. Rev. Lett.* **1983**, *51*, 1884–1887. [CrossRef]
31. Levy, M.; Perdew, J.P.; Sahni, V. Exact Differential Equation for the Density and Ionization Energy of a Many-Particle System. *Phys. Rev. A* **1984**, *30*, 2745–2754. [CrossRef]
32. Winter, B.; Weber, R.; Widdra, W.; Dittmar, M.; Faubel, M.; Hertel, I.V. Full Valence Band Photoemission from Liquid Water Using EUV Synchrotron Radiation. *J. Phys. Chem. A* **2004**, *108*, 2625–2632. [CrossRef]
33. Gaiduk, A.P.; Pham, T.A.; Govoni, M.; Paesani, F.; Galli, G. Electron Affinity of Liquid Water. *Nat. Commun.* **2018**, *9*, 1–6. [CrossRef]
34. Miliaieva, D.; Djoumessi, A.S.; Čermák, J.; Kolářová, K.; Schaal, M.; Otto, F.; Shagieva, E.; Romanyuk, O.; Pangrác, J.; Kuliček, J.; et al. Absolute energy levels in nanodiamonds of different origins and surface chemistries. *Nanoscale Adv.* **2023**, *5*, 4402–4414. [CrossRef]
35. Nordlund, D.; Ogasawara, H.; Bluhm, H.; Takahashi, O.; Odelius, M.; Nagasono, M.; Pettersson, L.G.M.; Nilsson, A. Probing the Electron Delocalization in Liquid Water and Ice at Attosecond Time Scales. *Phys. Rev. Lett.* **2007**, *99*, 217406. [CrossRef]
36. Jesus, A.L.; Redinha, J. Molecular insight into the amine–water interaction: A combined vibrational, energetic and NBO/NEDA study. *Comput. Theor. Chem.* **2013**, *1023*, 74–82. [CrossRef]
37. Yanai, T.; Tew, D.P.; Handy, N.C. A New Hybrid Exchange–Correlation Functional Using the Coulomb-Attenuating Method (CAM-B3LYP). *Chem. Phys. Lett.* **2004**, *393*, 51–57. [CrossRef]
38. Dunning, T.H., Jr. Gaussian Basis Sets for Use in Correlated Molecular Calculations. I. The Atoms Boron Through Neon and Hydrogen. *J. Chem. Phys.* **1989**, *90*, 1007–1023. [CrossRef]

39. Kendall, R.A.; Dunning, T.H., Jr.; Harrison, R.J. Electron Affinities of the First-Row Atoms Revisited: Systematic Basis Sets and Wave Functions. *J. Chem. Phys.* **1992**, *96*, 6796–6806. [CrossRef]
40. Grimme, S.; Antony, J.; Ehrlich, S.; Krieg, H. A consistent and accurate ab initio parametrization of density functional dispersion correction (DFT-D) for the 94 elements H-Pu. *J. Chem. Phys.* **2010**, *132*, 154104 . [CrossRef]
41. Barone, V.; Cossi, M. Quantum calculation of molecular energies and energy gradients in solution by a conductor solvent model. *J. Phys. Chem. A* **1998**, *102*, 1995–2001. [CrossRef]
42. Neese, F. Software Update: The ORCA Program System, Version 4.0. *Wiley Interdiscip. Rev. Comput. Mol. Sci.* **2018**, *8*, 1327. [CrossRef]
43. Lu, T.; Chen, F. Multiwfn: A Multifunctional Wavefunction Analyzer. *J. Comput. Chem.* **2012**, *33*, 580–592. [CrossRef]
44. Martin, R.L. Natural Transition Orbitals. *J. Chem. Phys.* **2003**, *118*, 4775–4777. [CrossRef]
45. Hirshfeld, F.L. Bonded-Atom Fragments for Describing Molecular Charge Densities. *Theor. Chim. Acta* **1977**, *44*, 129–138. [CrossRef]
46. Abraham, M.J.; Murtola, T.; Schulz, R.; Páll, S.; Smith, J.C.; Hess, B.; Lindahl, E. GROMACS: High performance molecular simulations through multi-level parallelism from laptops to supercomputers. *SoftwareX* **2015**, *1*, 19–25. [CrossRef]
47. Wang, J.; Wang, W.; Kollman, P.A.; Case, D.A. Automatic atom type and bond type perception in molecular mechanical calculations. *J. Mol. Graph. Model.* **2006**, *25*, 247–260. [CrossRef]
48. Jorgensen, W.L.; Chandrasekhar, J.; Madura, J.D.; Impey, R.W.; Klein, M.L. Comparison of simple potential functions for simulating liquid water. *J. Chem. Phys.* **1983**, *79*, 926–935. [CrossRef]
49. Parrinello, M.; Rahman, A. Polymorphic transitions in single crystals: A new molecular dynamics method. *J. Appl. Phys.* **1981**, *52*, 7182–7190. [CrossRef]
50. Bennett, L.; Melchers, B.; Proppe, B. *Curta: A General-Purpose High-Performance Computer at ZEDAT, Freie Universität Berlin*; Freie Universität: Berlin, Germany, 2020.

MDPI AG
Grosspeteranlage 5
4052 Basel
Switzerland
Tel.: +41 61 683 77 34

Molecules Editorial Office
E-mail: molecules@mdpi.com
www.mdpi.com/journal/molecules